SEMICONDUCTOR
DEVICES
AND CIRCUITS

SEMICONDUCTOR DEVICES AND CIRCUITS

CHARLES L. ALLEY
KENNETH W. ATWOOD
Electrical Engineering Department
University of Utah
Salt Lake City, Utah

John Wiley & Sons, Inc.
New York · London · Sydney · Toronto

Library of Congress Catalogue Card Number: 70-136708

ISBN 0-471-02330-2

Printed in the United States of America

10   9   8   7   6   5   4   3   2   1

PREFACE

Most technical institutes feel that their graduates need the ability to design electronic circuits in order to gain the greatest satisfaction for themselves and their employers. However, the available textbooks stress circuit description or analysis and stop short of design. To meet this need, we have prepared a design-oriented semiconductor electronics text that is suitable for junior colleges and technical institutes.

Although an understanding of calculus is helpful, the student whose mathematical background is limited to trigonometry and algebra can understand the material in this textbook. In addition, a background in high school chemistry and physics will be most useful. The reader will also require some circuit theory. This background should include the ac steady-state analysis of RLC circuits. This book, in essentially its present form, has been used for two years as a text in the Technical Institute, Division of Continuing Education, at the University of Utah. The critiques of the students and instructors there were very helpful in the final revision.

After a brief introduction, the first chapter reviews some basic electron physics, which may be omitted if the student is already familiar with these concepts. Chapter 2 briefly treats basic semiconductor physics with the thermistor as an application. In chapter 3, the reader is introduced to several types of diodes and typical diode applications. Chapters 4–6 describe the operation of transistors (including FET's and IGFET's) and discuss typical one-stage amplifier circuit applications. Chapters 7–13 consider multistage amplifiers and the various types of coupling systems that can be used. The effects of noise and negative feedback are also treated in depth.

v

In Chapter 14, the linear integrated circuit (IC) is introduced, and design techniques are given for various types of linear IC's, including operational amplifiers. Chapter 15 treats the principles and design of power supplies, including voltage regulator circuits; some examples utilize linear IC's. Switching and pulse applications, as well as the use of digital IC units are presented in Chapter 16.

The material is current and includes such devices as MOSFET's, unijunction transistors, SCR's, and triacs. The description of each device is followed by various applications of these devices. The applications increase in complexity from single-stage amplifiers to fairly complex systems.

We appreciate the support and encouragement of our wives and children during the preparation of this book. We also thank Mrs. Marian Swenson and the secretarial staff of the Electrical Engineering Department at the University of Utah for their excellent and willing service in the preparation of the manuscript. Also, we are grateful to Russell Fraser, formerly of John Wiley, for so carefully surveying the field of technical education to establish the need for this text, and for inviting us to participate with him in the preparation of this book.

We hope this book will convey some of the excitement that we find in electronics.

CHARLES L. ALLEY
KENNETH W. ATWOOD

CONTENTS

SEMICONDUCTOR
DEVICES
AND CIRCUITS

CHAPTER 1

INTRODUCTION

During the twentieth century, a phenomenal growth of technology has occurred. The fields of electricity and electronics have shared in this fantastic growth. At the present time radio, television, radar, sonar, hi-fi, stereo, and computers are familiar to most people. While the electronic devices are well known and widely used, the general public considers them too complicated to be understood by anyone but an expert. Actually, most electronic devices can be reduced to a group of interconnected basic electronic building blocks or circuits. This text will describe the operation and design considerations for most of these basic circuits.

1.1 TYPICAL BLOCK DIAGRAMS

When an engineer begins the design of an electronic device, he usually begins by laying out a *block diagram*. This block diagram is then used as a guide while he designs the circuits represented by the various blocks in the diagram.

For example, let us consider a block diagram of an audio amplifier. Assume the input signal will come from a microphone or a phonograph pickup. The output signal must be large enough to drive (or to activate) a loudspeaker. A typical block diagram for this device might be sketched as shown in Fig. 1.1.

The preamplifier is used to amplify the electrical signal from the microphone until it is about as large as the signal from the phonograph pickup. The signal mixer combines the two signals (microphone and phonograph)

1

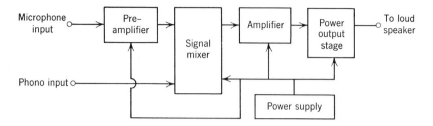

Fig. 1.1.   The block diagram of an audio amplifier.

and is usually designed so that an adjustment of one signal will not effect the other signal. The combined signals are then sent through an amplifier to increase the signal amplitude. This amplified signal is then used to drive the power output stage. The power output stage must produce enough electrical signal power to drive the loudspeaker at the required signal level. The power supply furnishes enough dc power to activate each of the amplifier stages. In a simple portable circuit, this power supply may be a single battery. However, in a high-quality, high-power amplifier, this power supply may be quite a sophisticated and complicated circuit.

The block diagram for a simple radio receiver is given in Fig. 1.2. The radio signal is received by the antenna and amplified in the radio frequency (RF) amplifier. This RF amplifier is tuned to amplify only one radio signal and to reject all the other radio signals that may be present. The detector converts the radio signal to an electrical audio-frequency signal. The audio signal is then amplified in the audio amplifier. The amplified audio signal is used to drive the power amplifier stage, which furnishes enough power to drive the loudspeaker. Again, a power supply is required to furnish dc power to the various amplifiers in the radio.

When the engineer has drawn the block diagram of the system and has determined the required specifications for this system, he is then ready to

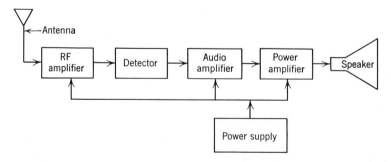

Fig. 1.2.   The block diagram of a radio receiver.

design a circuit for each of the blocks. Some of these circuits may be available as integrated-circuit modules, in which case, the entire block (or sometimes several blocks) may be available in a single tiny package.

To use integrated circuits or to design the various stages of a system, one must understand how these circuits operate. Although most of this book is devoted to the understanding of electronic devices and circuits, as was previously mentioned, these devices cannot be well understood without some insight into the characteristics of semiconductor materials, which will be provided in Chapter 2. Also, the basic laws that govern the flow of electric charges must be understood as a prerequisite to the study of semiconductor materials. Therefore, these basic electrical laws are briefly treated in the remainder of this chapter for the benefit of those who need to review them.

1.2   COULOMB'S LAW

One of the basic laws of electricity, *Coulomb's law*, states, in simplest terms, that the total number of electric flux lines that emanate from an electric charge Q is equal to the charge in coulombs (MKS system of units). For example, assume that a positive charge Q is placed at the center of a sphere of radius r in Fig. 1.3. If the charge inside the sphere is 10 coulombs (C) the total electric flux emerging from the surface of the sphere is $\Phi_E = 10$ lines.

Electric flux *density D* may be defined as the number of electric flux lines passing through a unit-area surface that is oriented normal to the direction of the flux flow. Then, referring to Fig. 1.3, the flux density D at the surface of the sphere of radius r (assuming uniform density) is

$$D = \frac{Q}{A} = \frac{Q}{4\pi r^2} \text{ lines/meter}^2 \text{ (lines/m}^2\text{)} \tag{1.1}$$

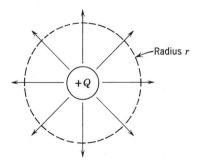

Fig. 1.3.   A positive charge Q enclosed in a sphere of radius r.

PROBLEM 1.1   If the radius of the sphere is 10 centimeters (cm), and $Q = 10$ C, what is the flux density D at the surface of the sphere?

Answer: $D = 79.5 \text{ lines/m}^2$.

1.3   ELECTRIC FIELD

Whenever electric flux lines exist, there is said to be an *electric field*. In fact, the electric-field intensity $\mathscr{E}$ is proportional to the flux density D. The constant of proportionality ϵ is known as the *permittivity* or *dielectric constant*.

$$D = \epsilon\mathscr{E} \tag{1.2}$$

Then, using Eq. 1.1,

$$\mathscr{E} = \frac{D}{\epsilon} = \frac{Q}{4\pi\epsilon r^2} \text{ volts/meter (V/m)} \tag{1.3}$$

The permittivity of free space (vacuum) is $\epsilon_v = 8.854 \times 10^{-12}$ farads/meter (F/m). The permittivity of other materials is usually given as a ratio to the permittivity of free space. This ratio is known as *relative* permittivity or relative dielectric constant ϵ_r. Then

$$\epsilon = \epsilon_v\epsilon_r = 8.854 \times 10^{-12} \epsilon_r \tag{1.4}$$

For example, the 10-C charge that provided a flux density $D = 79.5 \text{ lines/m}^2$ at a radius of 10 cm produces an electric field $\mathscr{E} = D/\epsilon = 9 \times 10^{12}$ V/m in a vacuum at $r = 10$ cm. A frightening number, isn't it?

PROBLEM 1.2   If the 10-C charge is immersed in a large sea of glass with relative permittivity $= 7$, what would be the field intensity at $r = 10$ cm?

Answer: $\mathscr{E} = 1.286 \times 10^{12}$ V/m.

Whenever two charges are at a finite distance, a force is exerted on each charge. In other words, whenever a charge is placed in an electric field, there is a force exerted on the charge. The force in newtons is

$$F = Q\mathscr{E} \tag{1.5}$$

One newton (N) is equal to 0.2248 pound-force (lbf). The electron has a negative electrical charge and would be attracted *toward* the positive charge in Fig. 1.3.

PROBLEM 1.3   If an electron with $q = 1.6 \times 10^{-19}$ C is placed in a vacuum, 10 cm from a 10-C charge, determine the force on the electron in newtons.

Answer: 1.44×10^{-6} N.

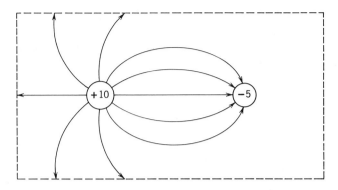

Fig. 1.4.   A 5-C negative charge in the vicinity of a 10-C positive charge.

The force on electric charges may be expressed in terms of the charges instead of the electric field. Using Eqs. 1.3 and 1.5 and designating the charges Q_1 and Q_2

$$F = \frac{Q_1 Q_2}{4\pi\epsilon r^2} \qquad (1.6)$$

The direction of this force is such that like charges repel each other and unlike charges attract each other.

The equation for the gravitational force on a mass is identical to Eq. 1.6 except that charge is replaced by mass and the dielectric constant (permittivity) is replaced by the gravitational constant. Therefore, charge in an electric field behaves like mass in a gravitational field, and we can call on our experience with mass and gravity to help us visualize problems involving charge in an electric field.

By definition, electric flux lines always begin on a positive charge and terminate on a negative charge. Figure 1.4 shows a 5-C negative charge in the vicinity of a 10-C positive charge. The net charge in the figure is 5 C and there are 5 flux lines passing through the boundaries of the figure, which terminate on charges outside the figure.

PROBLEM 1.4   If the charges in Fig. 1.4 are placed 1 kilometer (km) apart, find the force on each charge.   *Answer:* $F = 4.5 \times 10^5$ N.

1.4   POTENTIAL ENERGY AND POTENTIAL

Potential is the ability to do work because of position. For example, a lake of water on top of a hill has potential energy because as it flows down the hill it can turn a water wheel or turbine. The mass of water has potential energy when it is in a gravitational field. Similarly, a charge has potential energy when it is in an electric field. The work that can be done is equal to the product of the force and the distance through which the force can act (assuming constant force).

$$w = F(r_2 - r_1) \tag{1.7}$$

where w is the work done when the force F acts along the path r from point r_1 to point r_2, as shown in Fig. 1.5. If the force is not constant but changes with the location r, calculus must be used to determine the work done on the rectangular box in Fig. 1.5. The unit of work or energy is the joule if force is in newtons and distance is in meters. Also, power is the rate of doing work, and if one joule of work is accomplished in one second, the power is one watt, or watts = joules/sec.

Potential is the energy which a *unit mass* or *unit charge* has because of its position. For example, a one pound mass at the top of a 200-ft cliff has 200 foot-pounds (ft-lb) of potential energy, and the potential at the top of the cliff is said to be 200 ft with respect to the bottom of the cliff. The zero reference for potential may be assigned arbitrarily. Potential (or altitude) in the earth's gravitational field is usually given with respect to sea level. Therefore, the altitude at the top of the cliff might be 8500 ft and the altitude at the bottom of the cliff 8300 ft. The height of the cliff is the altitude difference, or 200 ft. Similarly, a unit charge (coulomb) has the capability, or potential, for doing work when it is placed in an electric field. The electric potential is the product of the electric field and the distance through which the field can act. Therefore, since the electric field is expressed in volts per meter, the potential is expressed in volts. Of course, calculus must be used to obtain the potential from the electric field intensity if the field is not constant. However, potential difference is readily measured with a voltmeter. Potential may be expressed with respect to an arbitrary reference such as the earth, ground, a metal chassis, or a point at infinity. Potential difference is, of course, the voltage between any two points.

PROBLEM 1.5   If a resistor is 1 cm long and has a uniform electric field $\mathscr{E} = 500$ V/m from end to end, what is the potential difference, or voltage across the resistor?   *Answer:* 5 V.

Fig. 1.5.   Work done by a force acting on a body.

1.5   BEHAVIOR OF AN ELECTRON IN AN ELECTRIC FIELD

The similarity between the behavior of an electric charge in an electric field and the behavior of a mass in a gravitational field was previously mentioned. Therefore, we can call on our personal experiences with gravity to assist us in understanding the basic laws of electricity. For example, we can visualize the acceleration of an electron placed in an electric field as being like the acceleration of a rock dropped from the top of a cliff. In either situation, newtons law of motion, $F = ma$, may be used to determine the acceleration, providing the velocities are small in comparison with the velocity of light, so the mass can be considered constant. Then

$$a = F/m \tag{1.8}$$

However, the mass of the electron is so very small (9.11×10^{-31} kg) that its acceleration is amazingly large even in weak fields. For example, if an electron is placed in a field strength $\mathscr{E} = 1$ V/m, it will experience an acceleration $a = q\mathscr{E}/m = 1.76 \times 10^{11}$ meters/second^2 (m/s^2) and its velocity will approach the velocity of light if it remains in the field for about one thousandth of a second. The precise velocity and the distance traveled in this millisecond can be calculated with the aid of calculus, but will not be pursued here.

Very often, the potential difference through which an electron has fallen, rather than the field intensity, is known. Then the velocity of the electron may be determined from basic energy relationships. For example, if a rock is dropped from the top of a cliff, the rock gains kinetic energy $= \frac{1}{2} mv^2$ as it loses potential energy. Except for energy loss due to air friction, the kinetic energy gained is equal to the potential energy lost. This is the principle known as conservation of energy. In this example, the potential energy lost is the product of the force on the rock and the distance through which it has fallen. Similarly, when an electron "falls" through a potential difference V, it loses potential energy $= qV$ and gains kinetic energy $= \frac{1}{2} mv^2$. Since the kinetic energy gained must be equal to the potential energy lost,

$$\tfrac{1}{2} mv^2 = qV \tag{1.9}$$

and the electron velocity is

$$v = \left(\frac{2qV}{m}\right)^{1/2} \tag{1.10}$$

where q and m are the charge and mass of the electron. When numerical values are substituted for q and m, the velocity becomes

$$v = 5.93 \times 10^5 \, (V)^{1/2} \text{ m/s} \qquad (1.11)$$

In the preceding discussion it was assumed that the mass remains constant. However, Einstein has shown that mass is a function of velocity as given in Eq. 1.12

$$m = \frac{m_o}{(1 - v^2/c^2)^{1/2}} \qquad (1.12)$$

where m_o is the rest (or zero velocity) mass and c is the velocity of light $= 3 \times 10^8$ m/s. From Eq. 1.12 it can be seen that the mass m differs from m_o by one percent or less if v is $0.1c$ or 3×10^7 m/s or less. This value of velocity will be acquired when the electron has fallen through 2560 V in a vacuum. However, the mass can be considered to be essentially constant until it increases by about 10 percent or $v \simeq 0.4c$. This velocity would be acquired by an electron in falling through about 40,000 V. This value is above the upper limit of accelerating voltages used in cathode ray or television picture tubes.

PROBLEM 1.6 Determine the velocity of an electron that has fallen through a potential difference of 1000 V.   *Answer:* $v = 1.87 \times 10^7$ m/s.

When electrons have been accelerated into the velocity where their mass begins to change, the equation which determines their velocity is

$$v = c \left[1 - \frac{1}{(1 + 1.966 \times 10^{-6} \, V)^2} \right]^{1/2} \text{ m/s} \qquad (1.13)$$

where V is the voltage through which the electrons have been accelerated. The mass of the electrons in this velocity region is given by the equation

$$m = m_o (1 + 1.966 \times 10^{-6} \, V) \qquad (1.14)$$

Table 1.1   Some Physical Constants

Electron charge q, C	1.6×10^{-19}
Electron rest mass m, Kilogram (kg)	9.11×10^{-31}
Permittivity of free space ϵ_v, F/m	8.854×10^{-12}
Speed of light, m/s	3.00×10^8
Planck's constant h, Joule-second (J-s)	6.62×10^{-34}
Boltzmann constant k, J/°K	1.38×10^{-23}
Avagadro's number, molecules/kg-mole	6.02×10^{26}

The physical constants used in this chapter and a few additional ones are listed in Table 1.1 for ease of future reference.

PROBLEM 1.7   If the mass of an electron at rest is 9.1066×10^{-31} kg, what is the mass of an electron that is traveling 2.7×10^8 m/s?

PROBLEM 1.8   To put a satellite into orbit, it must have a velocity of 18,000 miles per hour (mi/h). Through what potential difference must an electron be accelerated to achieve this velocity?   *Answer:* 1.84×10^{-4} V.

PROBLEM 1.9   What is the velocity and the mass of an electron that has been accelerated through a potential difference of 100,000 V?
Answer: $v = 1.64 \times 10^8$ m/s, $m = 1.09 \times 10^{-30}$ kg.

PROBLEM 1.10   An electron has been accelerated through a potential of 28,400 V when it enters the space between deflection plates as shown in Fig. 1.6.
   (a) Determine the velocity of this electron (assume all this velocity is in the x direction) when it enters the space between the 2 deflection plates.
   (b) How long does it take the electron to pass between the 2 plates?
   (c) What is the magnitude of the electric field between the two deflection plates?
   (d) What is the velocity in the y direction when the electron leaves the area between the deflection plates?

Given:
$m_e = 9.1066 \times 10^{-31}$ kg
$q = 1.6019 \times 10^{-19}$ C
$v = 5.93 \times 10^5 \, (V)^{1/2}$ m/s
Answer: $v_x = 10^8$ m/s, $t = 2 \times 10^{-10}$ s, $\mathscr{E} = 4000$ V/m, $v_y = 1.4 \times 10^5$ m/s.

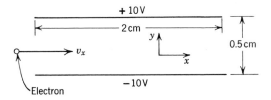

Fig. 1.6.   The configuration for Problem 1.10.

CHAPTER 2

SEMICONDUCTORS

This chapter presents the elementary semiconductor principles that are essential to the intelligent use of diodes and transistors in electronic circuits. Some devices such as thermistors and integrated circuits will be used to illustrate the characteristics of semiconductor materials.

Physical models and logical reasoning will be used whenever possible to establish rules and relationships instead of reliance on mathematics, which may become foreboding. Therefore, the Bohr model of the atom will be used because of the ease with which it can be visualized. The inaccuracy of this model, as compared with the quantum mechanics model, will not cause erroneous conclusions in the situations in which the Bohr model is used.

2.1 ATOMIC MODELS

You may recall from basic chemistry that the world is composed of atoms that have a positively charged nucleus and a set of planetary, negatively charged electrons. In the Bohr model, the atom resembles a miniature solar system with the electrons orbiting around the nucleus in either circular or elliptical orbits, as sketched in Fig. 2.1.

Hydrogen is the simplest atom, having only one orbital electron and a single proton as a nucleus. The next element in order of complexity is helium, with two orbital electrons. These two electrons *fill* the first, or lowest energy orbit, which is usually called the *K shell*. In the next few atoms of increasing complexity electrons are added to the second or *L*

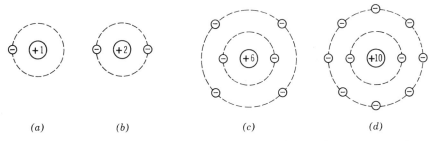

Fig. 2.1. Symbolic sketches of a few simple atoms. (*a*) Hydrogen, (*b*) helium, (*c*) carbon, and (*d*) neon.

shell until this shell becomes filled with eight electrons. Neon has both its *K* and *L* shells filled, and therefore has ten orbital electrons.

If an atom were given a personality, we could say that an atom likes to have its outer shell filled and is only content when this occurs. In fact, when the opportunity arises, atoms combine in a manner that will permit them to share their electrons and effectively fill their outer shells. For example two hydrogen atoms, each having one orbital electron, readily join with an oxygen atom, which has eight orbital electrons and which thus needs two more to fill its *L* shell. The result, of course is water. Actually, the desire to share electrons is not a peronality quirk, but the electron sharing results in a lower total energy for the molecule. The energy released by the combination of hydrogen and oxygen, known as burning, has been used to place a good number of astronauts into orbit. Also, two hydrogen atoms join together to form diatomic molecules because their *K* shells are effectively filled and their total energy is lower than that of two single atoms. Helium and neon will not combine with anything and are known as inert gases because their outer shells are filled.

The chemical properties of an atom depend almost entirely on the number of electrons in the outer shell, which is known as the valence shell. The atoms that have four valence electrons, such as carbon, are known as tetravalent atoms and are of special interest to us because they include the semiconductors. These elements may form crystals in which each atom shares its valence electrons with its four adjacent neighbors as illustrated schematically in Fig. 2.2. The actual crystal structure is three-dimensional, of course. The nucleus and the filled inner shells are represented by the circle with the +4, which is the net positive charge of this group. The valence electrons are represented by the negative signs and the bonds, which result from the reduced energy and which hold the crystal together, are represented by the curved lines. These bonds are known as covalent bonds. Carbon in this crystalline form is known as diamond.

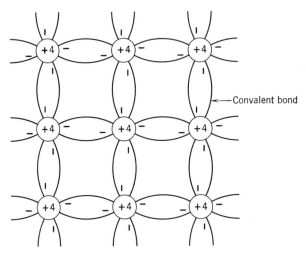

Fig. 2.2.   Two-dimensional representation of covalent bonds in a tetravalent crystal.

2.2 CONDUCTORS, INSULATORS, AND SEMICONDUCTORS

An electrical conductor is a material that will permit electric charge to flow (current) when a potential difference is applied to the material. Therefore, conduction relies on the availability of charge carriers, such as electrons, which can move freely about in the material. A good conductor has many free electrons but a semiconductor has comparatively few electrons at normal temperatures. Of course, an insulator does not have free electrons. At 0°K (absolute zero) all the electrons in a diamond crystal are bound by the covalent bonds as shown in Fig. 2.2. Therefore, diamond is an insulator at $T = 0°K$.

Diamond can become a conductor, however, if its temperature is raised high enough to give a sufficient number of electrons enough energy to break their covalent bonds and drift freely through the crystal. This electron liberation process compares roughly with the liberation of a rocket from the earth by transferring the energy of the rocket fuel, first primarily to kinetic energy and then, as the rocket gains altitude, to potential energy, until the rocket gains an altitude equal to many diameters of the earth. The gravitational pull of the earth is then negliglible and the rocket is free to drift through the solar system.

The tetravalent atoms such as carbon, silicon, germanium, and tin do not form crystals of equal conductivity at a given temperature because the energy required to break a covalent bond depends on the closeness of the atomic spacing in the crystal. The smaller the atom, the closer the spacing and the greater the energy required to break the covalent bonds. Figure 2.3 shows the relative atomic spacing and the relative energy required to break a covalent bond for carbon, silicon, germanium, and tin. Compared with carbon, silicon has one more filled shell, germanium has two more, and tin has three more filled shells between the valence shell and the nucleus; therefore, the atom size and spacing increases in that order. The energy represented by the forbidden band is the minimum energy required by a valence (band) electron to break a covalent bond and become a conduction (band) electron. Smaller amounts of energy cannot free the electron from the valence band. Therefore, electrons can pass through the forbidden band, but cannot remain there. At room temperatures, germanium is neither a good insulator nor a good conductor; hence it is a semiconductor.

PROBLEM 2.1   Of those tetravalent crystals shown in Fig. 2.3, which would you expect to be:

(a) The best conductor?

(b) The best insulator?

(c) The best semiconductors?

Diamond is an insulator and tin is a fairly good conductor at normal room temperatures. At very high temperatures, assuming the crystal to remain

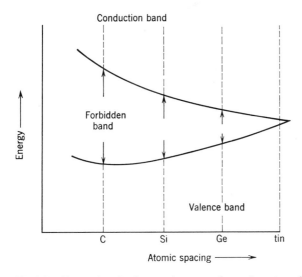

Fig. 2.3.   Energy bands of tetravalent crystals as a function of lattice spacing.

intact, they all become conductors, while at 0°K they all become insulators. Photons, or light energy, and strong electric fields may also break covalent bonds and elevate electrons to the conduction band, thus increasing their conductivity.

2.3   CONDUCTION IN A SEMICONDUCTOR

As was mentioned, heat or other sources of energy cause valence band electrons to break their covalent bonds and become free electrons in the conduction band, thus producing modest conductivity in a semiconductor at normal temperatures. As the electron leaves the valence band it creates a missing covalent bond, which is known as a *hole*. This hole permits charge movement and hence conduction by the valence electrons as illustrated by Fig. 2.4. A nearby valence-band electron can fill a hole and thus create another hole with practically no exchange of energy. If an observer could see electrons and holes he would see the hole move instead of the electrons. The hole acts as a positive (polarity) charge since it moves in the direction opposite to that of the electron. Thus electric conduction is caused by free electrons and free holes, which are known as *charge carriers*.

The conduction due to the electrons in the conduction band is a different process from the conduction due to the holes left in the valence band. In the intrinsic or pure semiconductor material, there are as many holes as there are free electrons because the free electron leaves a missing covalent

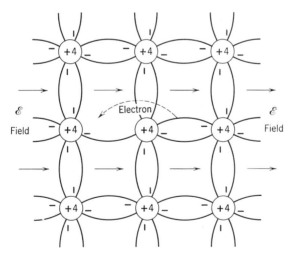

Fig. 2.4.   Illustration of conduction resulting from broken covalent bonds or holes.

bond or hole. An analogy (which is attributed to Shockley) might be used to illustrate the conduction processes. A parking garage with two floors has the lower floor completely filled with automobiles and the upper floor completely empty. Under these conditions there can be no movement of automobiles on either floor. If one automobile is elevated from the lower to the upper floor, there can be motion of automobiles on each floor. The auto on the upper floor may move freely over comparatively large distances. In contrast, the motion on the lower floor is accomplished by moving one car at a time into the available space. Hence, an observer near the ceiling of the first floor would see the "hole" move rather than the automobiles. The hole would have less mobility than the auto on the upper floor. Nevertheless, both would contribute to the total motion.

We now know that electrical conductivity depends on free charge carriers in a material and that semiconductors have two different charge carrier mechanisms known as electrons and holes. Our next goal is to determine the conductance (or resistance) of a piece of semiconductor material. But before that can be done we need to know how many of each type of carrier there are (or the carrier densities) in the material, and how the conductivity is related to the carrier densities. Since the latter problem is the easiest, it will be considered first.

As you have probably learned in basic physics or electricity,

$$\text{resistance } R = \frac{V}{I} = \rho \frac{l}{A} \tag{2.1}$$

where ρ is the resistivity of the material, l is the length and A is the cross-sectional area of the conductor.

Similarly

$$\text{conductance } G = \frac{I}{V} = \sigma \frac{A}{l} \tag{2.2}$$

where $\sigma = 1/\rho$ is the conductivity of the material.

Then, from Eq. 2.2,

$$\sigma = \frac{I/A}{V/l} = \frac{J}{\mathscr{E}} \tag{2.3}$$

where $I/A = J$ is the current per unit cross sectional area, or *current density*. The term $V/l = \mathscr{E}$ is known as the electric-field intensity. The electric-field intensity $\mathscr{E}$ is constant in a uniform conductor of constant cross section (area A). Equation 2.3 is therefore Ohm's law on a *per unit* basis.

Since current density is the rate of flow of charge through a unit cross-sectional area,

$$J = qn\bar{v} \tag{2.4}$$

where n is the number of carriers (electrons, for example) per cubic meter, q is the charge per carrier (1.6×10^{-19} C for electrons or holes) and $\bar{v}$ is the average drift velocity of the carriers due to the electric field. Substituting the expression for J in Eq. 2.4 into Eq. 2.3,

$$\sigma = \frac{qn\bar{v}}{\mathscr{E}} \tag{2.5}$$

One problem remains. Neither the drift velocity $\bar{v}$ nor the electric field $\mathscr{E}$ are usually known. Fortunately, however, their ratio $\bar{v}/\mathscr{E}$ is known and is constant at a given temperature for a given material. This constant is known as the carrier mobility and has been given the symbol μ. Thus

$$\mu = \frac{\bar{v}}{\mathscr{E}} \tag{2.6}$$

As was previously mentioned, the mobility of electrons is greater than the mobility of holes.

Since the total conductivity is the sum of the conductivities due to electrons and holes,

$$\sigma = qn\mu_n + qp\mu_p = q(n\mu_n + p\mu_p) \tag{2.7}$$

where n is free electron density, p is free hole density, μ_n is electron mobility, and μ_p is hole mobility.

The mobilities of electrons and holes in both silicon and germanium at 300°K are given in Table 2.1. With this information, the conductivity or conductances (or resistance) can be determined. Thus, a piece of pure germanium crystal has about 2.4×10^{19} free electrons per cubic meter at 300°K (room temperature). Then, from Eq. 2.7, $\sigma = q(n\mu_n + p\mu_p) = 1.6 \times 10^{-19}$ $(2.4 \times 10^{19} \times 0.39 + 2.4 \times 10^{19} \times 0.19) = 2.22$ mho-meter. The resistivity is equal to the inverse of the conductivity. Thus, $\rho = 1/\sigma = 1/2.22 = 0.45$ *ohm-meter* at room temperature.

PROBLEM 2.2   Determine the resistance of a germanium bar that has a cross section of 2 mm $\times$ 1 mm and is 1 cm in length.   *Answer:* $R = 2250\,\Omega$.

Table 2.1   Carrier mobilities, 300°K

Material	μ_n	μ_p	Units
Germanium	0.39	0.19	m²/Vsec
Silicon	0.135	0.05	m²/Vsec

2.4   CHARGE CARRIER DENSITY IN A SEMICONDUCTOR

The next problem to be treated is that of determining the free carrier density as a function of temperature and forbidden band energy in a semiconductor. A formal derivation, or even an informal one, is beyond the scope of this work. However, an analogy will be used that will make plausible the carrier density formula to be presented later.

As was pointed out in Chapter 1, charge in an electric field behaves very much like mass in a gravitational field. Therefore, the air molecules, or atmosphere, above the earth may be used as a model to predict the density of electrons in the conduction band of a semiconductor. Figure 2.5 is a sketch of the surface of the earth. An instrumented baloon is released vertically to measure the density of the atmosphere or air molecules.

As a beginning point, we should recall that the atmosphere exists only because of its temperature, which provides kinetic energy and mobility to the air molecules. At 0°K there would be no atmosphere, only a thin film of solid oxygen and nitrogen on the surface of the earth. As the temperature is increased the atmospheric density increases, particularly at higher altitudes, because the air molecules gain sufficient kinetic energy, which can be exchanged for potential energy, to place them high above the earth. If the temperature becomes sufficiently high, many molecules can reach escape velocity and go into outer space. This escape compares with

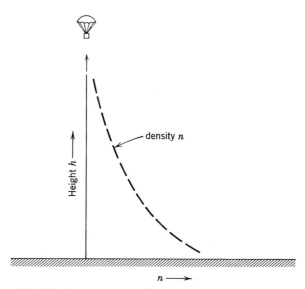

Fig. 2.5.   An arrangement for measuring air density as a function of altitude or height.

thermionic emission, where the electron is ejected from the surface of the emitting material.

Boltzmann has shown mathematically that if the temperature of a gas, such as the atmosphere, is constant, the density of the gas is an exponential function of the altitude h. His equation is

$$n = Ae^{-(mgh/kT)} \tag{2.8}$$

where n is the number of molecules per cubic meter, A is a constant, m is the mass of a molecule, g is the gravitational constant, and k is Boltzmann's constant, which relates temperature T to its equivalent energy. Note that mgh is the potential energy wh of a molecule at altitude h. Therefore, Eq. 2.8 can be written as

$$n = Ae^{-(wh/kT)} \tag{2.9}$$

This exponential density distribution, known as the Boltzmann distribution, has been theoretically and experimentally verified. However, the atmospheric temperature and composition vary with altitude, so that the Boltzmann distribution given by Eq. 2.9 is more easily applicable to semiconductors than it is to the atmosphere.

Notice two things about the ideal atmospheric model.

1. The Boltzmann equation gives the density at any altitude h, providing the density at any given reference $h = h_1$ is known. The reference may be at sea level, on a mountain top, or at the center of the earth, assuming a hole could be dug to the center of the earth, and the temperature is uniform. For example, if the density of air is 5×10^{25} molecules/m³ at sea level, the constant $A = 5 \times 10^{25}$ if the reference for $h(h = 0)$ is at sea level.

2. The Boltzmann distribution predicts the density only if the molecules are not restricted or forbidden for some reason. For example, the density of air molecules in Moffat tunnel in Colorado may be 4×10^{24} molecules/m³, but the air density 10 m above the tunnel inside the mountain is zero if the soil and rocks inside the mountain completely exclude the air.

When the gas molecule analogy is applied to the free electron density problem in a semiconductor, the availability of occupiable electron energy levels or states must be considered, and must weigh, or multiply, the Boltzmann density distribution function. For example, electrons cannot exist in the forbidden band. Also, as was previously mentioned, each electron energy level, or shell, in an atom will accept only a limited number of electrons. When these factors are included and a summation or integration is performed over the range of conduction band energy levels, the total number of electrons in the conduction band can be obtained for a unit volume of the material as follows

$$n_i = AT^{3/2} \, e^{-(W_g/2kT)} \qquad (2.10)$$

where T is temperature in °K, k is the Boltzmann constant $= 1.38 \times 10^{-23}$ J/°K, W_g is the forbidden band or *gap* energy (the energy required to break a covalent bond) in joules, and A is a constant.

The constant A may be either theoretically or experimentally determined. The values given in Table 2.2 for silicon and germanium are usually used. The gap energies are also given for $T = 300$°K. The gap energies are functions of temperature because the lattice spacing increases with temperature due to thermal expansion. Thus the gap energy W_g decreases as the temperature increases. However, in this work we shall use only the 300°K values given.

Table 2.2   Constant A and Gap Energy at 300°K

Material	W_g Electron Volts	Constant A
Silicon	1.12	7.96×10^{21}
Germanium	0.67	1.83×10^{21}

Since the gap energies are given in electron volts, these values must be multiplied by the electric charge $q = 1.6 \times 10^{-19}$ to convert them to joules.

The free electron density can now be calculated for temperatures near 300°K, which is the normal operating range. For example, the density of free electrons in pure, or intrinsic, silicon at 300°K is 1.7×10^{16} electrons/ m³. The conductivity of pure silicon at 300°K is 5.0×10^{-4} mho/m, and the resistivity is 2000 ohm-meter.

PROBLEM 2.3   Using the value of resistivity determined for germanium (0.45 ohm-meter) at 300°K, what is the approximate ratio of the resistivity of silicon to that of germanium at this temperature?   *Answer:* 4450.

As one would expect, free carriers are continually being generated in a semiconductor and their density would continually increase if recombinations did not also occur. Since the opportunities for recombination are proportional to the product of the number of free electrons and the number of holes, the recombination rate is proportional to this product pn. Thus, at a given temperature, the recombination rate must equal the generation rate so that the free carrier density remains essentially constant at the value predicted by Eq. 2.10. Therefore, if the temperature of an intrinsic semiconductor were increased enough to cause the rate of charge carrier generation to double, the carrier density would increase until the rate of recombination would also double. However, each carrier density would increase by only a factor of $\sqrt{2}$.

The variation of resistivity of a semiconductor with temperature is utilized in a temperature sensitive resistor known as a *thermistor*.

PROBLEM 2.4 If a thermistor is made of intrinsic (pure) silicon, calculate its resistivity at $T = 400°K$.   *Answer:* 3.94 ohm-meter.

PROBLEM 2.5 By what factor is the resistance of the silicon thermistor decreased if its temperature is raised from 300°K to 400°K?   *Answer:* 508.

In addition to its resistivity, the resistance of a thermistor depends on its length and cross-sectional area.

2.5 DOPED SEMICONDUCTORS

The conductivity of a semiconductor may be greatly increased if small amounts of specific impurities are introduced into the crystal. For example, if a few pentavalent atoms, such as arsenic or antimony, which have five valence electrons, are added to each million semiconductor atoms, the impurity atoms form covalent bonds with their neighbors as shown in Fig. 2.6. The pentavalent atom, after furnishing four electrons

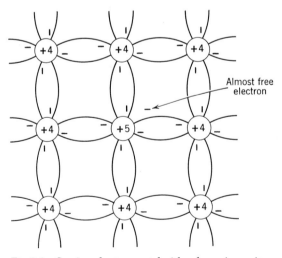

Fig. 2.6.   Semiconductor crystal with a donor impurity.

for the covalent bonds, has an extra valence electron that will not fit into the lattice arrangement. This electron is loosely bound, or almost free, at absolute zero temperature, and will almost certainly be free at temperatures above about 100°K. Observe that, essentially, each impurity atom provides a free electron *without* creating a hole. Thus the pentavalent atom is known as a *donor* because it donates or provides a free electron, and the semiconductor is said to be *doped* or extrinsic. Also note that the donor provides a *fixed* positive charge in the crystal lattice and the crystal is electrically neutral. The donor-doped crystal is known as *n*-type crystal because most of the charge *carriers* are negative.

EXAMPLE 2.1 The effectiveness of the donor impurity in increasing the conductivity of a semiconductor will be emphasized by an example. Let us consider a silicon crystal that has approximately 1.7×10^{16} free electrons resulting from thermally broken covalent bonds, per cubic meter at 300°K, according to our previous calculation. Then let us assume that one part arsenic per million parts silicon was added when the crystal was formed. Since there are approximately 5×10^{28} silicon atoms/m³ the density of arsenic atoms is 5×10^{22} atoms/m³. Then, assuming all the donors to be activated (electrons are free), the conductivity of the doped silicon is about 10^3 mhos/m, which is 2×10^6 times as high as the 300°K conductivity of intrinsic silicon that we calculated to be 5×10^{-4} mho/m.

PROBLEM 2.6 Verify that the conductivity of the doped silicon is as given in Example 2.1.

The loosely bound electrons of the donor atoms become activated at about 50°K, and for the 5×10^{22} atoms/m³ doping concentration of the preceding example, the thermally generated carriers are few compared with the donors until the temperature reaches about 450°C. Figure 2.7 is a sketch of the free electron and hole densities in this doped silicon as functions of temperature.

In a donor-doped crystal, the free electrons are known as the *majority* carriers and the holes are known as the *minority* carriers. The minority carrier density is usually so small at normal operating temperatures that it can be neglected in conductivity calculations. The conductivity equation, Eq. 2.7, can then be simplified to

$$\sigma = qn\mu_n \tag{2.11}$$

Since the majority carrier density is essentially equal to the doping concentration, N_d (for useful operating temperatures and normal doping), the

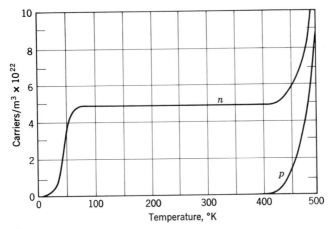

Fig. 2.7. Charge-carrier density in a donor-doped semiconductor as a function of temperature.

conductivity equation can also be written as

$$\sigma = qN_d\mu_n \qquad (2.12)$$

The minority carrier density is much lower than you may have suspected. As was previously discussed, the rate of carrier recombination is proportional to the product of the free electrons and holes, but the rate of carrier generation is dependent only on the temperature. Therefore, when additional free electrons are introduced into the crystal by doping, the rate of recombination increases until the population of holes is reduced to the point where the recombination rate again equals the rate of thermal generation. For example, if the doping suddenly increased the free electron density by a factor of 100, the recombination rate would initially increase by a factor of 100, but would then decrease as the hole population decreases until the normal rate of recombination, which is equal to the rate of thermal generation, is reached. This normal rate is reached when the hole density reaches 1 percent of its value for the intrinsic material. An analogy of the reduction of minority carrier density may be the reduction of free girl density in San Diego after the U.S. Navy comes ashore.

Since the rate of carrier generation is the same for both intrinsic and doped semiconductors, assuming the temperature to be the same, it should be apparent from the above discussion that the steady-state recombination rates must be the same. Therefore, since the recombination rate is proportional to the product of the free electrons and holes, the following relationship holds true.

$$pn = p_i n_i = p_i^2 = n_i^2 \qquad (2.13)$$

Therefore, in a donor-doped crystal, the minority carrier density p may be written

$$p = \frac{p_i^2}{n} = \frac{n_i^2}{n} \qquad (2.14)$$

Since n is very nearly equal to the doping concentration, or density N_d,

$$p = \frac{n_i^2}{N_d} = \frac{p_i^2}{N_d} \qquad (2.15)$$

PROBLEM 2.7 In Example 2.1 the doping concentration was assumed to be 5×10^{22} donors/m³ and the free electron density in the intrinsic crystal is 1.7×10^{16}. Determine the minority carrier, or hole, density in the doped silicon of Example 2.1.   *Answer:* 5.8×10^9 holes/m³.

P-doped semiconductors can be produced by adding a trace of trivalent material such as indium or gallium to the molten tetravalent germanium or silicon. Then, as the crystal forms, each impurity atom fits into the crystal structure but lacks one electron in forming covalent bonds with its neighbors, as shown in Fig. 2.8. Therefore, each trivalent atom provides a hole in the semiconductor crystal. At temperatures above about 50°K, the valence electrons have enough energy to transfer from a neighboring atom to fill a hole. Therefore, the hole moves freely through the crystal while a fixed negative charge remains with the trivalent atom. This atom is therefore known as an acceptor, since it accepts an electron from its

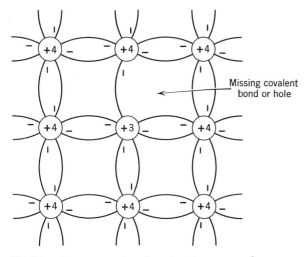

Fig. 2.8.   An acceptor-doped semiconductor crystal.

neighbors, and the symbol for acceptor doping concentration or density is N_a. The trivalent-doped crystal is called p-doped because most of the charge carriers are holes that act like positive charges.

Additional insight into the behavior of a doped semiconductor may be gained from the energy level diagram of Fig. 2.9. The energy level of the donor electrons is only about 0.01 electronvolt (eV) below the conduction band, as indicated. Therefore, at normal temperatures, the percentage of donor atoms that retain their extra electron is about the same as the percentage of donor atoms as compared with total atoms, which is of the order of one in a million, since the doping is of that order. Thus, nearly all the donor electrons are free charge carriers. Similarly, the energy level of the holes (acceptor level) in an acceptor or p-doped semiconductor is very near the valence band. Therefore the probability that the acceptor holes are filled at normal temperatures is about the same as the probability that the valence band holes are filled. Thus at normal doping levels of a few parts per million, the holes associated with an acceptor atom are nearly all filled.

From the preceding discussion, the following relationships can be written for a p-doped semiconductor:

$$p \simeq N_a \tag{2.16}$$

$$n = \frac{n_i^2}{N_a} = \frac{p_i^2}{N_a} \tag{2.17}$$

$$\sigma = qN_a\mu_p \tag{2.18}$$

Since the conductivity of a semiconductor, especially silicon, can be controlled over such wide limits by doping, a complete circuit including

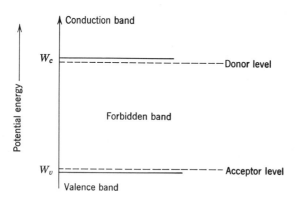

Fig. 2.9.   Energy-level diagram for a doped semiconductor.

resistors, capacitors, and conductors can be made in a single piece or chip of material by controlling the area and concentration of doping. In the following chapters you will learn how diodes and transistors are formed from *p*-doped and *n*-doped semiconductors, and thus how they may be an integral part of the circuit formed in the single or monolithic chip.

PROBLEM 2.8 If you were making an integrated or monolithic circuit, how would you dope for:

(a) A conductor?

(b) An insulator?

(c) A resistor?

(d) A capacitor?

If a semiconductor is doped with equal concentrations of donor and acceptor atoms, the free electrons from the donors fill the holes created by the acceptors, and the material behaves as though it were intrinsic. Also, a semiconductor may be doped with unequal concentrations of donors and acceptors. Then the effective doping is the difference between the two concentrations and has the characteristic of the higher concentration.

2.6 GALLIUM ARSENIDE SEMICONDUCTORS

A semiconductor crystal can be formed from a compound composed of trivalent and pentavalent atoms. Gallium arsenide is the prime example of this type compound and the only one in common use at the time of this writing. The gallium arsenide forms a crystal very similar to the silicon or germanium crystal except the gallium and arsenic atoms alternate in the crystal lattice, so that each gallium atom is surrounded by four arsenic atoms and each arsenic atom is surrounded by four gallium atoms. Covalent bonds are formed as in a germanium or silicon crystal. The extra electrons from the pentavalent arsenic atoms fill the holes produced by the trivalent gallium atoms so that the crystal has the same general properties as intrinsic semiconductors composed of tetravalent atoms.

The gallium arsenide (GaAs) crystal can be *p*-doped by adding small amounts of group II atoms, such as zinc, with two-valence electrons. These atoms replace the trivalent gallium atoms and provide an extra hole, in addition to accepting the extra electron from an arsenic atom neighbor. Also, the crystal can be *n*-doped by adding a small quantity of group VI atoms, such as selenium, with six-valence electrons. These atoms replace arsenic atoms and provide a free electron in addition to donating an electron to fill the hole of a neighboring gallium atom.

The GaAs crystal is used to make semiconductor devices that are superior to either germanium or silicon in some respects. These advantages arise from the following characteristics of GaAs.

1. The forbidden band or gap energy is 1.40 eV at 25°C compared with about 1.1 eV for silicon and 0.67 eV for germanium. This higher gap energy provides satisfactory performance of GaAs devices up to about 300°C with present technology; this limit could be increased to 400°C or above if some aging problems can be overcome. The upper temperature limit for silicon devices is about 200°C, and germanium is useful up to about 100°C. The intrinsic GaAs crystal is a much better insulator at normal temperatures than silicon or germanium because of the higher gap energy. This characteristic is desirable for making high-quality integrated circuits and low-loss varactor diodes, which are discussed in Chapter 3.

2. GaAs has a much higher electron mobility than either silicon or germanium. Therefore, the upper frequency limit of a transistor, which is proportional to charge carrier mobility, can be much higher if the transistor is made of GaAs. The theoretical electron mobility in a pure, perfect GaAs crystal is 1.1 m²/V-sec; about three times as high as that of germanium and six times as high as electron mobility in silicon. Present refining processes can produce GaAs with about 0.8 m²/V-sec electron mobility, about twice that of germanium.

3. Electrons that return from the conduction band to the valence band in a GaAs crystal produce light, in contrast with the heat produced in silicon and germanium. Therefore, GaAs can be used in low-voltage solid-state display devices. An example of such a device is the light diode discussed in Chapter 3.

GaAs has other interesting properties that are useful in microwave devices, which will not be discussed in this book. Some other semiconductor compounds with interesting properties have also been produced, but are not yet available commercially. These compounds, including GaAs will undoubtedly become increasingly important in the world of semiconductors as improved techniques of refining and metallurgy are developed.

PROBLEM 2.9   Calculate the end-to-end resistance of a rectangular bar of intrinsic GaAs crystal at 300°K if $w = 2$ millimeters (mm), $t = 1$ mm, and $l = 1$ cm.   *Answer:* $R = 1.94 \times 10^{10}$ ohms (Ω).

PROBLEM 2.10   There are 4.43×10^{22} atoms/cm³ in gallium arsenide at

Table 2.3   Some Fundamental Properties of Germanium, Silicon, and GaAs

Specific Property	Silicon	Germanium	GaAs
Atomic number	14	32	
Atomic weight	28.06	72.6	
Density (25°C) kg/m³	2.33×10^3	5.33×10^3	5.32×10^3
Melting point, °C	1420	936	1238
Relative dielectric constant ϵ/ϵ_v	12	16	11
Intrinsic carrier density n_i (300°K)/m³	1.7×10^{16}	2.4×10^{19}	1.4×10^{12}
Gap energy, W_g, electron volts: 0°K	1.2	0.782	1.63
300°K	1.12	0.67	1.40
Carrier mobility (300°K) m²/V-sec μ_n	0.135	0.39	1.10
μ_p	0.05	0.19	0.05
Diffusion constant (300°K) m²/sec D_n	33.8×10^{-4}	98.8×10^{-4}	43×10^{-3}
D_p	13×10^{-4}	46.6×10^{-4}	13×10^{-4}

300°K. If donor type atoms are added until the total number of free electrons is equal to $2 n_i$, what will be the free hole density? What will be the ratio of intrinsic atoms to donor atoms?   *Answer:* $p_i = 0.5 n_i, 3.16 \times 10^{16}$.

PROBLEM 2.11   Would impurity atoms (either donors or acceptors) either increase or decrease the rate of change of resistance with temperature? Why?

CHAPTER 3

JUNCTION DIODES

A diode is an electronic device that readily passes current in one direction, but does not pass appreciable current in the opposite direction. Such a device is formed when a donor-doped or n-type semiconductor is connected to an acceptor-doped or p-type semiconductor. In actual production a single crystal of semiconductor is formed with part of the crystal n-doped and the other part p-doped.

Diodes are used to change alternating current to direct current, detect radio signals, provide reference voltages and perform many other useful functions, some of which will be discussed in this chapter. Also, the theory of a p-n junction is basic to the understanding of transistors, which will be presented in Chapter 4.

3.1 THE *p-n* JUNCTION

The basic mechanism of a p-n junction is illustrated in Fig. 3.1. In the p-type material the acceptor atoms with their fixed negative charges are represented by the circles with the negative sign inside. The positive signs represent the holes that are free to move through the crystal. Also the thermally generated holes and free electrons are represented by the positive and negative signs in the squares. Similarly, in the n-type crystal the donor atoms with their fixed positive charges are represented by the circles, the free electrons are represented by the negative signs and the thermally generated carriers by the charges in the rectangles. Note that both pieces of crystal are electrically neutral.

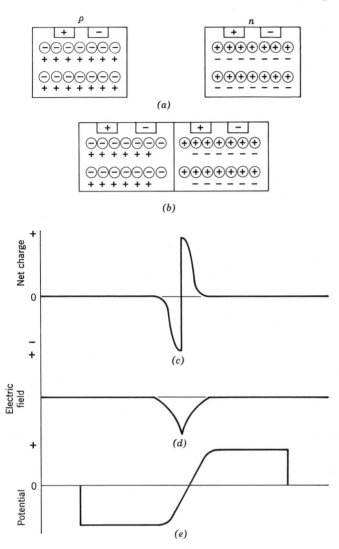

Fig. 3.1.   The *p-n* junction including charge and potential distribution.

When the *p*-type and *n*-type crystals are joined together, assuming their crystal structure to be continuous, a charge redistribution occurs as shown in Fig. 3.1*b*. Some of the free electrons from the *n*-material migrate across the junction and combine with the free holes in the *p*-material. Similarly, free holes from the *p*-material migrate across the junction and combine with free electrons in the *n*-material. Therefore, as a result of the redistribution, the *p*-material acquires a net negative charge and the *n*-material acquires a net positive charge (Fig. 3.1c). This electric charge produces an electric field (Fig. 3.1d) and a potential difference between the two types of material (Fig. 3.1e). As was previously mentioned, the diode manufacturing process does not actually employ the joining technique, but usually employs selective doping of a continuous crystal.

The electric field and potential difference, or *hill*, is in a direction that inhibits the migration of majority carriers across the junction. Also, the recombination of carriers near the junction leaves only fixed charges in this region that is known as the (carrier) depletion region. The junction therefore resembles a capacitor with a layer of positive charge in the n-material and a negative charge layer in the p-material (Fig. 3.2). However, there are some important differences between a diode and a capacitor. In the diode, the fixed charges are distributed throughout the depletion region, which acts as an insulator because of the depletion of free charge carriers. Also, the minority carriers that migrate or diffuse to the junction are swept across the junction by the electric field because the minority carriers are of opposite polarity to the majority carriers and are therefore aided instead of inhibited by the field.

The potential hill at the junction builds up because of the kinetic energy of the majority carriers. Therefore, the height of this hill is just sufficient to inhibit the migration of the highest energy carriers across the junction when there are no external connections to the diode. However, each minority carrier that is swept across the junction lowers the height of the potential hill a bit and allows a high energy majority carrier to surmount the hill in the opposite direction. Thus very small currents flow across the junction in opposite directions and these currents must be equal when no external voltage is applied to the diode as shown in Fig. 3.3b. The current due to the majority carriers is known as either *majority current* or *injection current*, I_i and the current due to minority carriers is called either *minority current* or *saturation current*, I_s. Potential differences, known as *contact potential*, also exist between the semiconductor and connecting leads as indicated in

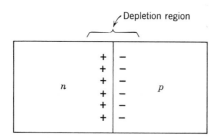

Fig. 3.2.   Net charge distribution in a junction diode.

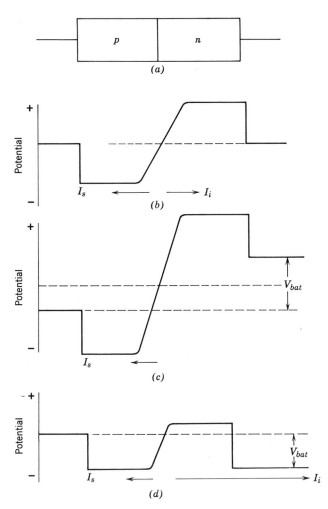

Fig. 3.3. (*a*) Junction diode, (*b*) potential distribution with no external voltage, (*c*) potential distribution with reverse bias, and (*d*) potential distribution with forward bias.

the potential diagram. Thus the connecting leads are at the same potential when no external voltage is applied.

When an external voltage, such as a battery is connected with the negative terminal to the *p*-side and the positive terminal to the *n*-side of the diode, the battery voltage adds to the potential barrier across the junction, as shown in Fig. 3.3*c*. Therefore, the majority carrier, or injection current is reduced or stopped, but the minority carriers that diffuse to the junction are still swept across by the field. Thus the saturation current remains essentially constant. Since the minority carriers are few in number, the saturation current is very small, microamperes or less, and the diode resistance is very high. This battery polarity is known as *reverse bias*.

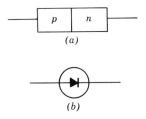

(a)

(b)

Fig. 3.4.   The diode symbol. (a) Actual diode and (b) symbol of the diode.

If the battery polarity is reversed, the height of the potential hill is reduced, as shown in Fig. 3.3d. Then a large number of majority carriers have enough kinetic energy to surmount or to be *injected across* the junction and the majority current is comparatively large. The minority current still flows but is very small compared with the majority current if the battery voltage is as much as 0.1 V. This polarity is known as *forward* bias because large currents may flow even though the applied voltage is small. Thus, the diode is forward biased when the positive terminal of the battery is connected to the p-type material and the negative terminal is connected to the n-type material and reverse bias occurs with the opposite polarity.

The diode symbol is shown in Fig. 3.4. The arrow points in the direction of positive current flow when forward bias is applied to the diode.

3.2   THE DIODE EQUATION

A relationship between the voltage across a diode and the current through it is quite easily derived. The exponential relationship between carrier density and potential energy, known as the Boltzmann distribution, will be used (Eq. 2.9). If n_p is the number or density of electrons in the n-material that has enough energy to surmount the potential hill, then n_p is related to the majority carrier density N_d in the n-material by the following equation

$$n_p = N_d e^{-qV_h/kT} \tag{3.1}$$

where V_h is the difference of potential, or height of the potential hill, between the two materials. Similarly, the density of holes p_n in the p-material that has enough energy to cross into the n-material is

$$p_n = N_a e^{-qV_h/kT} \tag{3.2}$$

But the injection current I_i is proportional to the total number of majority carriers that have enough energy to pass through the potential barrier V_h. Thus, letting K be the proportionality constant,

$$I_i = K(n_p + p_n) = K(N_d + N_a)e^{-qV_h/kT} \tag{3.3}$$

But $K(N_d + N_a)$ is another constant K', since the doping concentrations are constant for a given diode. Then,

$$I_i = K'e^{-qV_h/kT} \tag{3.4}$$

The constant K' can be evaluated in terms of the saturation current I_s, since the injection current is of equal magnitude but opposite direction to I_s when the externally applied voltage is zero. In this zero bias condition, we shall let V_h be V_{h0}. Then,

$$I_i = -I_s = K'e^{-qV_{h0}/kT} \tag{3.5}$$

and

$$K' = -I_s e^{qV_{h0}/kT}$$

Substituting this value for K' into Eq. 3.4,

$$I_i = -I_s e^{q(V_{h0} - V_h)/kT} \tag{3.6}$$

But $V_{h0} - V_h = V$, the bias voltage or change in barrier height resulting from the external voltage source. Thus,

$$I_i = -I_s e^{qV/kT} \tag{3.7}$$

The diode current I is the algebraic sum of the injection current I_i and the saturation current I_s. Therefore

$$I = I_i + I_s = -I_s(e^{qV/kT} - 1) \tag{3.8}$$

At normal temperatures, around 300°K, the factor $q/kT = 38.7$. This number is usually rounded to 40, and used in Eq. 3.8 to obtain a simple equation that is adequately accurate for normal temperatures. Making this approximation,

$$I = -I_s(e^{40V} - 1) \tag{3.9}$$

The theoretical relationship, Eq. 3.9, between diode current and applied voltage is plotted in Fig. 3.5, assuming normal temperature, for a value of

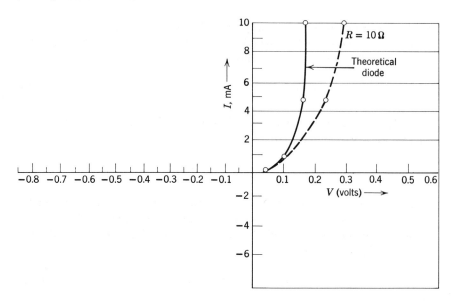

Fig. 3.5. Theoretical and actual current as a function of diode voltage for typical germanium diodes.

$I_s = -10$ microamperes (μA), which is typical for a modest size germanium diode. It is negative because the minority current flows in the opposite direction to the arrow in the symbol.

PROBLEM 3.1 A typical value of saturation current for a modest size silicon diode is $I_s = -10^{-8}$ A. Sketch a current versus voltage curve for this silicon diode by calculating values of diode current for forward-bias voltages of 0.1 V, 0.2 V, 0.4 V, and 0.6 V, respectively, and for reverse-bias voltages of -0.1 V and -1.0 V.

The voltage-current characteristics of an actual diode differ somewhat from the theoretical curves of Fig. 3.5 for two reasons.

1. The forward-bias voltage V appearing across the junction is never as large as the voltage applied to the diode terminals because of the IR drop in the n-doped and p-doped crystals. If this voltage is added to the junction-bias voltage, a correct value may be obtained for the terminal voltage. For example, let us assume the total ohmic resistance of the germanium diode of Fig. 3.5 is 10 Ω. The voltage versus current, or characteristic, curve is then altered as shown by the dashed line in Fig. 3.5.

2. The reverse current may exceed the minority current because of surface leakage and carrier multiplication in the depletion region, which will be discussed later.

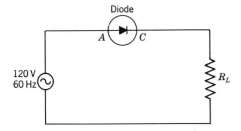

Fig. 3.6. An elementary half-wave rectifier.

3.3 DIODE RECTIFIER CIRCUITS

One common use of diodes is to change alternating currents to unidirectional currents. This process is known as rectification and can be accomplished very simply by a diode as shown in Fig. 3.6. The diode symbol, as was previously mentioned, indicates that current can flow essentially in only one direction. The arrow head points in the direction of positive current flow. Therefore, the diode terminal marked A (for anode) is connected to the p-type material. The terminal labeled C is the cathode and is usually identified by either a color dot or color coding bands on that end of the diode case.

Let us assume that the diode in Fig. 3.6 is the germanium diode of Fig. 3.5 and the maximum current through the diode is 10 mA. This diode current, which is also the current through R_L, is sketched as a function of time on the coordinates of Fig. 3.7.

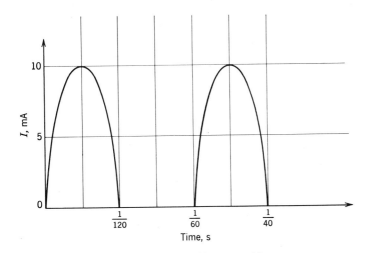

Fig. 3.7. Current waveform for the half-wave rectifier.

PROBLEM 3.2   Determine the maximum, or peak voltage across the load resistor R_L in the circuit of Fig. 3.6.   *Answer:* 169.4 V.

PROBLEM 3.3   What is the maximum reverse, or inverse, voltage across the diode?   *Answer:* 169.5 V.

The circuit of Fig. 3.6 is called a half-wave rectifier because only half of the input voltage wave appears across the load resistor R_L.

A full-wave rectifier circuit is shown in Fig. 3.8. This rectifier is known as a *bridge* circuit because of its similarity to a wheatstone bridge. Let us first assume the upper side of the 120 V source is positive. The current then flows through diode D_1, *down* through the load resistor, and through diode D_3 to the lower side of the voltage source. Next, assume the lower side of the 120 V source is positive. The current then flows through diode D_2, *down* through the load resistor and through diode D_4 to the upper side of the source. The full-wave load current is sketched as a function of time in Fig. 3.9 assuming the peak current is 10 mA. The average current in the load is 6.36 mA, which is twice as great as the average current in the half-wave circuit.

Rectifier circuits frequently include a capacitor in parallel with the load to reduce the voltage variations, or ripple, and provide a nearly smooth dc output voltage. This type of rectifier circuit is shown in Fig. 3.10.

In this circuit the capacitor charges to the peak value of the input voltage V_{max} minus the forward drop across the conducting diodes. But this forward drop is usually small in comparison with the input voltage v_S and can then be neglected. As the input voltage v_S decreases from its maximum value, the voltage across the capacitor reverse biases the rectifier diodes

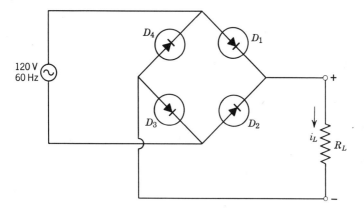

Fig. 3.8.   A full-wave bridge rectifier.

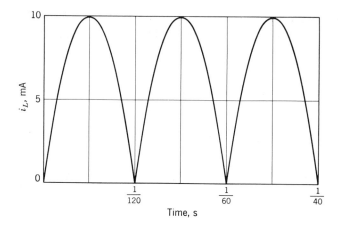

Fig. 3.9.   Load current in a full-wave rectifier circuit.

and thus the capacitor can discharge only through the load resistance R_L. Therefore the capacitor provides the load current while the diodes are not conducting. However, the load voltage v_L decreases as the capacitor discharges. The rate of this decrease depends on the time constant $R_L C$. If the discharge were to continue at the initial rate, the discharge current would cease and the load voltage would reduce to zero in time $t = R_L C$. However, the capacitor becomes charged again as the input voltage approaches its maximum value one-half period later at time $T/2$. The variation of the load, or output, voltage from v_{max} to v_{Lmin} is called the peak-to-peak ripple voltage. This variation will be small if the time constant $R_L C$ is long in comparison with the period of one-half cycle, $T/2$. Then the load voltage is maintained at almost the peak value V_{max} of the input voltage and the capacitor charges during brief intervals near the peaks of the input wave, so that the point v_L' becomes very nearly v_{Lmin} (Fig. 3.10).

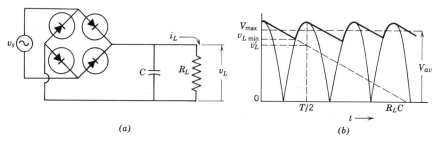

(a)

(b)

Fig. 3.10.   (a) A full-wave rectifier circuit, which includes a filter capacitor and (b) voltage waveform across the load.

When the approximation $v'_L = v_{Lmin}$ is made, the peak-to-peak ripple voltage can be easily determined in terms of the peak voltage V_{max}. From the similar triangles in Fig. 3.10b, we can see that

$$\frac{V_{max} - v'_L}{V_{max}} = \frac{T/2}{R_L C} \simeq \frac{\text{Peak-to-peak ripple}}{V_{max}} \tag{3.10}$$

and

$$\text{Peak-to-peak ripple} = \frac{TV_{max}}{2R_L C} \tag{3.11}$$

Let us define a peak ripple voltage v_{Rmax} which is one-half of the peak-to-peak ripple voltage. Then, using Eq. 3.11,

$$v_{Rmax} = \frac{\text{Peak-to-peak ripple}}{2} = \frac{TV_{max}}{4R_L C} \tag{3.12}$$

As a design procedure, it is preferable to specify the ripple ratio $v_{Rmax}/V_{max} = \eta_r$, and to calculate the capacitance required to produce this ripple ratio. Using Eq. 3.12,

$$C = \frac{TV_{max}}{4R_L v_{Rmax}} = \frac{T}{4R_L \eta_r} \tag{3.13}$$

The average value, or dc component, of load voltage is $V_{max} - v_{Rmax}$, as seen in Fig. 3.10b.

PROBLEM 3.4   A rectifier and filter are needed to provide 20 V dc to operate a transistor amplifier. If the amplifier requires or draws 100 mA at this voltage and the tolerable ripple ratio η_r is 0.05, determine the required filter capacitance and the root-mean-square (rms) value of input voltage V_s required. Assume 0.5 V forward drop across each diode.

PROBLEM 3.5   Can a half-wave rectifier be used for the rectifier-filter or power supply application in Problem 3.4? If so, what value of capacitance and what rms input voltage V_s is required?

Another common rectifier application is the *detector*, which recovers the modulating signal from an amplitude-modulated wave in a radio receiver. The main difference between a *diode* detector and the rectifier-filter circuit previously considered is the tuned, or resonant, circuit which is usually used in the detector circuit to accept or select the desired radio station and to discriminate against the undesired stations. A typical diode-detector

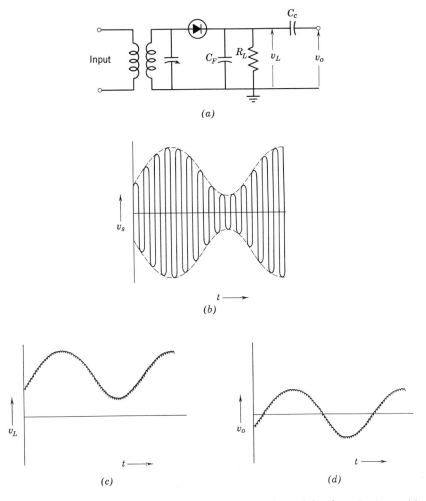

Fig. 3.11. (a) A diode detector circuit, (b) the amplitude modulated carrier input, (c) the voltage across the detector load resistor, and (d) the output voltage after the dc load voltage has been blocked by capacitor C_c.

circuit is shown in Fig. 3.11. The rectifier and filter operate precisely as described above. However, the radio frequency filter, or bypass, capacitor C_F must not be too large or it will eliminate or seriously reduce the amplitude of the relatively slow modulating frequency which is the desired output. Thus the time constant $C_F R_L$ must be large compared with the period of one cycle of the radio frequency, but it must be small compared with one cycle of the modulating frequency. Another way of stating this relationship is that the reactance of the capacitor C_F must be small in comparison with R_L at the radio frequency, but large in comparison with R_L at the highest modulating frequency. The coupling capacitor C_c is usually used to block the dc component across the load resistor and permit only the ac component to pass on to the headset, amplifier or whatever follows the detector. Therefore, C_c should have small reactance at the frequency of the modulation. This diode-detector circuit can be used as a complete radio receiver if a large antenna is connected to its input. Unfortunately, this receiver will have poor sensitivity and selectivity.

PROBLEM 3.6   The input carrier frequency of the detector of Fig. 3.11 is 10^6 hertz (Hz) and the highest modulating frequency is 5 kilohertz (kHz). If $R_L = 10$ kilohm (kΩ), determine C_F so that $\eta_r = 0.01$. How does the time constant $R_L C_F$ compare with the period of the highest modulating frequency? How does the X_{CF} compare with R_L at 5 kHz?

3.4   JUNCTION CAPACITANCE

The p-n junction has previously been compared with a capacitor because of its stored charge and depletion region. In this section we shall investigate the magnitude of the junction capacitance and consider a voltage-controlled capacitor, which is an interesting application of this capacitance. Circuit limitations resulting from this capacitance will be included in Section 3.7.

An expanded view of the junction region is shown in Fig. 3.12a. As was previously noted, the stored charge in the depletion region results from the removal of free electrons from the n-material, leaving fixed positive donors and the filling of free holes in the p-type material, leaving fixed negative acceptors. Therefore, the charge density in the depletion region on the p-side of the junction is the acceptor atom density times the charge per atom or qN_a. This fixed negative charge extends an effective distance l_p from the junction into the p-material as seen in Fig. 3.12. Similarly, the charge density in the depletion region on the n-side of the junction is qN_d and extends a distance l_n into the n-material. The total charge on the p-side of the junction is the charge density times the volume of the depletion region on the side in question. But we shall assume the crystal to have unit cross sectional area so that the volume will be equal to the length l_p. Then

$$Q_p = -qN_a l_p \tag{3.14}$$

Similarly, the charge in the depletion region in the n-material is

$$Q_n = qN_d l_n \tag{3.15}$$

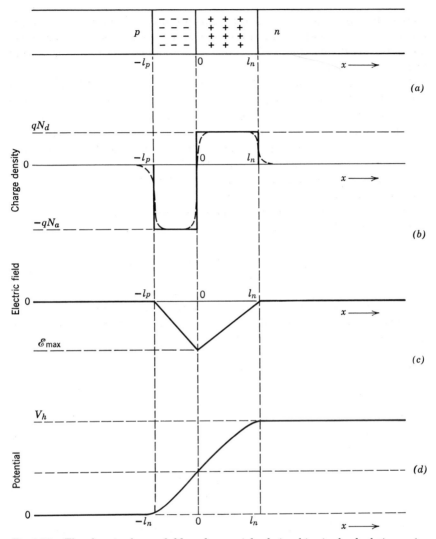

Fig. 3.12.   The electric charge, field, and potential relationships in the depletion region associated with a *p-n* junction. (*a*) semiconductor junction, (*b*) electric charge, (*c*) electric field, and (*d*) electric potential.

But the total charge must be equal to zero, or $Q_n = -Q_p$. Therefore, from Eqs. 3.14 and 3.15,

$$N_d l_n = N_a l_p \tag{3.16}$$

or

$$l_n = \frac{N_a}{N_d} l_p \tag{3.17}$$

Notice that the depletion region lengths are inversely proportional to the doping concentrations. The excess charge regions do not terminate abruptly, as shown, but little error is introduced by assuming abrupt termination at l_p and l_n.

The electric-field intensity (Fig. 3.12c) can be determined by the use of Gauss' law. As the junction region is viewed from the left-hand side of l_p, no net charge is seen since $Q_n = -Q_p$. Therefore there is no electric field to the left of l_p or, by the same reasoning, to the right of l_n. However, as one walks past l_p toward the junction, the net charge to the left increases linearly with distance from l_p. The total electric flux therefore increases linearly and the flux density is equal to the total flux because of the unit cross-sectional area and because the flux is all in the $-x$ direction, since it originates on the positive charges and terminates on the negative charges. The field intensity is maximum at the junction and is

$$\mathcal{E}_{max} = \frac{-Q_p}{\epsilon} = \frac{qN_a l_p}{\epsilon} = \frac{-qN_d l_n}{\epsilon} \tag{3.18}$$

Potential difference can be obtained by integrating or summing the electric field over the range of interest. Therefore, using the p-material to the left of l_p as a reference

$$V_h = -\int_{l_p}^{l_n} \mathcal{E} \, dx \tag{3.19}$$

But the integral is the area under the $\mathcal{E}$ versus x curve, Fig. 3.12c. Since this area is a triangle, its area is one half times the product of the base $(l_p + l_n)$ and the altitude $\mathcal{E}_{max}$. Therefore

$$V_h = \frac{\mathcal{E}_{max}(l_p + l_n)}{2} \tag{3.20}$$

But from Eq. 3.17, $l_n = N_a l_p / N_d$. Also using the value of $\mathcal{E}_{max}$ from Eq. 3.18,

$$V_h = \frac{qN_a l_p^2 (1 + N_a/N_d)}{2\epsilon} \tag{3.21}$$

The depletion region width l_p can be found in terms of the junction voltage V_h by solving for l_p in Eq. 3.21.

$$l_p = \left[\frac{2\epsilon V_h}{qN_a(1 + N_a/N_d)}\right]^{1/2} \tag{3.22}$$

Also l_n can be easily found from l_p by using Eq. 3.17. Observe, from Eq. 3.22, that the depletion region width is proportional to the square root of the potential difference V_h across the junction.

PROBLEM 3.7   If a silicon diode has $N_a = 10^{22}$ acceptors/m³ and $N_d = 10^{21}$ donors/m³ and the potential difference across the junction is 1.0 V, determine the depletion region widths l_p and l_n. From Table 2.3 we note that ϵ for silicon is $12\,\epsilon_v = 12 \times 8.855 \times 10^{-12} = 1.062 \times 10^{-10}$.

Answer: $l_p = 1.1 \times 10^{-7}$ m, $l_n = 1.1 \times 10^{-6}$ m.

The initial goal was to determine the junction capacitance as a function of the diode parameters. Since capacitance is the stored charge per volt,

$$C = \frac{Q}{V} = \frac{Q_p}{V_h} \tag{3.23}$$

Substituting the expression for Q_p in Eq. 3.14 and the expression for V_h in Eq. 3.21 into Eq. 3.23, but disregarding the negative sign of Q_p,

$$C = \frac{qN_a l_p(2\epsilon)}{qN_a l_p{}^2(1 + N_a/N_d)} = \frac{2\epsilon}{l_p(1 + N_a/N_d)} \tag{3.24}$$

But we would like to have an expression for C in terms of V_h instead of l_p; substituting the value of l_p in Eq. 3.22 into Eq. 3.24,

$$C = \left[\frac{2\epsilon qN_a}{(1 + N_a/N_d)V_h}\right]^{1/2} \tag{3.25}$$

Usually a dynamic, or ac, value of capacitance $C = dQ/dV$ or $\Delta Q/\Delta V$ is used, which is one-half the static or dc value given by Eq. 3.25. In either case, all the terms affecting C are constant except V_h. Therefore, Eq. 3.25 may be written

$$C = \frac{K}{V_h{}^{1/2}} = KV_h{}^{-1/2} \tag{3.26}$$

where K is a constant.

PROBLEM 3.8   Determine the static capacitance per square meter for the junction described in Problem 3.7 if the potential difference across the junction is 1 V.   *Answer:* $C = 1.76 \times 10^{-4}$ F/m².

PROBLEM 3.9   Determine the static and dynamic capacitance of a diode that has the junction characteristics and voltage of Problems 3.7 and 3.8 and has a rectangular junction 1 mm on each side.

Answer: C (static) $= 1.76 \times 10^{-10}$ F, C (dynamic) $= 8.80 \times 10^{-11}$ F.

PROBLEM 3.10 Determine the dynamic capacitance of the diode of Prob. 3.6 if $V_h = 4$ V.   *Answer:* 4.4×10^{-11} F.

The dynamic capacitance of the diode described in the preceding problems is sketched as a function of the potential difference across the junction in Fig. 3.13.

You may recall that $V_h = V_{h0} - V$, where V_{h0} is the zero-bias potential hill and V is the bias resulting from the external bias source. A typical value of V_{h0} for a silicon diode is 0.6 V. Thus values of junction capacitance given in Fig. 3.13 for values of V_h less than about 0.6 V occur when forward bias is applied to the diode, and values of C_j corresponding to V_h greater than about 0.6 V occur with reverse bias applied to the diode.

The voltage-variable capacitance of a junction may be utilized in tuning a radio receiver or controlling the frequency of an oscillator. Figure 3.14 shows how the diode could be used to control the resonant frequency of a tuned circuit. Diodes designed especially for this application are known as *varactor diodes*. In order to avoid loading the tuned circuit (seriously decreasing its Q) the diode must be reverse biased and R must be large. Also, the ac signal voltage across the diode should be small in comparison

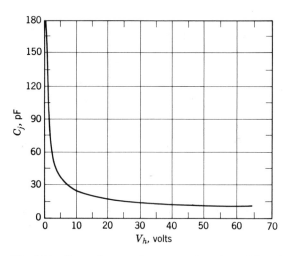

Fig. 3.13. Dynamic capacitance as a function of junction voltage in a typical abrupt-junction diode.

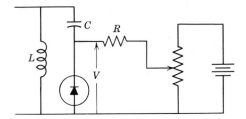

Fig. 3.14.   A circuit that can be tuned by a variable voltage V.

with the control voltage V so that the diode capacitance appears essentially constant, at the dynamic value, to the ac signal.

3.5   JUNCTION BREAKDOWN AND SPECIAL DIODES

When a sufficiently high reverse bias is applied to a junction diode a phenomenon known as junction breakdown occurs, as shown in Fig. 3.15. In this figure the diode is assumed to be a silicon diode with a breakdown voltage of 100 V. Breakdown occurs because of the very high field intensities in the depletion region. For example, the silicon diode of Prob. 3.7 had $N_a = 10^{22}$ acceptors/m^3, $N_d = 10^{21}$ donors/m^3, and a depletion region width $l_p = 1.1 \times 10^{-7}$ m when $V_h = 1$ V. This diode, therefore, has $\mathscr{E}_{max} = qN_a l_p/\epsilon = 1.66 \times 10^6$ V/m. As the reverse bias is increased the field intensity increases until breakdown occurs. A more heavily doped diode has a thinner depletion region and therefore breaks down at a lower

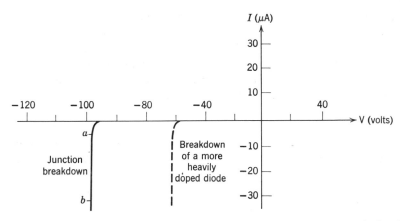

Fig. 3.15.   Junction breakdown and the effect of doping concentration on the breakdown voltage.

voltage as indicated in Fig. 3.15. Thus, the breakdown voltage may be controlled by the doping concentrations. One theory attributes the rather abrupt current increase to the high potential gradient (or electric field), which exists at the junction. According to this theory, the high electric field is able to disrupt the covalent bonds and therefore greatly increase the minority carriers. This effect is known as zener breakdown, and would require field intensities so high that about 1 V potential difference (equal to the gap energy) would appear across the diameter of an atom. Since an atomic diameter is about 10^{-8} m, this would require a field intensity $\mathscr{E} = 10^8$ V/m.

Another accepted theory for the voltage breakdown is the *avalanche* breakdown. According to the avalanche theory, a few carriers are generated in the intrinsic semiconductor material owing to thermal action, as previously discussed. These carriers are accelerated by the high electric field near the junction until high velocities are acquired. A carrier with sufficient energy can produce an electron-hole pair when this carrier collides with a neutral atom. The new carriers so produced are free to be accelerated and in turn to produce additional carriers. The origin of the term avalanche can thus be seen. Since the electrons are much more mobile than the holes, most of the carriers are produced by electron collisions. The electrons have many random collisions as they travel through the semiconductor. In order for the avalanche effect to manifest itself, the electrons must obtain sufficient energy in traveling one mean free path (distance between collisions) to produce ionization of the atoms in the semiconductor. Hence, the electrons must have a kinetic energy equal to or greater than the gap energy of the semiconductor for an avalanche to be produced.

Many diodes are designed and constructed for operation on the avalanche portion of the characteristic curve. These diodes are known as *zener diodes* or *reference diodes*. They operate in a region (from *a* to *b* on the curve of Fig. 3.15) where the current is essentially independent of voltage.

References diodes are available with breakdown voltages ranging from about 3 V to well over 100 V. The avalanche appears to be the primary breakdown mechanism in diodes with reference voltages above about 7 V because the breakdown voltages of these diodes increase with temperature. The reduced mean free path at elevated temperatures would account for this positive temperature coefficient. In contrast, diodes with breakdown voltages less than about 6 V have a negative temperature coefficient, which indicates that the zener breakdown mechanism is predominant in this range. The increased kinetic energy of the valence electrons would aid the high field in producing carriers and thus cause a negative temperature coefficient. Diodes that break down at about 6 to 7 V have essentially zero-temperature coefficient.

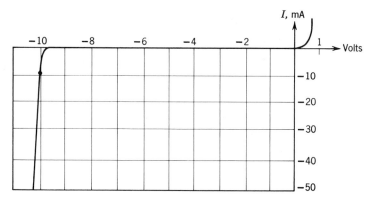

Fig. 3.16.   Characteristics of a typical 500-mW, 10-V reference diode.

The characteristic curve of a typical 10 V (at 10 mA) reference diode is given in Fig. 3.16. This diode is used in the voltage regulating circuit shown in Fig. 3.17. Note the modification of the diode symbol for identification of the reference voltage function. An example will help illustrate the regulating action of a reference diode.

EXAMPLE 3.1   A diode whose characteristics are given in Fig. 3.16 is connected as shown in Fig. 3.17. Let us determine the value of R_S if the minimum value of the input voltage V_{in} is 18 V. The diode is capable of dissipating 500 milliwatts (mW) at normal ambient temperatures.

The voltage across the 1-kΩ load is approximately 10 V. Therefore, the load current is $i_L \simeq 10\,\text{V}/1\,\text{k}\Omega = 10$ mA. To maintain the load voltage essentially constant, the diode current should not drop below approximately 3 mA as seen in Fig. 3.17. With $V_{in} = 18$ V, the value of R_S should be $R_S = (18-10)V/(10+3)$ mA $= 615\ \Omega$.

The maximum permissible current through the diode is approximately 500 mW/10 V $= 50$ mA. As the input voltage increases, the additional current through R_S flows through the diode. Thus the maximum per-

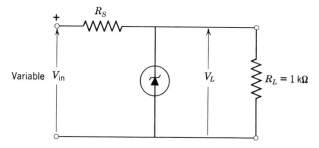

Fig. 3.17.   A reference diode in a voltage regulating circuit.

missible value of V_{in} is $10\,V + (50 + 10) \times 10^{-3} \times 615\,\Omega = 46.9\,V$. The input voltage can increase from 18 to 46.9 V and the voltage across the 1-kΩ load will only change from 9.9 V to 10.2 V.

PROBLEM 3.11   Repeat 3.1 with R_L changed to 500 Ω.

Answer: $R_S = 347\,\Omega$, max $V_{in} = 34.3$ V.

PROBLEM 3.12   A diode whose characteristics are given in Fig. 3.16 is connected as shown in Fig. 3.17. The input voltage V_{in} is maintained at 20 V but R_L is permitted to change from 1 kΩ to a lower value. Determine the value of R_S for this problem and determine the minimum value R_L can have.   *Answer:* $R_S = 167\,\Omega$, min $R_L = 176\,\Omega$.

A diode may be so heavily doped that the depletion region experiences zener breakdown with only V_{h0} (no external bias) across the junction. The characteristics of this diode are shown in Fig. 3.18. This diode is known as a *backward* diode because it conducts more readily in the reverse-bias direction than in the forward-bias direction. For example, the diode of Fig. 3.18 would be an effective rectifier of small ac voltages providing the peak amplitude does not exceed approximately 0.5 V.

If a diode is more heavily doped than a backward diode, the depletion region is in breakdown even with small amounts of forward bias applied. Thus, as the forward bias is increased, the current rises rapidly until the potential hill is reduced to the breakdown voltage, as shown at point B in Fig. 3.19. The current then falls rapidly until current again begins to rise due to normal majority carrier injection through the depletion region. This diode, known as a tunnel diode, or Esaki diode after its inventor, is useful because of its negative resistance or conductance that occurs between

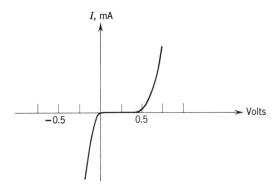

Fig. 3.18.   Characteristics of a backward diode.

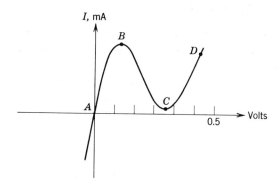

Fig. 3.19.   Characteristics of a tunnel, or Esaki, diode.

points B and C on the characteristic curve. This characteristic may be used in conjunction with a tuned circuit to produce very high-frequency oscillations.

As was previously mentioned, light (or photons) can give sufficient energy to a bound electron to break the covalent bond and produce a free electron and a hole. Conversely, an electron can fall from the conduction band into a hole and give up its energy in the form of a photon, or light. However, the momentum and energy relationships in silicon and germanium are such that the electron gives up its energy as heat when it returns from the conduction band to the valence band. On the other hand, the electron in a gallium arsenide crystal *does* produce a photon when it returns from the conduction band to fill a hole in the valence band. This event does not occur often enough to produce a useful light intensity in the intrinsic crystal. However, gallium arsenide is used to make light-emitting or *light* diodes. When forward bias is applied to these diodes large numbers of electrons are injected from the n-material to the p-material. Most of these electrons combine with the holes in the p-material. Since these holes are at the valence-band energy level, photons are given off as the electrons combine with these holes. The light intensity is proportional to the rate of recombination of electrons, and thus is proportional to the diode current. The GaAs diode emits wavelengths in the near infrared region. To obtain visible light, a mixture of GaAs and gallium phosphide must be used.

A photo diode is sort of the inverse of a light diode. Reverse bias is applied to the photo diode, and the saturation (reverse) current is controlled by the light intensity that shines on the diode and generates electron-hole pairs.

3.6 GRAPHICAL ANALYSIS

Section 3.3 included rectifier circuits which were easily solved for load currents and voltages because the supply voltage was so large in comparison with the forward drop across the diode that the latter could be neglected. Also the reverse current could be assumed to be zero. Sometimes the forward voltage across the diode cannot be neglected, but since the diode current is not linearly related to the diode voltage, an accurate circuit solution is difficult unless a graphical solution is used. For example, the rectifier circuit in Fig. 3.20a has such a low-voltage power source that the forward-diode voltage must be considered. Unfortunately, the diode voltage cannot be determined until the circuit current is known and vice versa. However, both of these parameters can be easily determined for a given input voltage if a straight line known as a load line is drawn on Fig. 3.20b, which includes the diode characteristic. In other words, if the load characteristic is also included on the graph, the intersection of the two characteristics represents a simultaneous solution. First, an equation for the load line is needed. We can see from Fig. 3.20a that

$$v_S = v_D + iR_L \tag{3.27}$$

Solving for i,

$$i = \frac{v_S}{R_L} - \frac{v_D}{R_L} \tag{3.28}$$

This equation is in the form

$$y = b + mx \tag{3.29}$$

where b is the y-axis intercept, or value of y when $x = 0$ and m is the slope of the straight line that this equation represents. Thus, in Eq. 3.28, the i-axis intercept is v_S/R_L and the slope is $-1/R_L$. Also note from Eq. 3.27 that when $i = 0$, $v_D = v_S$, which is the x- or v_D-axis intercept. The load line can be

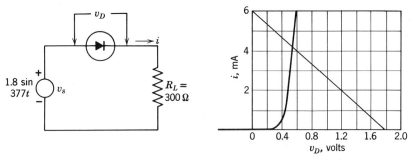

Fig. 3.20.   (a) A diode circuit that calls for graphical solution and (b) a method of solution.

drawn most easily by using the v_D- and i-axis intercepts which, for a value $v_S = 1.8$ V and $R_L = 300\ \Omega$ are $v_D = 1.8$ V when $i = 0$ and $i = 1.8/300 = 6$ mA when $v_D = 0$. The load line for this value of v_S has been drawn in Fig. 3.20b. Thus the circuit current at this value of v_S is 4.4 mA, the voltage across the diode is 0.54 V and the voltage across the load is $1.8{-}0.54 = 1.26$ V.

The waveform of the load current, or load voltage, can be determined as a function of time for any given input waveform. For example, the sinusoidal input voltage of Fig. 3.20a will be used. This input voltage is sketched in Fig. 3.21a. The circuit current corresponding to $v_S = 1.8$ V was found to be 4.4 mA by drawing the load line in Fig. 3.20, as was previously discussed. This value occurred at $t = 1/240$ s. The circuit current can be determined at any other time by noting the source voltage at that time and drawing a load line on the characteristic curve to determine the current value at that source voltage. This method was used to determine the current at each $1/960$ s time division as shown in Fig. 3.21b. A smooth curve was drawn through these points to provide the current wave form.

Another technique that is useful in determining the current-voltage characteristics of a diode circuit is the graphical addition of the diode characteristics and the load resistor characteristics as illustrated in Fig. 3.22. The curves in Fig. 3.22 represent the circuit components given in

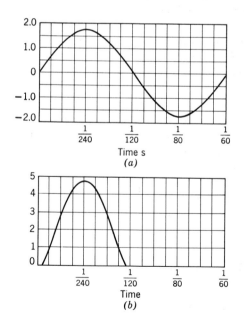

Fig. 3.21.   Source voltage and circuit current as function of time for the rectifier circuit of Fig. 3.20a.

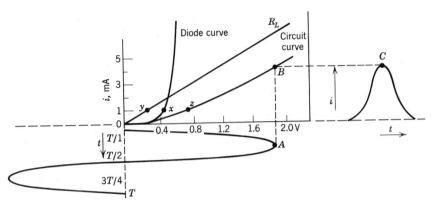

Fig. 3.22.   Circuit solution by the graphical addition of diode and resistor characteristics.

Fig. 3.20. The resultant curve (labeled circuit) is found by adding the voltage across the diode to the voltage across the load resistor at a given current. Thus, point x on the curve ($i = 1$ mA) is added to point y on the load resistance curve ($i = 1$ mA) to obtain point z on the circuit curve ($i = 1$ mA). In Fig. 3.22, the addition was made at 1 mA current increments, as shown. These points were then connected to form a smooth curve.

The source voltage is plotted along the voltage axis. To find the current that would flow for a given voltage, the following procedure is used. The voltage (point A) is projected vertically to the *circuit* curve (point B). The value of current at point B is projected horizontally to give the required current on the current curve (point C). Enough points are projected to complete the current plot.

PROBLEM 3.13   In Fig. 3.22, the source voltage $v_S = 1.5$ V and the load resistance is increased to 500 Ω. Draw a load line for these values on the diode characteristic given and determine the circuit current, the voltage across the diode and the voltage across the load.

$Answer:$ $i = 2.1$ mA, $v_D = 0.5$ V, $v_L = 1.0$ V

PROBLEM 3.14   Solve Problem 3.13 using the technique given in Fig. 3.22.

3.7   DIODE-EQUIVALENT CIRCUITS OR MODELS

The characteristics of a (nonlinear) diode may be approximated by a combination of linear elements and a switch, or ideal diode. This combination is known as an *equivalent circuit* or *model*. The characteris-

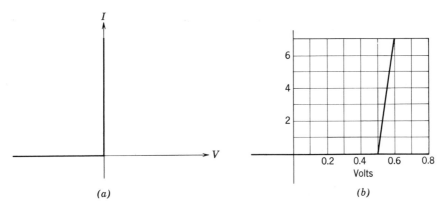

Fig. 3.23. (a) The characteristics of an ideal diode, and (b) approximation of actual diode characteristics (Fig. 3.20) by linear segments or elements.

tics of an ideal diode are shown in Fig. 3.23a. Observe that the ideal diode has zero voltage drop in the forward-bias direction and zero current flow in the reverse-bias direction. Also, the characteristics of an actual diode have been approximated by two straight-line segments in Fig. 3.23b. This approximation is represented by the equivalent circuit or model shown in Fig. 3.24. As seen in this equivalent circuit, no current flows through the diode until the forward-terminal voltage equals V_0 because the ideal diode is reverse biased otherwise. The ideal diode is enclosed in a rectangle to identify it.

PROBLEM 3.15   Determine the values of V_0 and r_d in the model of Fig. 3.24, using the approximate diode characteristics of Fig. 3.23b.

Answer: $V_0 = 0.5 \text{ V}, r_d = 14 \, \Omega.$

The equivalent circuit of a diode given in Fig. 3.24 is used in the rectifier circuit of Fig. 3.25. The current in this circuit can be readily determined as a function of time if v_S and R_L are known. In fact, the circuit can be simplified for specific ranges of v_S and R_L. For example, the diode resistance r_d can be neglected if R_L is large in comparison with r_d. The actual magnitude of R_L required for this approximation depends on the accuracy required by the user. For example, if the current calculations are to be accurate within one percent, r_d can be neglected in Fig. 3.25 only if R_L is

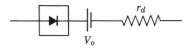

Fig. 3.24.   An equivalent circuit or model for a junction diode.

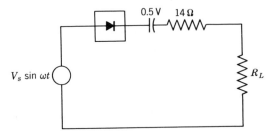

Fig. 3.25.   A rectifier circuit incorporating the diode equivalent circuit of Fig. 3.24.

$14 \times 100 = 1400 \ \Omega$ or greater. Similarly, the voltage V_o can be neglected if it is very small in comparison with v_S. Again, if a one percent error is permissible, V_o may be neglected if $v_S = 0.5 \times 100 = 50$ V or higher. Note that the errors introduced by neglecting both r_d and V_o are additive.

The value of dynamic resistance r_d can be determined analytically if the ohmic resistance of the doped semiconductor is neglected. As seen in Fig. 3.26, the slope of the diode characteristic curve is the dynamic conductance g_d, which is the reciprocal of r_d. Thus $g_d = \Delta I / \Delta V = dI/dV$, is the slope of the diode curve at a given point. Therefore, using calculus, the diode equation $I = -I_s(e^{qV/kT} - 1)$ can be differentiated with respect to V to give g_d. (If you are not acquainted with calculus, skip to Eq. 3.32.)

$$g_d = \frac{dI}{dV} = -\frac{q}{kT} I_s e^{qV/kT} \tag{3.30}$$

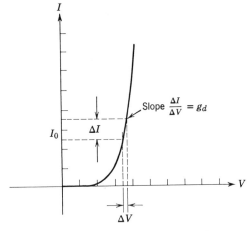

Fig. 3.26.   Dynamic diode conductance determined from the slope of the characteristic curve.

But, using the diode equation, the exponential term can be expressed in terms of the diode currents.

$$I - I_s = -I_s e^{qV/kT} \tag{3.31}$$

Making this suggested substitution into Eq. 3.30,

$$g_d = \frac{q}{kT}(I - I_s) \tag{3.32}$$

When forward bias is applied to the diode, I is much larger than I_s, which can then be neglected. Also, at normal temperatures q/kT is approximately 40. Making these simplifications in Eq. 3.32,

$$g_d = 40I \tag{3.33}$$

Therefore

$$r_d = \frac{0.025}{I} = \frac{25}{I(\text{mA})} \, \Omega \tag{3.34}$$

Thus, if I_o is 2 mA in Fig. 3.26, the value of g_d at the point indicated is $40 \times 2 \times 10^{-3} = 0.08$ mho and the value of r_d is $25/2 = 12.5 \, \Omega$.

An equivalent circuit can also be used to represent a reference diode. For example, the voltage regulating circuit of Fig. 3.16 is redrawn in Fig. 3.27 with the reference diode replaced with an equivalent circuit. This diode is assumed to have the characteristics given in Fig. 3.15. Thus the value of V_z is 9.8 V and the value of r_z is $\Delta V/\Delta I = 0.2$ V/0.04 A $= 5 \, \Omega$ assuming the minimum current through the diode exceeds 3 mA.

The saturation current and leakage resistance have been neglected in the preceding equivalent circuits. In most circuits, particularly those employing silicon diodes, these reverse-bias currents may be neglected with very little error. But in high-resistance circuits the reverse-bias characteristics of a diode need to be considered. For example, the german-

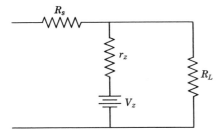

Fig. 3.27.   An equivalent circuit for a reference-diode regulator.

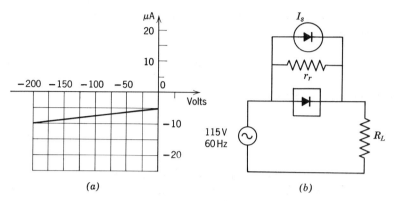

Fig. 3.28. (a) Reverse characteristics of a germanium diode and (b) a rectifier circuit incorporating an equivalent circuit for this diode.

ium diode with characteristics given in Fig. 3.28a is used in the rectifier circuit of Fig. 3.28b. The ideal diode is incorporated in the equivalent circuit. The value of I_s is 5 μA and the value of r_r is $\Delta V/\Delta I = 200/5 \times 10^{-6} = 4 \times 10^7 \, \Omega$. Therefore, the peak voltage across the load during the reverse-bias half cycle will not exceed one percent of the peak supply voltage if the value of load resistance R_L does not exceed $4 \times 10^5 \, \Omega$.

The preceding examples should lead one to the conclusion that the non-ideal reverse characteristics need to be included in an equivalent circuit only when the load resistance is comparatively high. On the other hand, the forward resistance r_d needs to be included only when the load resistance is comparatively low, and the offset voltage V_o needs to be included only when the supply voltage is a few volts or less. Therefore an equivalent circuit need not include all the nonideal diode characteristics. In fact, in many circuit applications, high accuracy may be obtained in circuit calculations if the diode is assumed to be ideal.

The junction capacitance has been neglected in the preceding equivalent circuits. When the frequencies involved are low, such as the 60 Hz heretofore used, the reactance of the junction capacitance is so high that it can be neglected. However this capacitance needs to be considered when frequencies in the kHz and MHz (megahertz) ranges are employed. An additional type of capacitance, known as diffusion capacitance, will be considered in Section 3.8.

Figure 3.29a shows a diode-equivalent circuit that includes the junction capacitance C_j. The effects of this capacitor on the circuit current of a rectifier can be seen from the simplified equivalent-diode circuit of Fig. 3.29b. The capacitor has essentially no effect during the forward-bias half cycle because it is essentially shorted out by the low resistance of the for-

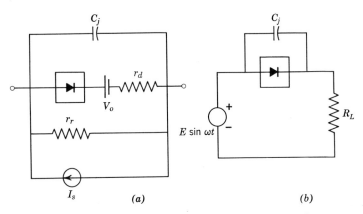

Fig. 3.29. (a) A fairly complete diode equivalent circuit including junction capacitance C_j and (b) a rectifier circuit that incorporates a simplified high-frequency equivalent diode circuit.

ward-biased junction. However, during the reverse-bias half cycle, the capacitor permits current to flow through the load resistance. The effectiveness of this reverse current in degrading the rectifier depends, of course, on the ratio of reverse current to forward current. This ratio will be low, as it should be, if the time constant $R_L C_j$ is short in comparison with the period $1/f$ of the input signal. Since C_j is a function of the diode voltage, an average value of C_j must be assumed for the easy determination of acceptable input frequencies. For example, an average value of $C_j = 40$ pF for the diode of Section 3.4 might be suitable. Then if $R_L = 10$ kΩ and the circuit time constant is to be no greater than ten percent of the input signal period, the maximum permissible input frequency is $f = 0.1/(4 \times 10^{-7}) = 250$ kHz. Special diodes with very small junction capacitances are required for efficient rectification or detection of signals in the megahertz region, or radio frequencies. These diodes are commonly known as *fast diodes.*

PROBLEM 3.16 A diode is needed to detect a 1.0-MHz signal from a radio station. If the load resistance is 5 kΩ, what is the maximum permissible diode capacitance if the circuit time constant should not exceed ten percent of the period of the input frequency?

3.8 DIFFUSION CURRENT AND DIFFUSION CAPACITANCE

In the preceding discussion no attention was given to the majority carriers after they were injected across the *p-n* junction in a diode. The

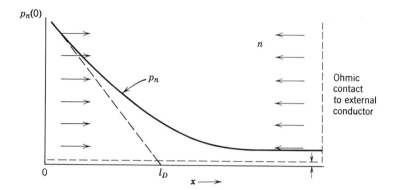

Fig. 3.30.   Density of holes in the n-material, p_n, as a function of distance x from the depletion region.

assumption was made that they somehow continued their journey through the doped semiconductor and caused the circuit current. A closer view will now be taken of these charge carriers for improved understanding of the diode and also the transistor, which will be discussed in Chapter 4.

Figure 3.30 shows the junction and the n-doped side of the diode into which holes are being injected. After being injected across the depletion region, the holes first diffuse away from the junction because of their velocity or kinetic energy. This diffusion of holes into the region causes two effects.

1. A net positive charge is produced in the conducting n-doped region. This charge produces an electric field which causes a flow of electrons in the p-material and terminal conductor in such a direction as to neutralize the charge. This is a capacitive effect and causes a *diffusion capacitance* C_D in the diode. As soon as the injected holes reach their steady state distribution in the n-material the neutralizing charge remains stationary and the current resulting from this charge ceases.

2. As the holes diffuse away from the depletion region some of them combine with electrons in the n-region. This reduces the population of holes and causes p_n to decrease as x increases as shown in Fig. 3.30. If p_n were to continue to decrease at the initial rate $\Delta p_n/\Delta x$ at $x = 0$, the hole density resulting from injected holes would reduce to zero at $x = l_D$ (the diffusion length). But the recombination rate is proportional to the hole density and therefore p_n decreases less rapidly as x increases. Actually, p_n decreases exponentially as indicated. The recombination of holes and electrons would cause a continual buildup of electric field in the n-material except for the fact that an electric field causes electrons to flow from the terminal conductor in such a direction as to neutralize the field. Thus, the

current at, or near, the *p-n* junction results almost entirely from the diffusion of holes into the *n*-material. This current is known as *diffusion current.*
But as the holes combine with electrons the diffusion current decreases
and the current due to electron flow in an electric field (*drift current*)
increases. The sum of these two currents is continuous or constant throughout the diode and is equal to the majority current due to holes. A similar
situation exists in the *p*-material, except, that the charge polarities are
reversed.

We shall now take a closer look at the diffusion current, including its
cause and magnitude. The current flows as a result of a net migration of
carriers, which in turn results from a nonuniform distribution of carriers,
as illustrated in Fig. 3.31. In this figure the holes with density $p(x)$ are
moving in random directions as a result of their kinetic energy. The two
slices of material on either side of x_0 are just thin enough so that all the
holes in the slice can pass out of the slice without having a collision in a
given time t. But the number of holes that will pass from the left to right
across the imaginary plane at $x = 0$, in time t, is proportional to the number
of holes in the left hand slice N_L. If the semiconductor is assumed to have
unit cross sectional area, the volume of the left hand slice is l, and since the
average density in that slice is $p_0 + \Delta p/2$, the number of holes in that slice
is

$$N_L = (p_0 + \Delta p/2)l \qquad (3.35)$$

Also, the number of holes that pass the plane at $x = 0$ from right to left in

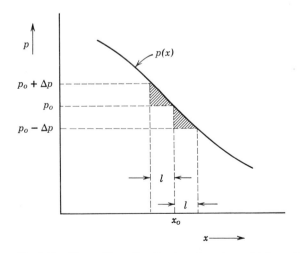

Fig. 3.31.   Nonuniform distribution of carriers, which results in a diffusion current.

time t is proportional to the number of holes in the slice on the right, which is

$$N_R = (p_0 - \Delta p/2)l \qquad (3.36)$$

The net flow of carriers across the plane at x_0 is

$$N = N_L - N_R = (\Delta p)l \qquad (3.37)$$

But $\Delta p/l$ is the slope of the $p(x)$ curve $= \Delta p/\Delta x$ at x_0. Then $\Delta p = l\Delta p/\Delta x$ and

$$N = l^2 \Delta p/\Delta x \qquad (3.38)$$

The number given in Eq. 3.38 is the number of holes that cross the plane at x_0 in time t. Then the net flow of carriers per second is

$$N/t = \frac{l^2}{t}\frac{\Delta p}{\Delta x} = l\bar{v}\frac{\Delta p}{\Delta x} \qquad (3.39)$$

where l is known as the mean free path and $\bar{v}$ is the average carrier velocity. This product $(l\bar{v})$ is known as the *diffusion constant D_p*. Then

$$N/t = D_p\frac{\Delta p}{\Delta x} \qquad (3.40)$$

But the rate of flow of holes times the charge per hole is the rate of flow of charge which is electric current. Therefore, if we multiply both sides of Eq. 3.40 by q, we obtain the current per unit cross section, or current density, due to hole diffusion.

$$J_p = -qD_p\frac{\Delta p}{\Delta x} \qquad (3.41)$$

The diffusion current can be obtained for any given cross-sectional area A if Eq. 3.41 is multiplied through by A.

$$i_p = -qAD_p\frac{\Delta p}{\Delta x} \qquad (3.42)$$

The negative sign appears because current flows in the positive x direction when $\Delta p/\Delta x$ is negative. Similarly the diffusion current for electrons is

$$i_n = qAD_n\frac{\Delta n}{\Delta x} \qquad (3.43)$$

Table 3.1

Constant	Silicon	Germanium	GaAs	Units
D_p	13.0×10^{-4}	46.8×10^{-4}	13×10^{-4}	m^2/sec
D_n	33.8×10^{-4}	98.8×10^{-4}	28×10^{-3}	m^2/sec

The negative sign does not appear in Eq. 3.43 because current flows in a direction opposite to electron flow. The total diffusion current is the sum of the hole diffusion current and the electron diffusion current, of course. The diffusion constants for holes and electrons in silicon, germanium, and gallium arsenide for $T = 300°K$ are given in Table 3.1.

These diffusion constants enable one to calculate the diffusion current if $\Delta p/\Delta x$ or $\Delta n/\Delta x$ is known. For example, if a forward-biased silicon diode has a diffusion length $l_D = 0.1$ cm. (Fig. 3.31) and an injected hole density $p_n(0) = 10^{22}$ holes/m³ at the junction, (Fig. 3.30), $\Delta p_n(x)/\Delta x$ near the junction is $-10^{22}/10^{-3} = -10^{25}$ holes/m. If the cross-sectional area of the diode is 10^{-6} m² at the junction, the current resulting from hole diffusion at the junction is 2.08 mA.

PROBLEM 3.17   If 5×10^{21} electrons/m³ are injected into the *p*-material of the diode above $(n_p(0) = 5 \times 10^{21})$ in addition to the given hole injection, and $l_D = 0.1$ cm for electrons as well as holes, determine the total diode current, due to both holes and electrons.   *Answer: i = 4.78 mA.*

Electron and hole diffusion are important concepts in the understanding of transistor operation and will be discussed further in Chapter 4.

A *diffusion capacitance* is associated with the diffusion of holes into the *n*-type crystal or the diffusion of electrons into the *p*-type crystal because of the stored charge which results. This stored charge, which is proportional to the forward-diode current, degrades the performance of the diode as a rectifier or switch, as shown in Fig. 3.32. In the circuit of Fig. 3.32, a square-wave voltage is applied to the rectifier circuit with load resistance R. The voltage V is assumed to be large enough so that the forward-bias diode drop can be neglected. Therefore, the forward current through the circuit is approximately V/R. Figure 3.32b is a sketch of the stored holes in the *n*-region at the instant of input polarity reversal and for some time thereafter. Before reversal, and at the instant of switching (reversal) at $t = 0$, the hole density distribution is $p_0(x)$. But for some time after switching the stored charges near the junction diffuse back toward the junction and are swept back across the junction, causing a large reverse current, limited only by the circuit resistance R as shown in Fig. 3.32c.

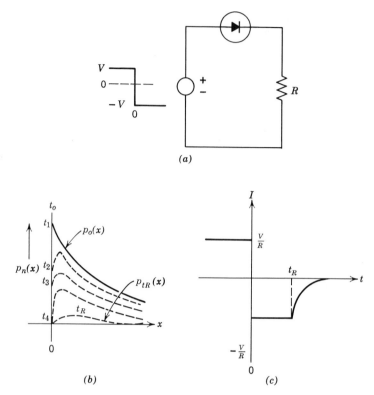

Fig. 3.32. (a) Diode circuit with square wave input, (b) sketch of hole density or stored charge on the n side of the junction as a function of time after input polarity reversal, and (c) sketch of circuit current before and after polarity reversal.

In fact, the diode is not actually reverse biased until the stored charge at the junction is reduced to essentially zero as indicated by the curve $p_{tR}(x)$. The time required to remove this stored charge is known as the recovery time t_R of the diode. The junction is then reverse biased and the reverse current decreases somewhat exponentially toward I_S. This exponential decrease results from the charging of the junction capacitance with time constant RC_j. This charge curve is not truly exponential because C_j is not a constant, but is a function of the diode voltage, as was previously discussed.

The stored minority carrier charge in the p- and n-regions are removed by recombination in addition to being swept back across the junction. Therefore the recombination rate, or recombination time, determines, to a large extent, the recovery time. Manufacturers can control the recombination rate and, therefore, the recovery time. Diodes designed for high-

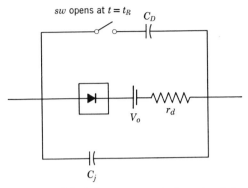

Fig. 3.33.   A high-frequency equivalent circuit for a diode.

frequency use usually have their recovery times given by the manufacturer. Some diodes, known as *hot carrier* diodes, which have a metal-semiconductor junction, have essentially zero recovery time. Efficient rectification may be obtained only if the recovery time is short in comparison with the period of the input cycle.

The diffusion capacitance C_D may be included in a diode-equivalent circuit as shown in Fig. 3.33. In this circuit, the capacitor C_D remains in the circuit and maintains forward bias across the diode until the switch opens at time t_R after the supply voltage reverses polarity. Diffusion capacitance will be quantitatively considered in Chapter 4 where the frequency limitations of a transistor will be treated.

PROBLEM 3.18   A detector (rectifier) is needed for a 2.0-MHz radio signal. What should be the maximum recovery time of the diode used in this application if efficient rectification efficiency is to be obtained? Assume $t_{R\max} = 0.05\ T$.   *Answer:* $t_R = 2.5 \times 10^{-8}$ s.

PROBLEM 3.19   A given junction diode has a 10^{-7}-A flow when 0.2 V reverse bias is applied. Determine the forward-bias voltage required to provide a 200 mA current if the internal ohmic resistance of the diode is 1 Ω. Assume $T = 300°$K.   *Answer:* 0.57 V.

PROBLEM 3.20   A reference or zener diode is used to provide essentially constant voltage to a transistor amplifier, as shown in Fig. 3.34. The current to the amplifier varies from 1 mA to 20 mA depending on the settings of the amplifier controls. The internal resistance R_b of the battery is 30 Ω when the battery is new but increases as the battery ages; thus, the battery voltage at terminals A-B decreases as the battery ages.

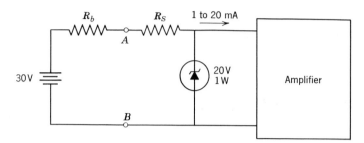

Fig. 3.34. A zener regulator.

(a) What maximum value of resistance can R_S have to maintain 20 V at the amplifier terminals when the battery terminal voltage has decreased to 24 V, if 1 mA is required through the zener diode to maintain 20 V across it?   *Answer:* 190 Ω.

(b) If $R_S = 190$ Ω, what maximum current will flow through the zener diode when the battery is new?   *Answer:* 44.5 mA.

(c) If the average dynamic resistance of the zener diode is 15 Ω, what will be the maximum voltage supplied to the amplifier?   *Answer:* 20.66 V.

(d) What will be the maximum power dissipation of the zener diode?
                                                                      Answer: 0.92 W.

(e) If the zener diode is rated 1.0 W at 25°C ambient (surrounding air) temperature and must be derated 5 mW for each degree ambient temperature above 25°C, what is the maximum ambient temperature at which the

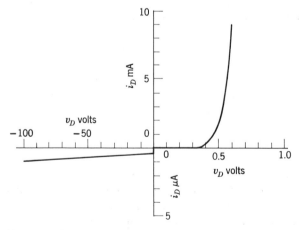

Fig. 3.35. Characteristics of the diode for Problem 3.20 (note the scale changes).

regulator can operate without exceeding its dissipation rating, assuming the battery is new?   *Answer:* 41°C.

PROBLEM 3.21   A given diode has the characteristics shown in Fig. 3.35. In addition it has C_i (dynamic) = 200 pF at $v_D = 0.5$ V and recovery time = 50 nanosecond (ns). Draw the circuit diagram, including an equivalent circuit for the diode for each of the following applications. Include only those diode elements that will alter the currents or voltages in the circuit by at least 2 percent of the maximum values.

(a) The diode is used to rectify (half-wave) a 50-V, 100-Hz signal. The load resistance is 10 kΩ.

(b) The diode is used to rectify a 10-V, 100-Hz signal. The load resistance is 2 kΩ.

(c) The diode is used as a detector for a 1-MHz, 5-V signal with $R_L = 5$ kΩ.

(d) The diode is used to rectify a 400-Hz, 10-V signal. The load resistance is 1 MΩ.

CHAPTER 4

JUNCTION
TRANSISTORS

Transistors are used in a wide variety of applications, including television, automatic control, satellite instrumentation, and medical electronics. The ability of the transistor to amplify electrical signals accounts for its wide use.

The amplifier is actually an energy converter. The input signal merely controls the current that flows from the power supply or battery. Thus the energy from the power supply is converted by the amplifier to signal energy.

4.1   *p-n-p* TRANSISTORS

When a semiconductor is arranged so that it has two *p-n* junctions as shown in Fig. 4.1*a*, it is known as a *p-n-p junction transistor*. The electrical potential in the transistor as a function of distance along the axis is shown in Fig. 4.1*b*. Note that this potential is the same as expected from two diode junctions. However, a basic difference exists between two diodes and a transistor (Fig. 4.1*c* and Fig. 4.1*d*). When the left-hand junction is forward biased the potential hill is lowered and carriers are injected into the center slice or *base*; but the base is made very thin, so nearly all the injected carriers diffuse across the base and are accelerated across the right-hand junction into the region known as the *collector*. The left-hand region is known as the *emitter* because it provides, or emits, the injected carriers. The current that flows out the base lead results from recombina-

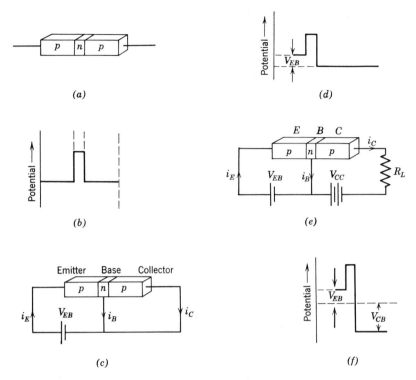

Fig. 4.1.   Potentials and currents in a *p-n-p* transistor. (*a*) *p-n-p* transistor, (*b*) idealized potential for (*a*), (*c*) forward-biased *E-B* junction, (*d*) idealized potential of (*c*), (*e*) load resistance and reverse-biased collector-base junction, and (*f*) idealized potential for (*e*).

tions in the base. Therefore, the base is lightly doped in addition to being thin in order to minimize recombinations. The ratio of carriers flowing into the collector to those injected across the emitter-base junction is known as alpha (α), which ranges from about 0.90 to 0.998 in modern transistors.

The transistor is useless as an amplifier unless a load resistance, such as a headset or relay, is placed in the collector, or output, circuit. But a resistance in the collector circuit of Fig. 4.1*c* would cause the collector terminal to become positive with respect to the base and would thus forward bias the collector-base junction. The injection current resulting from this bias would tend to cancel the initial collector current. However, if a battery is placed in the collector circuit as shown in Fig. 4.1*e*, the collector-base junction will remain reverse biased, providing the voltage drop across the load is less than the collector supply voltage V_{CC}. Then Fig. 4.1*f* applies. The supply voltage V_{CC}, and hence the voltage across the load resistor may be very large in comparison with the input voltage across the forward-biased emitter-base junction. Therefore, since the circuit has voltage amplification or *gain*, it is an *amplifier*.

The reverse bias across the collector-base junction increases the height of the potential hill (Fig. 4.1*f*). Therefore the electrons, or charge carriers, gain more kinetic energy as they are accelerated from the base into the collector region. This energy is dissipated as heat and causes the temperature of the transistor to rise.

From the preceding discussion, it would seem that either end of the transistor could be used as the emitter. That would be true for the transistor configuration of Fig. 4.1, but most transistors are made with collector junctions larger than emitter junctions for improved collector power dissipation and increased α (which will be discussed later) and thus perform poorly in the reverse direction.

Nevertheless, the transistor is sometimes used in switching applications that require that the normally intended collector serve as the emitter and vice versa. This mode is known as the *reverse* mode; and the *reverse alpha* α_R is only of the order of 0.7 or 0.8. The normal mode is known as the *forward* mode. The *forward alpha* α_F is specifically designated only when the transistor is used in these switching circuits. Otherwise the subscript is dropped and forward alpha is assumed.

The circuit diagram for the transistor amplifier of Fig. 4.1*e*, with an ac signal source added to the input circuit, is shown in Fig. 4.2*a*. Note that the emitter-base voltage v_{EB} is not the same as the bias-battery voltage V_{EE}, but is the sum of V_{EE} and the signal voltage v_i. Also observe the transistor symbol. The *n-p-n* transistor amplifier may be identical to the *p-n-p* amplifier, except that all the current and voltage polarities are reversed, as shown in Fig. 4.2*b*.

At this point a few words should be said about conventional current and

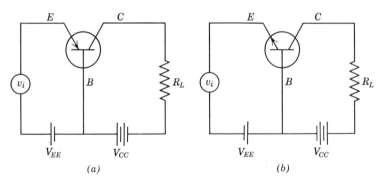

Fig. 4.2.   Circuit diagrams for (*a*) a *p-n-p* transistor amplifier and (*b*) an *n-p-n* transistor amplifier.

voltage directions. Actual current directions and potential polarities were shown in Fig. 4.1. However, IEEE standards require that current flow *into* a device be considered *positive*, and that currents that flow *out* of the device be considered *negative*. Therefore, the collector and base currents that usually flow out of the *p-n-p* transistor are negative currents because they flow in the opposite direction to the conventional positive currents. Also, a potential is considered positive if it is positive with respect to the common terminal or ground, which is the base terminal in Fig. 4.1. Thus the emitter terminal is positive and the collector terminal is negative in Fig. 4.2*a*. Either + and − signs, or an arrow pointing in the direction of positive potential are used to indicate voltage polarity.

An amplifier is frequently used to amplify ac signals. Capacitors are then used to couple the ac signal into the amplifier and out of the amplifier and to block (or prevent) the dc, or bias, components from entering the driving source or load as shown in Fig. 4.3. The capacitors are assumed to be large enough to have negligible reactance to the signal currents. The input signal voltage v_s varies the voltage across the forward-biased emitter junction and thus causes the emitter current i_E to vary. The collector current, being almost equal to the emitter current, varies in accordance with the emitter current, and thus causes a varying voltage across the collector circuit resistor R_C. This varying or signal component is passed through the capacitor C_2 and becomes the output voltage v_o. Note that the emitter-circuit resistor R_E should have high resistance compared with the input resistance of the transistor to avoid shunting an appreciable part of the signal current i_s to ground. Then V_{EE} must be larger than V_{EB} because of the $I_E R_E$ voltage drop across R_E. An example will be used to illustrate this concept.

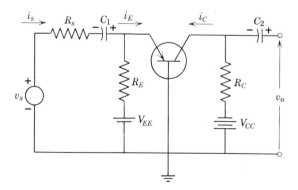

Fig. 4.3.   A common-base ac amplifier using a *p-n-p* transistor.

EXAMPLE 4.1   A transistor is connected as shown in Fig. 4.3. Let us determine the proper values of the bias elements in this circuit if the desired bias current I_E is 2.0 mA.

Since the emitter to base configuration is the same as a forward-biased diode, we can use Eq. 3.30 to determine the dynamic input resistance r_e. Thus, $r_e \simeq 25/I_E(\text{mA}) = 25/2 = 12.5\ \Omega$. If we choose $R_E = 100\ r_e$, then only 1 percent of i_s will be shunted through R_E. This value of R_E is $100 \times 12.5 = 1250\ \Omega$. The dc voltage drop across R_E is $i_E R_E = 2 \times 10^{-3} \times 1250 = 2.5\ \text{V}$. If the transistor is silicon, the voltage from emitter to base, V_{EB}, will be about 0.5 V. Then, the battery voltage V_{EE} should be $2.5 + 0.5 = 3.0\ \text{V}$.

A suitable value for the collector circuit resistor can be determined if we first choose a value for V_{CC}. To protect the transistor from damage, V_{CC} must be below the voltage breakdown rating of the collector junction. Let us assume $V_{CC} = -20\ \text{V}$ for this amplifier. Now, since $-I_C \simeq I_E$, the collector current $I_C = -2\ \text{mA}$. To permit a maximum collector voltage swing, we normally allow the voltage drop $-I_C R_C$ to be equal to $-V_{CC}/2$. Then, $-V_{CC}/2 = -10\ \text{V}$ and $R_C = -10\ \text{V}/I_C = -10\ \text{V}/2\text{mA} = 5\ \text{k}\Omega$.

PROBLEM 4.1   Repeat Example 4.1 with the desired emitter bias current, I_E, reduced to 1 mA. Let $R_E = 100\ r_e$ and assume $V_{EB} = 0.48\ \text{V}$ when $I_E = 1\ \text{mA}$.

Answer: $R_E = 2.5\ \text{k}\Omega$, $V_{EE} = 2.98\ \text{V}$, if $V_{CC} = -20\ \text{V}$, $R_C = 10\ \text{k}\Omega$.

A few words should be said here about voltage and current symbols. The IEEE standards are based on the system given in Table 4.1.

Table 4.1

Component	Symbol	Subscripts	Example
dc or average	Capital	Capital	V_{CB}, I_C
ac sinusoidal	Capital	Lower case	V_{cb}, I_c
total (dc + ac)	Lower case	Capital	v_{CB}, i_C
ac instantaneous	Lower case	Lower case	v_{cb}, i_c
bias-supply voltage	Capital	Double capital	V_{CC}, V_{EE}

The double subscripts associated with the voltage symbols (except the bias battery symbols) indicate the terminals between which the voltage is measured. For example V_{BC} is the dc or average value of potential difference between the base and the collector. Therefore $V_{BC} = -V_{CB}$. Sometimes a single subscript is used when the potential is with respect to ground or the chassis. For example, v_C is the total voltage between the collector and ground (or chassis or other ground reference) at any instant.

If the base terminal is grounded, $v_C = v_{CB}$. As was previously mentioned, the current flows into the transistor if its sign is positive; whereas a negative sign indicates that current is flowing out of the transistor. Current arrows always indicate the direction of positive current flow. Of course, ac or signal components may be arbitrarily (but consistently) assigned since they alternately change direction.

The transistor input voltage v_i due to the signal current i_s is

$$v_i = i_s r_e \tag{4.1}$$

Also, the output voltage v_o due to the signal current i_s is

$$v_o = \alpha i_s R_C \tag{4.2}$$

Then the voltage amplification or gain is

$$K_v = \frac{v_o}{v_i} = \frac{\alpha i_s R_C}{i_s r_e} = \frac{\alpha R_C}{r_e} \tag{4.3}$$

PROBLEM 4.2   If the transistor amplifier of Fig. 4.3 has the values previously used ($r_e = 12.5\ \Omega$ and $R_C = 5.0\ \mathrm{k\Omega}$) and $\alpha = 0.99$, calculate the voltage gain K_v of the amplifier.   *Answer: 396.*

PROBLEM 4.3   Draw the circuit diagram for an ac amplifier using an *n-p-n* transistor in the grounded base configuration. Also determine suitable values for the circuit elements (excluding capacitors) and calculate the voltage gain if $I_E = 1\ \mathrm{mA}$, $\alpha = 0.98$, and $V_{CC} = 20\ \mathrm{V}$. Assume $V_{BE} = 0.5\ \mathrm{V}$.
Answer: $R_E = 2.5\ \mathrm{k\Omega}$, (assuming $100\ r_e$), $V_{EE} = 3.0\ \mathrm{V}$, $R_C = 10\ \mathrm{k\Omega}$, and $K_V = 392$. (Any suitable set of answers is acceptable.)

Saturation currents due to minority carriers flow across the junctions in transistors as well as diodes. These currents, which were designated I_S in the diode, are known as I_{CO} for the collector junction and I_{EO} for the emitter junction in a transistor. Transistor data sheets usually list these currents, plus any leakage currents, as I_{CBO} and I_{EBO}. The first two subscripts designate the terminals between which the current flows, and the third subscript indicates that the third (other) terminal is open. The current I_{CO} (or I_{CBO}) must be included in the collector and base currents if highly accurate results are to be obtained. The relationships between these currents are shown in Fig. 4.4. Observe that I_E is a negative current

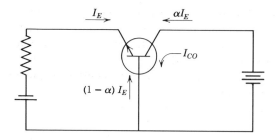

Fig. 4.4. Bias currents and the collector saturation (or thermal) current in an *n-p-n* transistor.

while I_C and I_{CO} are positive currents. Therefore

$$I_C = -\alpha I_E + I_{CO} \qquad (4.4)$$

and

$$I_B = -(1-\alpha)I_E - I_{CO} \qquad (4.5)$$

The thermal current I_{CO} is usually small in comparison with αI_E, but may not be small in comparison with $(1-\alpha)I_E$. In fact these currents may be equal at moderate values of I_E in a germanium transistor and the base current will then be zero. For example, if a germanium transistor with $\alpha = 0.98$ has $I_{CO} = 10\,\mu\text{A}$, the value of I_E, which will reduce I_B to zero, is -0.5 mA.

PROBLEM 4.4   A germanium transistor has $\alpha = 0.99$ and $I_{CO} = 12\,\mu\text{A}$. For what value of I_E will $I_B = 0$?   *Answer:* -1.2 mA.

4.2   GRAPHICAL ANALYSIS

Improved understanding of the common-base transistor amplifier can be gained from a graphical presentation of the input and output (or collector) characteristics of the transistor. The diagram of a circuit that might be used to obtain the common-base characteristics is given in Fig. 4.5. The batteries and potentiometers that are used as variable bias sources in this circuit are customarily replaced by ac operated power supplies with adjustable dc output voltages. The voltmeters shown should be highly sensitive, preferably electronic voltmeters, so that their operating currents do not make an appreciable contribution to i_C or i_E.

There are four important variables in the circuit, as indicated by the meters. Some of these variables must be held constant while the relationships between the others are found. For example, the emitter current must

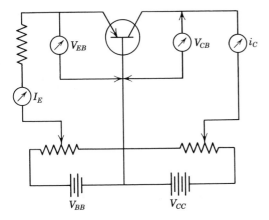

Fig. 4.5.   A circuit for determining transistor characteristics.

be held constant while the effect of the collector voltage on collector current is determined. When the collector current is plotted as a function of the collector-base voltage (with emitter current held constant), the resulting curve is known as a *collector-characteristic curve*. Each different value of emitter current will yield a different collector-characteristic curve. A set of several curves obtained from several representative values of emitter current is known as a *family* of collector-characteristic curves. A typical family or set of collector characteristics for a *p-n-p* transistor is shown in Fig. 4.6. The negative current indicates that the current is flowing out of the transistor where the reference direction is into the

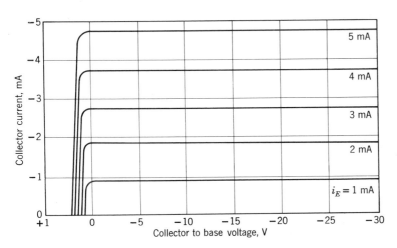

Fig. 4.6.   A family of collector characteristic curves.

transistor. A set of curves for an *n-p-n* transistor might be identical to the set shown except that the polarities of currents and voltages would be reversed.

From Fig. 4.6 it should be observed that:

1. The collector current is almost equal to the emitter current when reverse bias is applied to the collector junction.

2. The collector current is almost independent of the collector voltage when reverse bias is applied to the collector junction.

3. The collector current is rapidly reduced to zero and then reversed when forward bias is applied to the collector junction. This behavior occurs because the injection current across the collector junction opposes the injection current across the emitter junction. The collector current is essentially the algebraic sum of these two injection currents.

The emitter-base voltage does not appear in the set of collector characteristics. Therefore, a relationship is needed between emitter current and emitter-base voltage to determine the input resistance of the transistor. Experimentation reveals that the collector voltage has a slight influence on the emitter current. Therefore, a family of curves is needed to completely define the input characteristics of the transistor. Normally the emitter-base voltage is plotted as a function of the emitter current with the collector voltage held constant (Fig. 4.7). A curve is obtained for each different value of collector voltage. Notice that these curves are typical junction-diode curves. The voltage and current axes have been exchanged when compared with the diode characteristics of Chapter 3 because the emitter current is here considered as the independent variable. Fig. 4.7 shows that an increase of reverse collector bias increases the emitter

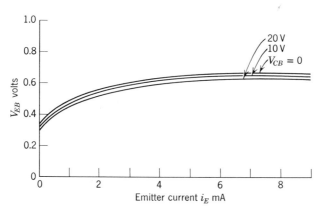

Fig. 4.7.   Input characteristics of a typical common-base transistor.

current slightly if v_{EB} is held constant. The reason for this collector voltage dependence will be discussed in Section 4.3.

The characteristic curves given above are known as *static* curves because at least one parameter is held constant while others are varied. However, all parameters vary with time when a signal is applied to a transistor amplifier. Therefore, it may seem that the static curves are of little use in determining the operating characteristics of an amplifier. However, this is not true because the dynamic characteristics can be obtained simply by drawing a *load line* on the static collector characteristics. A load line may be drawn on the collector characteristics in the same manner as for the diode characteristics described in Section 3.6. Since $v_{CB} = V_{CC} - i_C R_C$, the load line intercepts the v_{CE} axis ($i_C = 0$) at $v_{CE} = V_{CC}$ and the i_C axis ($v_{CB} = 0$) at $i_C = V_{CC}/R_C$. This load line is known as a dc load line because the direct current must flow through the collector load resistor R_C. For example, the dc load line for a value of $R_C = 5\,\text{k}\Omega$ has been drawn on a set of collector characteristics in Fig. 4.8a for a value of $V_{CC} = 20\,\text{V}$. The dc load will be considered to be the only load at this time.

A bias point or *quiescent point* must be selected some place along the load line. If the expected input signals are symmetrical about the quiescent or q-point, such as sinusoidal signals, a good selection for the q-point might be near the center of the load line.

The effect of the varying collector voltage on the input characteristics may be seen in Fig. 4.8b. The resulting curve is known as a *dynamic* input curve and can be obtained by determining, from Fig. 4.8a, the collector

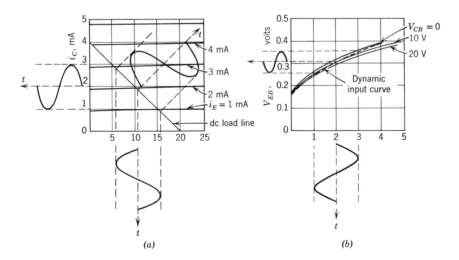

Fig. 4.8.   (a) A load line and (b) dynamic input characteristics for $R_L = 5\,\text{k}\Omega$.

voltage for each value of emitter current along the load line and spotting these points in the family of *static* input characteristics. For example, when $i_E = 1$ mA, $v_{CB} = 15$ V, approximately.

The voltage gain of the amplifier may be determined from the curves of Fig. 4.8. For example, if the q-point is selected at $i_E = 2$ mA and a sinusoidal input current of 1 mA peak value is applied, as indicated, the peak-to-peak output voltage (Δv_{CB}) is 9 V and the peak-to-peak input voltage (Δv_{EB}) is 0.1 V. Therefore, the voltage gain is $K_v = 90$.

The effect of the collector voltage on emitter current is so small in some transistors that a single input characteristic curve, instead of a family, will give adequately accurate results in the calculation of voltage gain. A single curve is usually provided by the manufacturer for these transistors.

PROBLEM 4.5   The common-base characteristics of a 2N3903 transistor are given in Fig. I-1 (appendix I). If this transistor is used in a common-base amplifier with $V_{CC} = 30$ V and $R_C = 10$ kΩ, determine the voltage gain of the amplifier when the peak-to-peak variation of the emitter current is 1.0 mA. R_C is the total load resistance. Draw the load line on the characteristics, choose a suitable quiescent operating point, and graphically determine the voltage gain.

4.3   BASE-WIDTH MODULATION

The collector voltage has an influence on the collector and emitter currents because of the dependence of the depletion-region width on the collector voltage, as illustrated in Fig. 4.9. In this figure the vertical line at $x = 0$ represents the emitter junction and the solid vertical line beyond $x = w$ represents the collector junction. Since the base is lightly doped compared with the collector, and since depletion-region width is inversely proportional to the doping density, the depletion region extends primarily into the base region. But the injected carriers that diffuse across the base

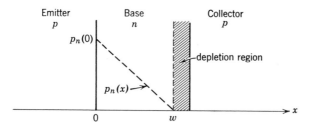

Fig. 4.9.   Diagrammatic sketch showing the effect of depletion region width and, hence, collector voltage, on base width.

region are swept into the collector as soon as they reach the depletion region; therefore, the effective base width (Fig. 4.9) is w. Now the current flow through the base is primarily by diffusion, as discussed further in Section 4.5. The diffusion equation was developed in Section 3.8 and is repeated here for hole diffusion.

$$J_p = -qD_p \frac{dp}{dx} \tag{4.6}$$

But the diffusion current through the base region must be continuous or constant, except for the small loss due to recombinations. Therefore, in the rectangular configuration assumed here, the slope of the $p_n(x)$ curve, dp/dx, must be essentially constant. Then the density of injected carriers $p_n(x)$ must decrease almost linearly from $p_n(0)$ at the emitter junction to essentially zero at the edge of the collector-junction depletion region. Figure 4.9 shows that this slope may be expressed as $p_n(0)/w$. The slope is negative since it is downward. Then, since the diffusion current is proportional to the slope $p_n(0)/w$ and the base width w decreases as the collector voltage increases, the emitter current increases as the collector voltage increases, assuming the emitter-junction voltage remains constant. Alpha (α) also increases with collector voltage because the narrower base region provides less opportunity for carrier recombination. This variation of effective base width with collector-base voltage is known as *base-width modulation*.

The effect of collector voltage on α becomes more pronounced as the collector voltage becomes high and approaches avalanche breakdown of the collector junction. Then, carrier multiplication occurs across

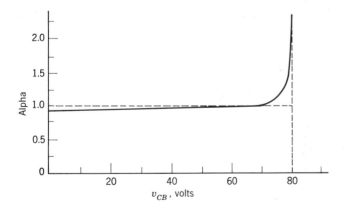

Fig. 4.10.   A sketch of α as a function of v_{CB} for a typical transistor.

the collector junction because high velocity carriers collide with atoms in the crystal and produce additional carriers. A sketch of α as a function of v_{CB} is given in Fig. 4.10 for a typical transistor with a collector-junction breakdown voltage of 80 V. Observe that α is unity at a collector voltage considerably below junction breakdown. In the transistor of Fig. 4.10, α is unity at approximately $v_{CB} = 65$ V.

4.4 TRANSISTOR EQUIVALENT CIRCUITS

Equivalent circuits or models of the transistor may be devised, as was done for the diode. One model, known as an *equivalent-T* circuit is shown in Fig. 4.11. This is a low-frequency model because it does not include the effects of junction capacitance or diffusion capacitance, which will be included later. The point b' can be imagined as being in the center of the base region. Then the resistance r_e accounts for the voltage-current relationship across the forward-biased emitter junction. This resistance is identical with the dynamic resistance of a forward-biased diode previously discussed. Thus its value at normal temperatures and at a given q-point emitter current I_E may be obtained from the relationship $r_e \simeq 25/I_E(\text{mA})\,\Omega$. The ideal-diode symbols are not included because the emitter junction is always forward biased and the collector junction is always reverse biased so that no switching occurs.

The resistance r_b represents the effective transverse ohmic resistance in the lightly doped base region. This resistance is typically 100 to 500 Ω in low-power transistors and varies somewhat with collector voltage because the base width varies with the collector voltage. The resistance r_c appears in the same position as the leakage resistance for the reverse-biased collector junction. However r_c is much smaller than normal values of surface leakage because it primarily accounts for the change in collector current

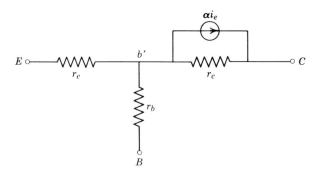

Fig. 4.11. An equivalent-T circuit for a common-base transistor.

that results from a change in collector voltage, assuming emitter current remains constant, because of base width modulation. Typical values of r_c for a low-power transistor range from 1 to 10 megohms (MΩ).

The current generator αi_e (Fig. 4.11) accounts for the current that diffuses across the base region from the emitter junction. This current source does not include the thermal current I_{CO} since the model is a small-signal, ac equivalent circuit and the dc components such as I_{CO} do not appear in the circuit diagram. The small-signal requirement results from the assumption that the components, such as r_e, do not vary over the signal cycle, which is a good approximation only if the signal currents are considerably smaller than the bias currents.

The equivalent-T circuit may be used to solve an amplifier problem instead of using graphical analysis, as illustrated by the following example.

EXAMPLE 4.2   We assume that the transistor with characteristics given in Table 4.2 is used in the circuit of Fig. 4.12a.

Table 4.2

Parameters for a typical low-power transistor $I_E = 1$ mA
$\alpha = 0.99$
$r_e = 25\ \Omega$
$r_b = 300\ \Omega$
$r_c = 1\ \text{M}\Omega$

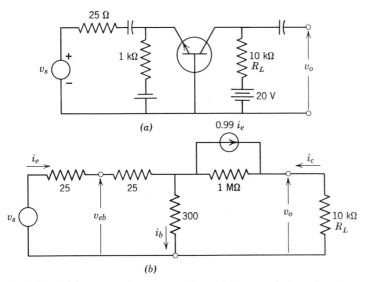

Fig. 4.12.   (a) A common-base ac amplifier and (b) an equivalent circuit.

The equivalent T circuit for the transistor is included in the equivalent circuit in Fig. 4.12b. The batteries do not appear in this circuit because it is an ac equivalent circuit, and the capacitors do not appear because they are assumed to have negligible reactance. The 1 kΩ resistor in the emitter-bias circuit has also been omitted, but it may not be obvious that its resistance is large compared with the transistor input resistance r_{in}, so we shall determine r_{in} by finding the input current in terms of the input voltage.

$$v_{eb} = i_e r_e + i_b r_b \tag{4.7}$$

But $i_b = i_e - i_c = (1-\alpha)i_e$ if the load resistance R_L is small compared with r_c, so that the percentage of αi_e shunted through r_c is small. Since $r_c = 100\,R_L$ in this example, we use $i_b = (1-\alpha)i_e$. Making this substitution into Eq. 4.7,

$$v_{eb} = i_e r_e + (1-\alpha)i_e r_b \tag{4.8}$$

Since $r_{\text{in}} = v_{eb}/i_e$

$$r_{\text{in}} = r_e + (1-\alpha)r_b \tag{4.9}$$

The value of input resistance for the amplifier of Fig. 4.12b is $r_{\text{in}} = 25 + (0.01)300 = 28\ \Omega$, which is very small compared with 1 kΩ, so that the latter may be neglected.

Either the voltage gain v_o/v_{eb} or the voltage gain v_o/v_s may be determined with the aid of the equivalent circuit (Fig. 4.12b). Again, neglecting the current which is shunted through r_c,

$$v_o = \alpha i_e R_L \tag{4.10}$$

and

$$v_{eb} = i_e r_{\text{in}} \tag{4.11}$$

Then

$$K_v = \frac{v_o}{v_{eb}} = \frac{\alpha R_L}{r_{\text{in}}} \tag{4.12}$$

Also

$$v_s = i_e(R_s + r_{\text{in}})$$

and

$$K_v' = \frac{v_o}{v_s} = \frac{\alpha R_L}{R_s + r_{\text{in}}} \tag{4.13}$$

The values of voltage gain for the amplifier of Fig. 4.12 are

$$K_v = \frac{0.99(10^4)}{28} = 354$$

and

$$K_v' = \frac{0.99(10^4)}{25 + 28} = 187.$$

The equivalent-T circuit shows, as was previously discussed, that the collector current is less than αi_e because of the shunting effect of r_c. However, when R_L is very small in comparison with r_c, the approximation may be made that $i_c = \alpha i_e$ and circuit calculations are simple. Otherwise, loop or nodal equations or similar techniques must be used to solve for the currents or voltages in the circuit, and the solution is tedious.

Many different models have been developed for the transistor. You may wonder why we need any more models (or equivalent circuits), since the equivalent T seems to yield satisfactory circuit solutions and gives additional insight into the characteristics of the transistor. In fact, you may wonder why models are necessary at all, since graphical analysis will also yield solutions to electronic circuits. This latter question will be answered first, since it is the easiest. Graphical analysis gives no indication concerning the frequency capabilities of a device. Graphical analysis is cumbersome and relies on the availability of suitable sets of characteristic curves, which are usually *not* available. Then to the first question; why not stick with one model. You may as well ask, "Why not use only one model automobile — why compacts, sedans, station wagons, and trucks?" The reasons are similar. One model will not perform all functions well. The equivalent T is a poor high-frequency model. It is not even a good low-frequency model because it is not easy to use. It *does* look like the skeleton of the physical model we have used for a transistor and therefore has some educational value in developing better models because its components can be related to the physical parts of a transistor.

At least two additional models are essential to an adequate understanding of semiconductor circuits and a reasonable dexterity in their solution. The first of these is the h-parameter circuit, which can yield easy *approximate* solutions to circuit problems, providing the high-frequency characteristics are not important. The z-parameter and y-parameter models are used as intermediate steps in the development of the h-parameter model in this chapter. The other popular and very useful model, particularly at high frequencies, is the hybrid-Π model, which will be developed in Chapter 7.

The general solution of the equivalent-T circuit could be obtained most easily by writing two loop or voltage equations for the circuit. In general form these two equations are:

$$v_1 = Z_{11}i_1 + Z_{12}i_2 \tag{4.14}$$

$$v_2 = Z_{21}i_1 + Z_{22}i_2 \tag{4.15}$$

These equations may also be written for the equivalent circuit of Fig. 4.13a, as may be seen by adding the voltages in the two loops. This circuit is known as a z-parameter circuit because all elements include either self impedances or transfer impedances. Note that the value of these impedances can be determined by making open-circuit measurements on the transistor, which may be difficult to do.

The voltage source, or Thevenin's equivalent, generators of Fig. 4.13a may be replaced by current source, or Norton's equivalent, generators (Fig. 4.13b). Nodal equations can be written for this circuit as follows:

$$i_1 = y_{11}v_1 + y_{12}v_2 \tag{4.16}$$

$$i_2 = y_{21}v_1 + y_{22}v_2 \tag{4.17}$$

This equivalent circuit is known as a y-parameter circuit because all the elements include either self or transfer admittances. Observe that the element values can be determined by making short-circuit measurements on the transistor.

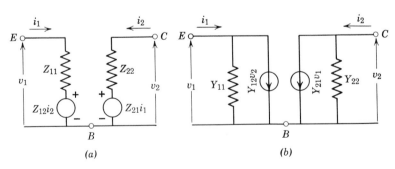

Fig. 4.13.   (a) A Z-parameter equivalent circuit and (b) a Y-parameter equivalent circuit.

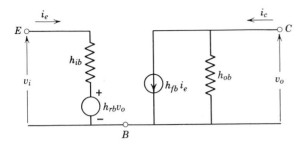

Fig. 4.14.   An *h*-parameter equivalent circuit.

An equivalent circuit, which is a cross between the *z*-parameter and *y*-parameter circuits, and is therefore called a hybrid or *h-parameter* circuit, is shown in Fig. 4.14. The basic equations for the *h*-parameter circuit are:

$$v_i = h_{ib}i_e + h_{rb}v_o \qquad (4.18)$$

$$i_c = h_{fb}i_e + h_{ob}v_o \qquad (4.19)$$

The subscript notation has been changed from that given to *z*- and *y*-parameter circuits to conform to standard transistor form. The second subscript, *b*, designates the common *base* configuration. The meaning of the first subscript can be seen from the following definitions of the *h*-parameters.

$h_i = input$ impedance with the output shorted (v_2 or $v_o = 0$).
$h_r = reverse$ voltage transfer ratio with input open (i_1 or $i_e = 0$).
$h_f = forward$ current transfer ratio with output shorted ($v_o = 0$).
$h_o = output$ admittance with input open ($i_e = 0$).

These parameters are easily obtained in terms of the equivalent-*T* parameters by inspection of the equivalent-*T* circuit of Fig. 4.11, as shown in the following discussion. The input resistance with R_L small compared with r_c was determined and is given by Eq. 4.9. This equation is precisely correct if R_L is zero, in which case $r_{in} = h_{ib}$ by definition. Then, using Eq. 4.9,

$$h_{ib} = r_e + (1-\alpha)r_b \qquad (4.20)$$

Also, note from Fig. 4.15 that a signal voltage v_c applied to the output terminals will cause current i to flow through r_c and r_b. With the input open ($i_e = 0$) this is the only current. Then the voltage appearing in the input

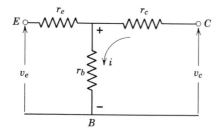

Fig. 4.15. Equivalent circuit used to determine $h_{rb} = v_e/v_c$.

circuit is the drop across r_b (Fig. 4.15). Then

$$h_{rb} = \frac{v_e}{v_c} = \frac{ir_b}{i(r_c + r_b)} = \frac{r_b}{r_c + r_b} \qquad (4.21)$$

From observing Fig. 4.11, or from the definition of α, it can be seen that

$$h_{fb} = -\alpha \qquad (4.22)$$

The negative sign appears because the current generators αi_e and $h_{fb} i_e$ have opposite polarities. From Fig. 4.15, it can also be seen that the output admittance with the input open is

$$h_{ob} = \frac{1}{r_c + r_b} \simeq \frac{1}{r_c} \qquad (4.23)$$

One main advantage of the h-parameter equivalent circuit is that the h-parameters may be obtained from the transistor input and the collector characteristics previously discussed. Since these curves are easily obtained from a transistor curve tracer, the h-parameters are readily available. They are also frequently given by the transistor manufacturer.

Input and collector characteristics are given in Fig. 4.16 for a typical low-power transistor. The (dynamic) input resistance with the output shorted h_{ib} is $\Delta v_{EB}/\Delta i_e$ with v_{CB} held constant (shorted for ac). This is the slope of an input characteristic curve at a given point. The point at which the h-parameters are desired is, of course, the q-point. In Fig. 4.16 the q-point has been selected at $i_E = 2.0$ mA, $v_{CB} = 10$ V. Therefore, h_{ib} at this point is the slope of the line AB, which has the value $h_{ib} = 0.03$ V/1 mA = 30 Ω.

The parameter h_{rb} is $\Delta v_{EB}/\Delta v_{CB}$ with i_E held constant (ac input current = 0). This value of h_{rb} may be determined from the line CD, which gives the

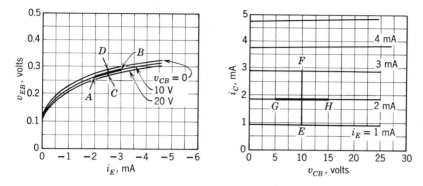

Fig. 4.16. Illustration of the technique used to obtain the *h*-parameters from characteristic curves.

change in v_{EB} produced by a 20 V change in v_{CB}. The value of h_{rb} at this point is 0.025 V/20 V $= 1.25 \times 10^{-3}$.

The forward current transfer ratio with output shorted, h_{fb}, can be determined at the q-point from line EF, which gives the collector current change Δi_C for an emitter current change $\Delta i_E = 2$ mA with v_{CB} constant (or zero ac voltage). Thus, the value of h_{fb} is -1.9 mA/2.0 mA $= -0.95$.

The output admittance with the input open, h_{ob}, is $\Delta i_C / \Delta v_{CB}$ with i_E held constant. Thus h_{ob} is the slope of the output characteristic curve at the q-point or the line GH. This slope is so small that it is difficult to determine it with satisfactory accuracy. However, h_{ob} for this q-point is roughly 0.01 mA/10 V $= 10^{-6}$ mho.

EXAMPLE 4.3 We now calculate the voltage gain of the transistor $\Delta v_{CB} / \Delta v_{BE}$ with $R_L = 10$ kΩ, using the *h*-parameters determined from Fig. 4.16. The *h*-parameter circuit with these values listed is given in Fig. 4.17. At this time we shall avoid a solution using simultaneous equations by

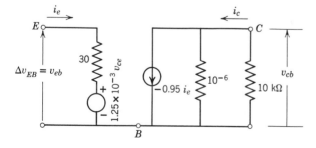

Fig. 4.17. An *h*-parameter circuit with values determined from Fig. 4.16.

assuming that $i_e = 1$ mA. Then $v_{cb} = -0.95$ mA$/(10^{-6} + 10^{-4}) = 9.4$ V. Then $v_{eb} = h_{ib}i_e + h_{rb}v_{cb} = 0.03 + 0.0117 = 0.0417$ V and the voltage gain is 9.4 V/0.0417 V $= 225$.

PROBLEM 4.6   A given transistor has $r_e = 15\ \Omega$, $r_b = 200\ \Omega$, $\alpha = 0.98$, and $r_c = 2\ \text{M}\Omega$ at a certain q-point. Determine the h-parameters, draw an h-parameter equivalent circuit, and calculate the voltage gain v_{cb}/v_{eb} if $R_L = 5\ \text{k}\Omega$.

PROBLEM 4.7   Determine the approximate common-base h-parameters of the transistor in Figs. 4.6 and 4.7 at the point $I_C = 3.0$ mA, $V_{CB} = 10$ V.

4.5   DIFFUSION CAPACITANCE AND ALPHA-CUTOFF FREQUENCY

The response time, or frequency response of a transistor, like a diode, is limited by diffusion capacitance and junction capacitance. The diffusion capacitance results from the carriers injected into the base region (Fig. 4.18). The injected carriers tend to produce an excess charge (positive for a p-n-p in the base), but the electric field that begins to build up causes electrons to flow through the base connecting lead to neutralize the charge. Notice that these neutralizing charges do not necessarily combine with the injected carriers. In fact, they do not appreciably increase the recombination rate as long as their density is small in comparison with the base doping concentration. The capacitive current that flows in the emitter-base circuit does not contribute to the collector current.

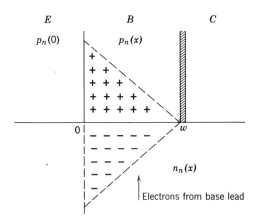

Fig. 4.18.   A representation of stored charge in the base, which results in diffusion capacitance.

Therefore, as the frequency increases and this capacitive current becomes significant, alpha decreases.

The dynamic or ac value of the diffusion capacitance can be determined as a function of the effective base width of the transistor with the aid of Fig. 4.19a. If the injected carrier density is increased by ΔP from P_0, the stored charge is increased by ΔQ, where ΔQ is the increase in average charge density times the effective volume of the base region. Then, since the average increase in charge density is $q \Delta P/2$,

$$\Delta Q = \frac{q \Delta P w A}{2} \qquad (4.24)$$

where A is the effective cross-sectional area of the base region. The diffusion capacitance is

$$C_D = \frac{\Delta Q}{\Delta v_E} \qquad (4.25)$$

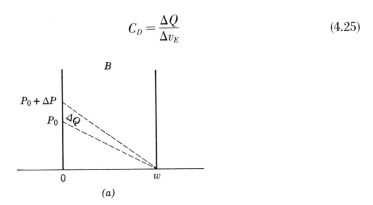

(a)

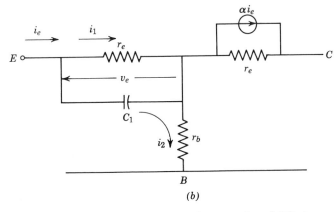

(b)

Fig. 4.19. (a) A sketch to assist in the determination of diffusion capacitance and (b) a high-frequency equivalent-T circuit.

where Δv_E is the change in voltage across the capacitor or the ac voltage v_e (Fig. 4.19b). But, from this figure, $v_e = i_1 r_e$. Then, using Eq. 4.24,

$$C_D = \frac{q\Delta PwA}{2i_1 r_e} \tag{4.26}$$

Since i_1 is the ac component of diffusion current, (from Eq. 4.6)

$$i_1 = AqD_p\frac{\Delta P}{w} \tag{4.27}$$

Substituting this value of i_1 into Eq. 4.26,

$$C_D = \frac{w^2}{2D_p r_e} \tag{4.28}$$

The value of diffusion capacitance may be calculated for a uniform base transistor at a given q-point if the effective base width is known. For example, a p-n-p silicon transistor with $w = 10^{-4}$ m and $I_E = 1$ mA has $C_D = (1 \times 10^{-8})/(2 \times 13 \times 10^{-4}(25)) = 1.54 \times 10^{-7}$ F. The capacitor C_1 in Fig. 4.19 is the diffusion capacitance C_D plus the emitter-junction capacitance C_{je} and is therefore somewhat larger than C_D.

Alpha-cutoff frequency f_α is defined as the frequency at which alpha decreases to $\alpha_0/\sqrt{2}$ where α_0 is the low-frequency value of α. This is the frequency at which the capacitive current i_2 is equal to the diffusion current i_1 (Fig. 4.19), since these currents are 90° out of phase and must be added vectorially to obtain i_e. Remember that $\alpha = i_c/i_e$.

Then r_e must equal X_c at f_α or ω_α, or

$$r_e = \frac{1}{\omega_\alpha C_1} \tag{4.29}$$

Thus

$$\omega_\alpha = 2\pi f_\alpha = \frac{1}{r_e C_1} \tag{4.30}$$

Note that if C_D is large compared with C_{je}, so that the latter can be neglected, the expression for C_D in Eq. 4.28 can be substituted for C_1 in Eq. 4.30. Then

$$\omega_\alpha \simeq \frac{2D_p}{w^2} \tag{4.31}$$

Thus the *p-n-p* silicon transistor with $w = 10^{-4}$ m would have $\omega_\alpha \simeq$ $(2 \times 13 \times 10^{-4})/10^{-8} = 2.6 \times 10^5$ rad/s and $f_\alpha \simeq 4.14 \times 10^4$ Hz. This value of alpha-cutoff frequency is disappointingly low. If the base width w is reduced to 10^{-5} m, f_α is increased to 4.14×10^6 Hz.

PROBLEM 4.8   If the transistor above were an *n-p-n* silicon with $w = 10^{-5}$ m, what would be the value of f_α?   *Answer:* 1.08×10^7.

PROBLEM 4.9   Determine f_α for an *n-p-n* germanium with $w = 10^{-5}$ m.
Answer: 3.15×10^7.

Gallium Arsenide is an especially attractive material for high-frequency transistors because of the very high electron mobility and consequently high value of D_n. Physically, the high carrier mobility and large diffusion constant provide a high value of f_α because the carriers diffuse through the base so quickly that the stored charge, or diffusion capacitance, is relatively small for a given value of I_E.

PROBLEM 4.10   Determine the value of f_α for an *n-p-n* GaAs transistor with an effective base width $w = 10^{-5}$ m if $D_n = 200 \times 10^{-4}$ m²/s for GaAs.
Answer: 6.37×10^7 Hz.

PROBLEM 4.11   Determine the common-base *h*-parameters approximately for the 2N3903 transistor at the *q*-point $I_C = 3.0$ mA, $V_{CB} = 6.0$ V. Curves are given in Fig. I.1 (Appendix I).
Answer: $h_{ib} = 12\,\Omega$, $h_{rb} \simeq 0$, $h_{fb} \simeq -0.99$, $h_{ob} \simeq 3 \times 10^{-7}$ mho (estimate).

PROBLEM 4.12   Assume that you want to build a sound amplifier to eavesdrop on your sister and her boyfriend. You have an 8-Ω (internal resistance) loudspeaker that will serve as a microphone. You find by measurement that moderate sound levels will generate 0.01-V peaks in the 8-Ω voice coil of this loudspeaker. You also have a 2,000-Ω (internal impedance) headset with which you can listen. In addition, you have a 2N3903 transistor (Fig. I.1), a 12-V battery, a 3-V battery, and a wide assortment of resistors and capacitors. Design a common-base amplifier, using these components and determine the peak voltage expected across the headset.   *Answer:* 1.2 V, if $I_C = 3$ mA.

CHAPTER 5

COMMON-EMITTER
AMPLIFIERS

In Chapter 4, the student may have surmised that a rearrangement of the input terminals of the common-base amplifier would reduce the input current requirement for a given output current and thus provide higher power gain. Such a configuration would employ the emitter as the common or grounded terminal and the base as the other input terminal as shown in Fig. 5.1.

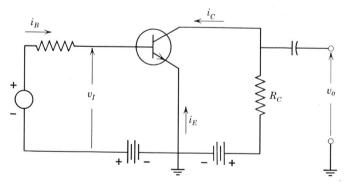

Fig. 5.1.   A common-emitter amplifier.

5.1 CURRENT GAIN OF THE COMMON-EMITTER AMPLIFIER

Since $i_B = (1-\alpha)i_E$, the input current is reduced by the factor $(1-\alpha)$ and the input impedance is increased by the factor $1/(1-\alpha)$, as compared with the common-base configuration, assuming the q-point remains the same. Since the base current is the input current, it is desirable to obtain an expression for the collector, or output, current as a function of the input current. This may be done by using Eq. 4.4 repeated below.

$$i_C = -\alpha i_E + I_{CO} \qquad (4.4)$$

But

$$-i_E = i_C + i_B \qquad (5.1)$$

Substituting this value of $-i_E$ into Eq. 4.4,

$$i_C = \alpha(i_C + i_B) + I_{CO} \qquad (5.2)$$

Then solving for i_C in terms of i_B and I_{CO},

$$i_C = \frac{\alpha}{1-\alpha}i_B + \frac{I_{CO}}{1-\alpha} \qquad (5.3)$$

The term $\alpha/(1-\alpha)$ is known as β (beta), the current gain of the common-emitter configuration when the output is shorted. A transistor that has $\alpha = 0.99$, has $\beta = 0.99/0.01 = 99$ and $1/(1-\alpha) = 1/0.01 = 100$. Note that $1/(1-\alpha) = \beta+1$, as can be shown by adding $\alpha/(1-\alpha)$ to $(1-\alpha)/(1-\alpha)$. Thus i_C may be written in terms of i_B, i_{CO}, and β, using Eq. 5.3.

$$i_C = \beta i_B + (\beta+1)I_{CO} \qquad (5.4)$$

Perhaps it is surprising to find that the thermally generated current I_{CO} is multiplied by $(\beta+1)$ if the base current is zero, or constant. For example, if $i_B = 10\ \mu A$, $I_{CO} = 10\ \mu A$, and $\alpha = 0.99$, $i_C = 99 \times 10\ \mu A + 100 \times 10\ \mu A = 1.99$ mA.

One way to visualize the control action of the base is to assume that the forward bias between the base and the emitter increases. This increased forward bias causes the height of the potential hill to be reduced and the majority current across the junction to be increased. Most of the carriers that comprise this current diffuse across the thin base region and are swept

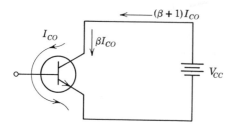

Fig. 5.2. An illustration of the process by which I_{CO} is multiplied in the common-emitter configuration.

into the collector. Thus, the increase of collector current may be much greater than the increase of base current. The collector current change is almost proportional to the base current change. Therefore, the base current is often assumed to be the control parameter. The thermal current I_{CO} is amplified because it is forced to flow across the emitter junction. As a result, a small forward bias is produced across that junction and the collector current is increased as though I_{CO} were added to the base current.

The process by which I_{CO} is amplified is shown in Fig. 5.2. With the base circuit open, the current I_{CO} is not free to flow out of the base lead to return to V_{CC}, but must flow across the emitter junction and, therefore, produces a forward bias across that junction; the forward bias, in turn, causes an additional current of magnitude βI_{CO} to flow in the collector circuit, thus making a total current of $(\beta + 1)I_{CO}$. The I_{CO} causes a similar increase in collector current when the base current is held at a constant nonzero value.

PROBLEM 5.1   A given transistor has $\alpha = 0.995$ and $I_{CO} = 5\,\mu\text{A}$. What is the value of collector current when $i_B = 20\,\mu\text{A}$?   *Answer:* 4.98 mA.

5.2 GRAPHICAL ANALYSIS OF THE COMMON-EMITTER AMPLIFIER

Collector characteristics and input characteristics may be obtained for the common-emitter configuration in a manner similar to that described for the common-base configuration. However, a curve tracer is recommended as the easy and accurate method of obtaining these characteristics.

A set of collector characteristics for a typical *n-p-n* germanium transistor is given in Fig. 5.3. Comparison of these curves with those given in Fig. 4.6 for the common base configuration show the following differences:

  1. The base current i_B is a parameter in the common-emitter set.
  2. The collector current is approximately equal to zero when v_{CE} equals

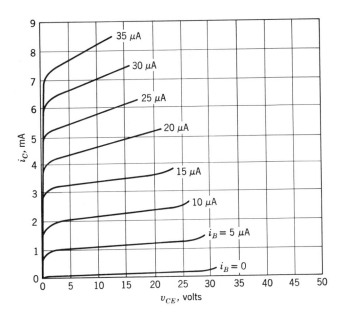

Fig. 5.3.   Collector characteristics for the common-emitter configuration.

zero, regardless of the value of i_B. The reason for this behavior is that the collector junction has the same forward-bias voltage as the emitter junction when $v_{CE} = 0$, as seen from Fig. 5.1. Therefore, an injection current flows across the collector junction into the base. This current is almost equal in magnitude and opposite in direction to the emitter-junction injection current and therefore the collector current is essentially the difference between these two currents. As v_{CE} is increased in the reverse-bias direction, the collector current increases very rapidly until the collector-junction bias is reduced to zero. Larger values of v_{CE} apply reverse bias to the collector junction and the collector current is then relatively independent of collector voltage, the collector characteristic is nearly horizontal.

3. The spacing between curves increases as v_{CE} increases because α and hence β increase with v_{CE}. The increase in β is much more pronounced than the increase in α as shown by the curves, because small changes in α produce large changes in β.

4. The slope of the collector characteristics are much steeper in the common-emitter set because the change in collector current that results from base-width modulation, like the thermal current I_{CO}, is forced to flow across the emitter junction, if base current is held constant. The change in forward bias thus produced causes β times as much change in collector current as that produced by the initial base-width modulation. Thus the

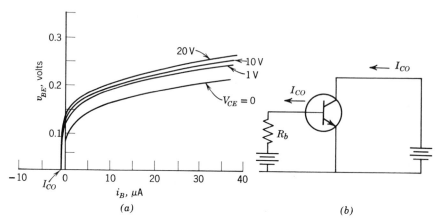

Fig. 5.4. (a) Common-emitter input characteristics and (b) illustration of reverse-base current required to eliminate the forward bias across the emitter junction.

slope of the output characteristic in the common-emitter set is increased by the factor $\beta + 1$.

A set of input characteristics for the transistor of Fig. 5.3 is given in Fig. 5.4a. The curve obtained with $v_{CE} = 0$ is widely spaced from the others because both junctions are forward biased when $v_{CE} = 0$, as was previously mentioned, and the majority currents across both junctions store charge in the base. This excess charge results in increased recombination and a larger base current for a given value of forward emitter bias. Except for switching applications, the transistor does not normally operate in this region of very small collector voltage, known as the *saturation region*.

Observe from Fig. 5.4a that, except for $v_{CE} = 0$, the emitter junction is forward biased when $i_B = 0$ because I_{CO} flows across the emitter-base junction and produces forward bias. Therefore, as shown in Fig. 5.4b, the emitter-junction voltage can be reduced to zero and consequently the collector current can be reduced to I_{CO} only if a reverse base current equal to I_{CO} flows in the base circuit.

The current gain, voltage gain, and permissible signal levels of a transistor amplifier may be determined graphically, using the characteristic curves, in the manner indicated for the common-base amplifier in Section 4.2. An example will be used to illustrate this procedure.

EXAMPLE 5.1   An *n-p-n* transistor is connected as shown in Fig. 5.5. The characteristics of this transistor are given in Fig. 5.6. Let us determine the current gain, voltage gain, and power gain for this transistor.

The collector supply voltage is $V_{CC} = 25$ V and $R_C = 6$ kΩ. Thus, a load

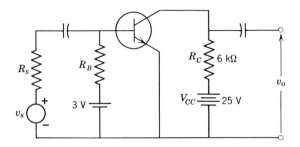

Fig. 5.5.   A circuit diagram for an ac common-emitter amplifier using an *n-p-n* transistor.

line can be drawn as shown in Chapters 3 and 4. In this example the inter-cept with the $v_{CE} = 0$ axis occurs at $i_C = 25$ V/6 kΩ = 4.17 mA. Thus, the load line is drawn on the collector characteristics as shown in Fig. 5.6. The q-point is chosen approximately at the midpoint of load line. Let us choose a q-point at $I_B = 10$ μA. The q-point collector current is then approximately 2.2 mA. To obtain this q-point, the base resistor R_B must have a value of $R_B = (V_{BB} - V_{BE})/I_B$. From the input characteristics, the value of V_{BE} for $I_B = 10$ μA and $V_{CE} = 12$ V (the q-point) is $V_{BE} \simeq 0.2$ V. Then, $R_B = (3 - 0.2)/10^{-5} = 280$ kΩ.

To obtain the current gain of this amplifier, we must assume a base-current variation. A base-current swing of a 5-μA peak looks reasonable for this circuit. Then, the base current varies along the load line from 5 μA to 15 μA. When $i_B = 5$ μA, the collector current $i_C \simeq 1.2$ mA and when $i_B = 15$ μA, $i_C \simeq 3.3$ mA. Thus, a change of 10 μA base current ($\Delta i_B = 10$ μA) produces a change of collector current equal to 2.1 mA ($\Delta i_C = 2.1$ mA). Then, $K_i = \Delta i_C/\Delta i_B = (2.1 \times 10^{-3})/10^{-5} = 210$.

From the collector characteristics we also note that when $i_B = 5$ μA, $v_{CE} = 17.5$ V and when $i_B = 15$ μA, $v_{CE} = 5.5$ V. Then, $\Delta v_{CE} = 12$ V. To find Δv_{BE}, we must use the v_{BE} versus i_B curves. Thus, when $i_B = 5$ μA (and $v_{CE} = 17.5$ V), $v_{BE} = 0.18$ V. Also when $i_B = 15$ μA ($v_{CE} = 5.5$ V), $v_{BE} = 0.21$ V. Then, $\Delta v_{BE} = 0.03$ V. The voltage gain is $K_v = \Delta v_{CE}/\Delta v_{BE} = 12/0.03 = 400$. The power gain $K_p = K_i K_v = 210 \times 400 = 8.2 \times 10^4$.

PROBLEM 5.2   The amplifier of Fig. 5.5 has $R_C = 10$ kΩ. Other compon-ents remain the same except R_B. Draw a load line, choose a suitable q-point and determine R_B, K_i, K_v, and K_p.

Answer: $R_B \simeq 560$ k, $K_i = 180, K_v = 270, K_p = 4.8 \times 10^4$.

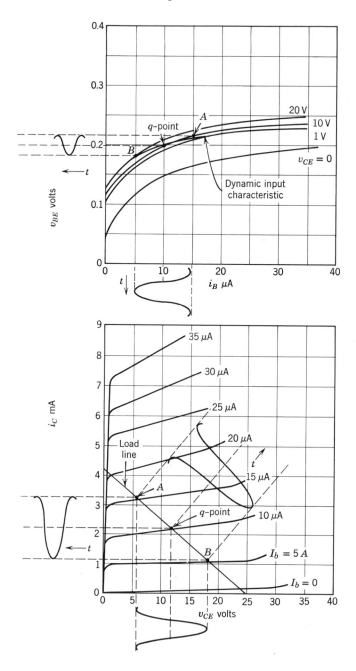

Fig. 5.6   Input and collector characteristics for the transistor of Fig. 5.5.

5.3   COMMON-EMITTER EQUIVALENT CIRCUITS

The common-base equivalent-T circuit could be used for the common-emitter configuration by exchanging the input terminals as shown in Fig. 5.7a. However, this equivalent circuit is not convenient for the common-emitter configuration because the current source in the collector circuit is given in terms of i_e, which is not the input current. The circuit of Fig. 5.7b is therefore useful. Note that the resistance $r_d = (1-\alpha)r_c = r_c/(\beta+1)$, as was previously discussed, because with $i_b = 0$, the output current that results from base-width modulation flows across the emitter junction and is therefore amplified by the factor $(\beta+1)$. This reduction in output resistance can be easily shown mathematically by applying a voltage v_{ce} to the output terminals of the circuit of Fig. 5.7a. Then the current flowing into the transistor is, assuming $r_e \ll r_c$,

$$i_c \simeq \frac{v_{ce}}{r_c} + \alpha i_e \tag{5.5}$$

But with $i_b = 0$ (or i_B constant since i_b is the signal component of i_B) $i_e = i_c$. Making this substitution for i_e in Eq. 5.5,

$$i_c = \frac{v_{ce}}{r_c(1-\alpha)} \tag{5.6}$$

Then

$$r_d = \frac{v_{ce}}{i_c} = r_c(1-\alpha) \tag{5.7}$$

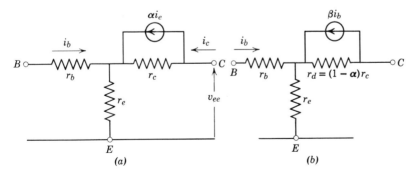

Fig. 5.7.   Equivalent-T circuits for the common-emitter configuration.

The h-parameter circuit for the common-emitter configuration is identical to the common-base h-parameter circuit (Fig. 5.8). The definitions of the h-parameters are the same for the common-emitter configuration as for the common base, although their values are different, as indicated by the second subscript. A more precise definition for h_{ie}, than $\Delta v_{BE}/\Delta i_B$ with v_{CE} constant, is the following derivative form.

$$h_{ie} = \frac{dv_{BE}}{di_B}\bigg|_{v_{CE} = \text{const.}} = \text{the input impedance with the output shorted}$$

The improved preciseness results because the derivative is the slope of the input curve at the desired point, whereas the ratio $\Delta v_{BE}/\Delta i_B$ is only the approximate slope and the goodness of the approximation depends on the size of Δv_{BE}. The other common-emitter h-parameters can be defined similarly.

$$h_{re} = \frac{dv_{BE}}{dv_{CE}}\bigg|_{i_B = \text{const.}} = \text{The reverse voltage transfer ratio with the input open}$$

$$h_{fe} = \frac{di_C}{di_B}\bigg|_{V_{CE} = \text{const.}} = \text{The forward current transfer ratio with the output shorted}$$

$$h_{oe} = \frac{di_c}{dv_{CE}}\bigg|_{i_b = \text{const.}} = \text{The output admittance with the input open}$$

The slope of the input characteristics will yield h_{ie} and the slope of the output characteristics will give h_{oe} at the prescribed point. Also, from the information given in these sets of curves, additional curves could be plotted for v_{BE} as a function of v_{CE}, with i_B held constant; i_C could be plotted as a

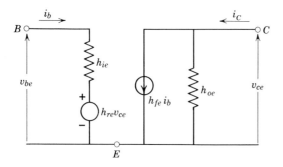

Fig. 5.8.   The common-emitter h-parameter circuit.

function of i_B with v_{CE} held constant. The slope of these curves would give h_{re} and h_{fe}, respectively, with improved accuracy. However, we assume that adequate accuracy will be obtained by using the Δ value technique employed in finding the common-base h-parameters h_{rb} and h_{fb}. Thus, all the common-emitter h-parameters can be determined from the curves of Fig. 5.6 for the transistor represented there. Notice that $h_{fe} = \beta$.

PROBLEM 5.3 Determine the common-emitter h-parameters for the transistor of Fig. 5.6 at the q-point $i_B = 10\ \mu\text{A}$, $v_{CE} = 10$ V.
   Answer: $h_{ie} = 4000\ \Omega$, $h_{re} = 8 \times 10^{-4}$, $h_{fe} = 230$, and $h_{oe} = 25\ \mu\text{mho}$.

As was previously noted, the input impedance (with output shorted) of the common-emitter transistor is $(\beta + 1)$ times as high as that of the common base. Therefore

$$h_{ie} = (\beta + 1)h_{ib} \tag{5.8}$$

Also, the output impedance (with input open) of the common-emitter configuration is reduced by the factor $(\beta + 1)$, as compared with the common base. Thus the admittance h_{oe} is greater than the admittance h_{ob} by the factor $(\beta + 1)$.

$$h_{oe} = (\beta + 1)h_{ob} \tag{5.9}$$

In addition, since $h_{fb} = -\alpha$, $h_{fe} = \alpha/(1-\alpha)$ and $(\beta + 1) = 1/(1-\alpha)$,

$$h_{fe} = -(\beta + 1)h_{fb} \tag{5.10}$$

As was shown above, the common-emitter and common-base h-parameters except h_{re} and h_{rb} are related by the factor $(\beta + 1)$ or $(h_{fe} + 1)$. In modern transistors h_{re} and h_{rb} are so small that they usually can be neglected at low frequencies without significant error. Therefore, sets of curves are not needed for both common-base and common-emitter configurations. The common-emitter curves are almost universally used because they provide better accuracy in obtaining the parameters.

PROBLEM 5.4 Using the h-parameters determined in Prob. 5.3, calculate the common-base h-parameters for the transistor of Fig. 5.6.
   Answer: $h_{ib} = 17.3\ \Omega$, $h_{fb} = -0.996$, and $h_{ob} = 1.1 \times 10^{-7}$ mho.

The current gain, voltage gain, input resistance, and output admittance (or impedance) can be determined for the transistor amplifier by use of the h-parameter equivalent circuit. A typical h-parameter circuit is shown in

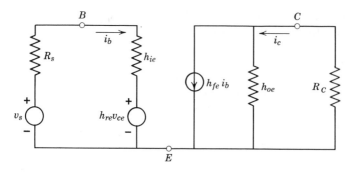

Fig. 5.9. A typical *h*-parameter circuit with input generator and output load.

Fig. 5.9. Let us assume we know i_b. Then, a current $h_{fe}i_b$ will flow through the parallel combination of h_{oe} and R_C. If $G_C = 1/R_C$, the admittance of this parallel combination is $G = (h_{oe} + G_C)$. Then, since $v = i/G$, we have the relationship

$$v_{ce} = -\frac{h_{fe}i_b}{h_{oe} + G_C} \tag{5.11}$$

The negative sign appears because v_{ce} becomes more negative as v_{be} becomes more positive, or there is a voltage polarity reversal across the transistor. From the input loop, we note that $v_{be} = h_{ie}i_b + h_{re}v_{ce}$. If we use the value of v_{ce} obtained in Eq. 5.11, this relationship becomes

$$v_{be} = h_{ie}i_b - \frac{h_{fe}h_{re}}{h_{oe} + G_C}i_b \tag{5.12}$$

We can now take the ratio of Eq. 5.11 to Eq. 5.12 to obtain the voltage gain of this amplifier.

$$K_v = \frac{v_{ce}}{v_{be}} = \frac{h_{fe}i_b/(h_{oe} + G_C)}{h_{ie}i_b - (h_{fe}h_{re}/h_{oe} + G_C)i_b} = \frac{-h_{fe}}{h_{ie}(h_{oe} + G_C) - h_{fe}h_{re}} \tag{5.13}$$

Both sides of Eq. 5.12 can be divided by i_b to give the input impedance R_{in}.

$$R_{\text{in}} = \frac{v_{be}}{i_b} = h_{ie} - \frac{h_{fe}h_{re}}{h_{oe} + G_C} \tag{5.14}$$

The collector current i_c is, by using Eq. 5.11,

$$i_c = -v_{ce}G_C = \frac{h_{fe}i_b G_C}{h_{oe} + G_C} \tag{5.15}$$

The current gain is i_c/i_b, so that Eq. 5.15 can be divided by i_b to yield

$$K_i = \frac{i_c}{i_b} = \frac{h_{fe}G_C}{h_{oe}+G_C} = \frac{h_{fe}}{1+h_{oe}R_C} \qquad (5.16)$$

If the driving source v_s is turned off and a voltage v_{ce} is applied between the collector and emitter, the ratio of i_c to v_{ce} can be taken to find the output admittance of the transistor. Then, $i_b = -h_{re}v_{ce}/(R_s + h_{ie})$ and

$$G_o = \frac{i_c}{v_{ce}} = h_{oe} - \frac{h_{re}h_{fe}}{R_s + h_{ie}} \qquad (5.17)$$

PROBLEM 5.5   Prove Eq. 5.17 is correct if $v_s = 0$.

It is better to develop skill in solving the equivalent circuit by inspection rather than attempting to memorize Eqs. 5.13, 5.14, 5.16 and 5.17.

EXAMPLE 5.2   To help clarify the use of the equivalent circuit, let us solve the circuit of Fig. 5.5 (Example 5.1) using the equivalent-circuit technique.

The h-parameters for this transistor were found in Problem 5.3 to be

$$h_{ie} = 4 \text{ k}\Omega$$
$$h_{re} = 8 \times 10^{-4}$$
$$h_{fe} = 230$$
$$h_{oe} = 25 \ \mu \text{ mho}$$

The value of R_C is 6 kΩ and let us assume that R_s is also 6 kΩ.

From Eq. 5.11, we have $v_{ce} = -h_{fe}i_b/(h_{oe}+G_C) = -i_b 230/(25 \times 10^{-6} + 167 \times 10^{-6}) = -1.19 \times 10^6 i_b$. From Eq. 5.12, $v_{be} = h_{ie}i_b - h_{fe}h_{re}i_b/(h_{oe} + G_C) = i_b[4 \times 10^3 - (2.3 \times 10^2 \times 8 \times 10^{-4})/(25 \times 10^{-6} + 167 \times 10^{-6})] = (4 \times 10^3 - 810)i_b = 3,1090i_b$. Then, the voltage gain is $-K_v = v_{ce}/v_{be} = -1.19 \times 10^6/3.19 \times 10^3 = -373$.

The value of R_{in} (Eq. 5.14) is $4 \times 10^3 - 810 = 3,190 \ \Omega$. Equation 5.16 can be used to find the current gain. Then, $K_i = 230/(1 + 0.15) = 200$. Finally, the output admittance of the transistor (Eq. 5.17) is $G_o = 25 \times 10^{-6} - [(2.3 \times 10^2 \times 8 \times 10^{-4})/(4 \times 10^3 + 6 \times 10^3)] = 25 \times 10^{-6} - 18.4 \times 10^{-6} = 6.6 \times 10^{-6}$ mhos.

PROBLEM 5.6   A given transistor has $h_{ie} = 1 \text{ k}\Omega$, $\beta = 100$, $h_{re} = 10^{-4}$, and $h_{oe} = 2 \times 10^{-5}$ mho at a given q-point. Calculate the voltage gain v_{ce}/v_{be}, current gain i_c/i_b, and input resistance of an amplifier that uses this transistor with $R_C = 5 \text{ k}\Omega$.   *Answer: $K_v = 476$, $K_i = 91$, $R_{in} = 955 \ \Omega$.*

Transistor types which have been developed recently have values of h_{re} much smaller than the value (8×10^{-4}) determined for the transistor of Fig. 5.6. Consequently, the value of h_{re} for these modern transistors can be assumed to be approximately equal to zero over the frequency range below beta-cutoff frequency (to be discussed later) where the h-parameter circuit is most useful. This assumption greatly simplifies the determination of the voltage gain, input impedance, and output impedance of the transistor in a given circuit configuration.

EXAMPLE 5.3   Let us consider the transistor amplifier of Example 5.2, assuming the input characteristics can be adequately represented by a single curve instead of the family of curves shown in Fig. 5.6; then $h_{re} \simeq 0$. Otherwise, the h-parameters have the same values as was previously determined. Then,

$$R_{\text{in}} \simeq h_{ie} \qquad = 4000 \ \Omega$$
$$R_{\text{out}} = 1/h_{oe} \qquad = 40 \ \text{k}\Omega$$

$$K_v = \frac{h_{fe}}{h_{ie}(h_{oe}+G_C)} = \frac{230}{4000(1.92 \times 10^{-4})} = 300$$

$$K_i = \frac{h_{fe}}{1+h_{oe}R_C} = 200 \ (\text{same as before})$$

Note that a further simplification can be made if $G_C \gg h_{oe}$, so that h_{oe} can be neglected. Then

$$K_v = \frac{h_{fe}}{h_{ie}G_C} = \frac{h_{fe}R_C}{h_{ie}} \tag{5.18}$$

$K_v = 345$ for this example.

$$K_i = h_{fe} = 230$$

PROBLEM 5.7   Calculate the voltage gain v_{ce}/v_{be}, the current gain i_c/i_b, the input resistance and output resistance (of the transistor) of the amplifier of Prob. 5.6, assuming $h_{re} \simeq 0$.
$\qquad\qquad$ *Answer:* $K_v = 454$, $\quad K_i = 91$, $\quad R_{\text{in}} = 1 \ \text{k}\Omega$, $\quad R_o = 50 \ \text{k}\Omega$.

PROBLEM 5.8   Calculate the voltage gain v_{ce}/v_{be}, current gain, and input resistance of the amplifier of Prob. 5.6, assuming $h_{re} \simeq 0$ and h_{oe} is negligible compared with G_C. $\qquad$ *Answer:* $K_v = 500$, $\quad K_i = 100$, $\quad R_{\text{in}} = 1 \ \text{k}\Omega$.

5.4   TRANSISTOR   RATINGS

Essentially the only cause of deterioration and destruction of a semiconductor device is heat, which may melt solder connections, deteriorate insulating materials, and produce changes in the crystal structure of the semiconductor. Therefore, manufacturers rate their semiconductor products in accordance with their power-dissipation capability and maximum permissible temperature. Also, maximum voltage ratings are given and sometimes maximum current is specified. These ratings and their application to circuit design will be discussed in this section.

The ratings of a typical low-power *n-p-n* germanium transistor are listed in Table 5.1.

Table 5.1   Absolute Maximum Ratings, 25°C

V_{CEO}	30 V
I_C	100 mA
Power dissipation, $P_d{}^a$	200 mW
Junction temperature, T_j	85°C

^aDerate 3.33 mW/°C for ambient temperatures above 25°C.

The safe operating area of this transistor may be marked on the collector characteristics as shown in Fig. 5.10. The transistor operation will be confined to this area if the dc load line remains *below* the maximum dissipation curve that is drawn through all points where $v_{CE}i_C = P_d$ max. For example, if $v_{CC} = 30$ V, as shown in Fig. 5.10, R_C min = 1.11 kΩ.

If the ambient, or surrounding, temperature is higher than 25°C, the maximum dissipation rating must be reduced, as specified by the manufacturer. For example, the transistor with the ratings given above must have its maximum dissipation rating reduced 3.3 mW for each degree C of ambient temperature above 25°C. Therefore, the maximum dissipation curve drawn on the collector characteristics should represent the maximum permissible dissipation at the highest expected ambient temperature. For example, if the transistor above is to be enclosed in a metal cabinet and used on the desert in the summer, the ambient temperature may rise to 55°C (131°F). The maximum dissipation rating is then P_d max = 200 − 3.33 × 30 = 100 mW.

PROBLEM 5.9   Draw the maximum dissipation curve for 100 mW dissipation on the collector characteristics of Fig. 5.10. Determine the minimum safe value of the collector load resistance for this dissipation if $V_{CC} = 20$ V.   *Answer:* $R_C = 1$ kΩ.

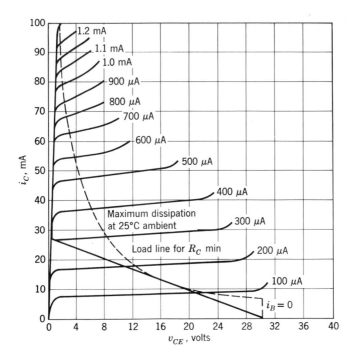

Fig. 5.10.   Safe operating area for a typical 200-mW transistor.

Sometimes a derating curve is given instead of a derating factor. A derating curve for the transistor of Fig. 5.10 is given in Fig. 5.11. Observe that the slope of the derating curve is equal to the derating factor. The slope of the curve in Fig. 5.11 (above 25°C) is −200 mW/60°C = −3.33 mW/°C. The negative sign results from the negative slope. However, the derating factor is usually given as a positive number.

Most of the power dissipation in a transistor occurs at the collector junction because of the relatively large voltage across that junction. The junction temperature rises above the ambient temperature because of this dissipation. The rise in junction temperature is equal to the power dissipation divided by the derating factor. For example, if the transistor of Fig. 5.10 is operating with $V_{CE} = 10$ V and $i_C = 10$ mA, the rise in junction temperature above the ambient is $v_{CE}(i_C)$/derating factor = 100 mW/3.3 mW per °C ≈ 30°C.

Heat flow, which results from a temperature difference, is analogous to current flow, which results from a potential difference. Therefore, a *thermal resistance* θ_T has been defined as the ratio of the temperature rise

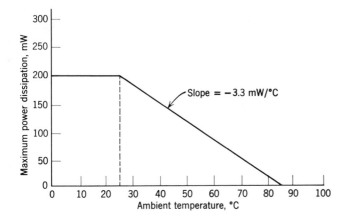

Fig. 5.11. Derating curve for the transistor of Fig. 5.10.

to the power dissipation, or

$$\theta_T = \frac{\Delta T}{P_d} \qquad (5.19)$$

and

$$\Delta T = \theta_T P_d \qquad (5.20)$$

These equations show that the thermal resistance is the reciprocal of the derating factor and is therefore the negative reciprocal of the slope of the derating curve. The thermal resistance between the junction and the ambient surroundings of the transistor of Fig. 5.10 is $\theta_T = 1/3.33 = 0.3°C/mW$.

The junction temperature is the ambient temperature T_a plus the temperature rise due to power dissipation. Therefore, by using Eq. 5.20

$$T_j = T_a + \Delta T = T_a + \theta_T P_d \qquad (5.21)$$

For example, if the ambient temperature is 40°C and the transistor of Fig. 5.10 is dissipating 100 mW, the collector-junction temperature $T_j = 40°C + 0.3 \times 100°C = 70°C$. Observe and verify that at any point on the derating curve, the ambient temperature plus the corresponding power dissipation times the thermal resistance θ_T gives the maximum permissible junction temperature.

PROBLEM 5.10   A given transistor has a 300-mW maximum dissipation rating at $T_a = 25°C$ and $T_j\text{max} = 175°C$. Determine (a) the derating factor, (b) the thermal resistance, and (c) the junction temperature when $T_a = 50°C$ and the average power dissipation is 100 mW.

Answer: (a) 2 mW/°C, (b) 0.5°C/mW, (c) 100°C.

The maximum current rating of a transistor, if given, usually indicates either the current at which the maximum dissipation curve crosses the saturation voltage (Fig. 5.10), or the current at which beta falls below the minimum specified by the manufacturer.

The maximum voltage rating is not as simply specified for a transistor as for a diode. The transistor may *appear* to break down at the voltage for which $\alpha = 1$, although this voltage may be considerable below the avalanche breakdown voltage shown in Fig. 4.10. The reason the transistor appears to break down is because the collector current approaches infinity as alpha approaches unity, as seen by Eq. 5.3, repeated below.

$$i_C = \frac{\alpha}{1-\alpha} i_B + \frac{I_{CO}}{1-\alpha} \tag{5.3}$$

Observe that i_C is equal to infinity for $i_B = 0$ or for any positive value of i_B when $\alpha = 1$. The *apparent* breakdown voltage when $i_B = 0$, is known as the sustaining voltage or *maximum* V_{CEO}. The first two subscripts indicate the electrodes to which the voltage is applied and the third subscript indicates the conditions at the third terminal (base), which in this case is open. Also observe from Eq. 5.3, however, that finite positive values of collector current may be obtained for values of α greater than unity if the base current is negative. In fact, if $i_B = -I_{CO}$, Eq. 5.3 shows that the collector current $i_C = I_{CO}(-\alpha/(1-\alpha) + 1/(1-\alpha)) = I_{CO}$ for any value of alpha. Of course, I_{CO} increases rapidly because of carrier multiplication as the avalanche breakdown voltage is approached. This avalanche breakdown voltage is known as the maximum V_{CBO}, because there is no current across the emitter junction when the emitter is open and the base current is automatically held to $-I_{CO}$.

When neither the base nor the emitter circuits are open and the base circuit has a finite resistance R as shown in Fig. 5.12a, part of I_{CO} flows as a negative base current i_B and the remaining part of I_{CO} flows across the emitter junction. This negative base current reduces the collector current, as compared with open base, and causes the apparent breakdown voltage, *maximum* V_{CER} to be higher than maximum V_{CEO}. The improvement in apparent breakdown voltage depends on the value of R. Maximum improvement occurs when $R = 0$, because this value gives the maximum

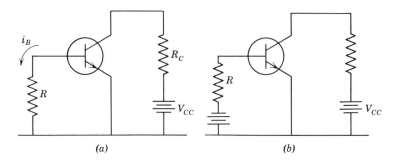

Fig. 5.12.   Negative i_B flows through resistance R in the base circuit.

negative i_B. The apparent breakdown voltage with $R = 0$ is known as maximum V_{CES} where the S means the base is shorted to the emitter. The value of base resistance must be specified when the maximum V_{CER} is given.

Not all of I_{CO} flows in the base circuit when $R = 0$ because of the internal resistance r_b. Therefore the negative base current can be increased and hence the apparent breakdown voltage increased if a reverse biasing voltage is included in the base circuit (Fig. 5.12b). The apparent breakdown voltage with this reverse voltage applied is known as *maximum V_{CEX}* and is essentially equal to V_{CBO}.

The collector characteristics for a typical *n-p-n* transistor, including the avalanche region, are given in Fig. 5.13a and the breakdown characteristics of this transistor are given in Fig. 5.13b. The various *maximum* voltage ratings for this transistor are listed along the voltage axis for a typical transistor.

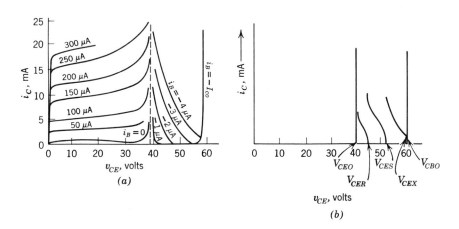

Fig. 5.13.   Voltage breakdown characteristics of a transistor.

5.5 BIASING CIRCUITS

The transistor circuits that have been considered previously have used a bias battery and a current-limiting resistor to provide forward bias to the emitter junction. However, in the common-emitter configuration the battery V_{BB}, which supplies forward bias to the base, has the same polarity, with respect to the emitter, as the battery V_{CC}, which supplies reverse bias to the collector. Therefore, a single battery or power supply can supply the proper bias to both junctions, as shown in Fig. 5.14. Since the dc voltage across the bias resistor R_B is $V_{CC} - V_{BE}$, the bias-circuit resistance can be determined, using Ohm's Law, from the relationship

$$R_B = \frac{V_{CC} - V_{BE}}{I_B} \tag{5.22}$$

EXAMPLE 5.4   If $V_{CC} = 20$ V, the desired q-point base current I_B is 20 μA and the q-point value of v_{BE} is $V_{BE} = 0.5$ V, then the proper value of R_B is 19.5 V$/(2 \times 10^{-5}) = 9.75 \times 10^5 \simeq 10^6$ Ω. Observe that V_{BE} may be neglected in Eq. 5.22 if V_{CC} is large in comparison with V_{BE} (factor of 10, at least).

This type of bias is known as *fixed* bias because the base bias current I_B is determined, or *fixed*, almost entirely by the values of R_B and V_{CC}. The idea of I_B being *fixed* may seem good at first thought. However, the thermally generated current I_{CO} approximately doubles for each 10°C temperature increase. Therefore, the q-point shifts up the load line with increasing temperature (Fig. 5.15) for a typical *n-p-n* germanium transistor. The reason for the large increase in i_C, with i_B held constant, is that I_{CO} is forced to flow across the emitter junction and is therefore amplified by the factor

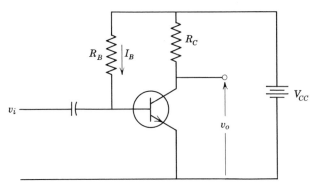

Fig. 5.14.   Fixed bias obtained from a single voltage source V_{CC}.

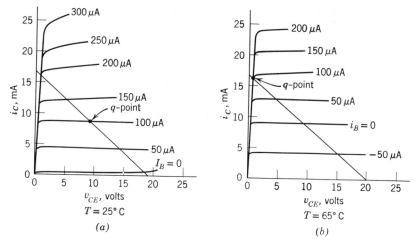

Fig. 5.15.   Collector characteristics of a typical *n-p-n* germanium transistor (*a*) at *T* = 25°C and (*b*) at *T* = 65°C.

($\beta + 1$), as was previously discussed. At the 65°C temperature, the amplifier of Fig. 5.15*b* is useless because the *q*-point is at the upper end of the load line and the transistor is in saturation.

The *q*-point shift, or collector current instability, can be reduced considerably if a resistor R_E is inserted in the emitter lead and the base-circuit resistor R_B is reduced from the fixed bias value. The reason for this improvement in *q*-point stability is shown in Fig. 5.16, where the thermally generated current I_{CO} is shown to divide between the base circuit and the

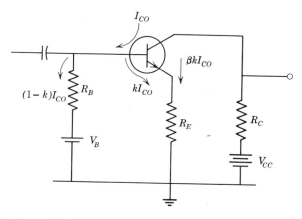

Fig. 5.16.   Circuit showing the division of thermal current I_{CO} between the emitter and base circuits.

emitter circuit. Only that portion, kI_{CO} that flows across the emitter junction is amplified by β; therefore, the collector current, which results from the thermal current, is $(I_{CO} + k\beta I_{CO}) = (1 + k\beta)I_{CO}$ and k can have values between 1 and 0. Since the shift in q-point is caused primarily by the increase of I_{CO} with temperature, the best stability is obtained when k is 0.

A current stability factor S_I has been defined as the ratio of collector current change to the change in I_{CO}, or

$$S_I = \frac{\Delta I_C}{\Delta I_{CO}} \tag{5.23}$$

Thus, if I_{CO} in Fig. 5.16 is increased by ΔI_{CO}, the collector current is increased by $(1 + k\beta)\Delta I_{CO}$ and the current stability factor $S_I = (1 + k\beta)$. Note that a small stability factor results in good q-point stability. If the base resistor R_B is very large and the emitter resistor R_E is zero, which occurs when fixed bias is used, the value of k is 1 and the $S_I = (1 + \beta)$. On the other hand, if $R_B = 0$ and R_E is large, which occurs when the common-base configuration is used, $k = 0$ and $S_I = 1$. Intermediate values of R_E and R_B give values of S_I that lie between one and $(\beta + 1)$. In fact, it can be shown that when S_I is large compared with 1, but small compared with $(\beta + 1)$.

$$S_I \simeq R_B / R_E \tag{5.24}$$

A desirable value for S_I can be determined after a transistor has been chosen for a given application, as illustrated by the following example.

EXAMPLE 5.5   A given n-p-n germanium transistor is to operate as a common-emitter amplifier with $V_{CC} = 25$ V and $R_L = 5$ kΩ. The load line is drawn on the collector characteristics in Fig. 5.17. Let us assume that the q-point should remain within the limits q_1 and q_2 as the ambient temperature changes from 25°C to 55°C. The transistor ratings are given in Table 5.2 for 25°C ambient temperature. As seen in Fig. 5.17, $\Delta I_C = 1$ mA. The value of I_{CO} depends on the junction temperature, not the ambient, so that the maximum junction temperature must be determined from the relationship $T_j = T_a + \theta_T P_d$, where $\theta_T = 0.5$°C/mW. Thus, the maximum junction

Table 5.2

V_{CEO}max	30 V
$P_d$max	140 mW
$T_j$max	95°C
I_{CO}	4 μA

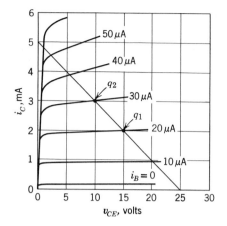

Fig. 5.17. Characteristics used to visualize the permissible shift in q-point.

temperature in this application is $T_a + \theta_T \times I_c V_c$ or $T_j \text{max} = 55° + 0.5 \times 3 \times \times 11 = 71°$.

We have no information from the manufacturer concerning the rate of increase of I_{CO} with temperature, so we assume that I_{CO} doubles for each 10°C junction temperature increase, which is typical for a germanium transistor. Then, at $T_j = 71°C$, the value of I_{CO} can be determined approximately from Table 5.3. We may estimate I_{CO} at $T_j = 71°$ to be about $100 \mu A$. Then, assuming $T_j \text{min}$ to be 25°C, $\Delta I_{CO} = 96 \mu A$ and $S_I = 1/0.096 = 10.4$. Note that $T_j \text{min}$ rises above 25°C as the transistor warms up, but if we want the q-point to *not* be below the lower limit q_1 during the warm-up period, we must use $T_j \text{min} = T_a \text{min} = 25°C$.

Table 5.3

T_j(°C)	I_{CO}
25	$4 \mu A$
35	$8 \mu A$
45	$16 \mu A$
55	$32 \mu A$
65	$64 \mu A$
75	$128 \mu A$

A more accurate method of predicting the maximum value of $I_{CO}(I_{CO_2})$, when a reference value I_{CO_1} is given, follows.

$$I_{CO_2} = I_{CO_1} 2^{(\Delta T_j/10)} \qquad (5.25)$$

Using Eq. 5.25, $I_{CO_2} = 97\,\mu A$. This value is so near the first estimate that we will not alter our value of S_I.

We now know the required S_I, which is approximately the ratio of R_B to R_E. But we need to determine a suitable value of R_E that will be bypassed (Fig. 5.18) to eliminate degeneration or loss of ac signal amplification. R_E is actually part of the dc resistance between the collector and emitter and therefore should be included in the dc load resistance and load line. Generally, R_E should be small in comparison with R_C to avoid unnecessary power loss in R_E. $R_E = 0.2\,R_C$ to $0.25\,R_C$ is usually adequate. In this example we choose $R_E = 1\,k\Omega$. R_C will then be $4\,k\Omega$ so that the total dc load resistance will be $5\,k\Omega$, as specified. The value of R_B required for proper stability is $R_B = S_I R_E = 10.4 \times 1\,k\Omega = 10.4\,k\Omega$ or approximately $10\,k\Omega$. Since the transistor input resistance is approximately $(\beta + 1)r_e = (100)(25)/2 = 1250\,\Omega$, the $10\,k\Omega$ value of R_B would shunt about 12 percent of the ac signal to ground, which is a small price to pay for the good stability achieved.

The two batteries shown in Fig. 5.18 are not required for a stabilized bias system. A single-battery system is shown in Fig. 5.19a. Observe that a dc Thevenins equivalent circuit looking into the base-bias circuit to the left of the points A and B is as shown in Fig. 5.19b, where R_b is the parallel combination of R_1 and R_2, or

$$R_b = \frac{R_1 R_2}{R_1 + R_2} \tag{5.26}$$

and the open-circuit voltage, or voltage between points A and B with the base disconnected, is

$$V_B = \frac{R_1}{R_1 + R_2}\,V_{CC} \tag{5.27}$$

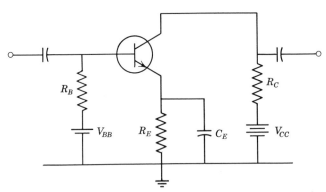

Fig. 5.18. A stabilized transistor amplifier with the emitter resistor bypassed.

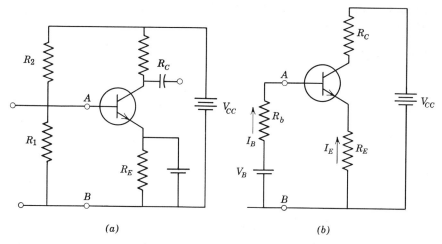

Fig. 5.19. (a) A stabilized bias system that uses a single battery and (b) the dc Thevenin's equivalent of this circuit.

Notice that this equivalent circuit is the same as the circuit of Fig. 5.18. The value of V_B can be found by adding the voltage drops around the base circuit at one of the q-points, preferably the lower one. Thus

$$V_B = I_B R_b + V_{BE} - I_E R_E \tag{5.28}$$

Notice that I_E is a negative current since it flows out of the transistor. Therefore the term $-I_E R_E$ will be positive for an n-p-n transistor. Input characteristics may not be available, in which case V_{BE} may be assumed to be about 0.2 V for a germanium transistor of 0.5 V for a silicon transistor. For example, in the preceding amplifier of Figs. 5.17 and 5.18, $V_B = 20 \times \times 10^{-6} \times 10^4 + 0.2 + 1.9 \times 10^{-3} \times 10^3 = 2.3$ V.

Since V_{CC} and V_B are now known, there are only two unknowns (R_1 and R_2) in the two equations, 5.26 and 5.27. Therefore, these equations can be solved simultaneously for R_1 and R_2 in terms of V_{CC}, V_B and R_b. This solution yields

$$R_1 = \frac{V_{CC}}{V_{CC} - V_B} \; R_b \tag{5.29}$$

$$R_2 = \frac{V_{CC}}{V_B} R_b \tag{5.30}$$

Therefore, the circuit of Fig. 5.19a can be used for the amplifier with characteristics shown in Fig. 5.17, in which case $R_1 = 25 \times 10.4 \,\text{k}\Omega/(25 -$

2.3) $= 11.5\,\mathrm{k}\Omega$ and $R_2 = 25 \times 10^4\,\mathrm{k}\Omega/2.3 = 113\,\mathrm{k}\Omega$. $R_1 = 12\,\mathrm{k}\Omega$ and $R_2 = 120\,\mathrm{k}\Omega$, which are stock size 10 percent resistors, would be suitable.

A more general and precise method of designing a bias circuit will be discussed in Section 5.6.

PROBLEM 5.11   The germanium transistor used in the preceding example, with characteristics given in Fig. 5.17 and Table 5.2, is used in the amplifier circuit in Fig. 5.19a. This amplifier must operate over the ambient temperature range $T_a = 5°\mathrm{C}$ to $T_a = 65°\mathrm{C}$ and the q-point may vary from $I_C = 1.5\,\mathrm{mA}$ to $I_C = 3.5\,\mathrm{mA}$. $R_C = 4\,\mathrm{k}\Omega$ and $R_E = 1\,\mathrm{k}\Omega$, as in the preceding example. Determine suitable values for R_1 and R_2.

Answer: $R_1 = 15\,\mathrm{k}\Omega, R_2 = 180\,\mathrm{k}\Omega$.

5.6 STABILIZED BIAS

In the preceding sections, I_{CO} was considered the only parameter that varies with temperature. If this were true, the bias stabilization technique discussed in the preceding section would be entirely adequate. However, two parameters in addition to I_{CO} may change with temperature and cause undesirable q-point shift. One of these is V_{BE}, which is primarily the forward bias voltage across the emitter junction. The diode equation (Eq. 3.8) shows that the diode current strongly depends on the saturation current I_S or I_{EO}, which is an exponential function of temperature, if the junction voltage is held constant. Similarly, if the emitter current is held constant, the junction *voltage* will vary with temperature as shown in Fig. 5.20.

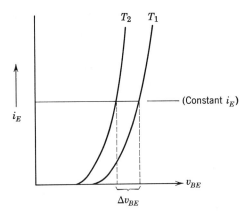

Fig. 5.20.   A sketch showing the effect of temperature on the input characteristics of a transistor.

When i_E is held constant and the temperature is increased from T_1 to T_2, v_{BE} decreases by the amount Δv_{BE}. The rate of change of v_{BE} with temperature is typically about $-2.5\ mV/°C$. Thus, if a typical silicon transistor has $v_{BE} = 0.55$ V at $i_E = 1$ mA and $T = 25°C$, it would have $v_{BE} = 0.425$ at $T = 75°C$ and $i_E = 1$ mA.

The other parameter which is temperature sensitive, particularly in a silicon transistor, is β. Figure 5.21 shows how a change in β causes a q-point shift, assuming base current is held constant. Generally, β increases with temperature.

The stabilized bias circuit of Fig. 5.19a can be used to control the q-point shift, which may result from a combination of changes in I_{CO}, v_{BE}, and β, as shown below. We begin with the base circuit voltage equation (Eq. 5.28), repeated below, which was written by inspecting Fig. 5.19b.

$$V_B = I_B R_b + V_{BE} - I_E R_E \qquad (5.28)$$

However, we specify the q-point in terms of I_C, not I_B and I_E. Therefore, the following relationships are needed. First, using Kirchoff's current law, $I_E + I_C + I_B = 0$.

$$-I_E = I_C + I_B \qquad (5.31)$$

Then, using the relationship $I_C = \beta I_B + (\beta + 1)I_{CO}$, which is Eq. 5.4 written for dc values,

$$I_B = \frac{I_C}{\beta} - \frac{\beta + 1}{\beta} I_{CO} \qquad (5.32)$$

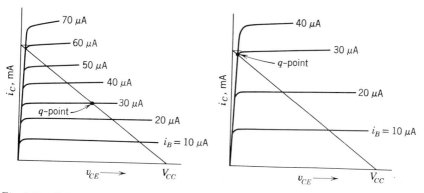

Fig. 5.21.   Collector characteristics showing q-point shift with changing β but i_B constant.

Substituting Eq. 5.31 into Eq. 5.28 and then substituting Eq. 5.32 into the result,

$$V_B = \left(\frac{I_C}{\beta} - \frac{\beta+1}{\beta} I_{CO}\right) R_b + V_{BE} + \left(I_C + \frac{I_C}{\beta} - \frac{\beta+1}{\beta} I_{CO}\right) R_e \qquad (5.33)$$

This equation can be simplified if β is assumed to be much larger than 1, then $(\beta+1) \simeq \beta$ and I_C is much larger than I_C/β in the right-hand term. Then

$$V_B \simeq \left(\frac{I_C}{\beta} - I_{CO}\right) R_b + V_{BE} + (I_C - I_{CO}) R_E \qquad (5.34)$$

Now, the objective is to restrict the q-point values of I_C to a predetermined range ΔI_C between I_{C_1} and I_{C_2} (Fig. 5.22). At I_{C_1} the collector-junction temperature is minimum, the thermal current I_{CO_1} is minimum, and β is minimum. Conversely, when the q-point is at I_{C_2}, the values of T_j, I_{CO_2}, and β_2 are maximum. To accomplish this, we select a suitable value for R_E, as was discussed in Section 5.5, and then calculate the required value of R_b, which may be determined by first writing Eq. 5.34 with values at I_{C_2} substituted and then writing Eq. 5.34 again with values at I_{C_1} substituted. The second equation is then subtracted from the first, letting $(I_{C_2} - I_{C_1}) = \Delta I_C$, $(I_{CO_2} - I_{CO_1}) = \Delta I_{CO}$, and $(V_{BE_2} - V_{BE_1}) = \Delta V_{BE}$, to give

$$0 = \left(\frac{I_{C_2}}{\beta_2} - \frac{I_{C_1}}{\beta_1} - \Delta I_{CO}\right) R_b + \Delta V_{BE} + (\Delta I_C - \Delta I_{CO}) R_E \qquad (5.35)$$

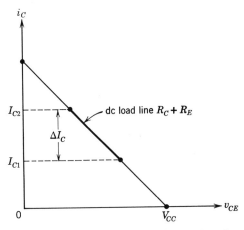

Fig. 5.22.   Confinement of q-point range from I_{C1} to I_{C2}.

Solving Eq. 5.35 explicitly for R_b,

$$R_b = \frac{\Delta V_{BE} + (\Delta I_C - \Delta I_{CO})R_E}{\dfrac{I_{C_1}}{\beta_1} - \dfrac{I_{C_2}}{\beta_2} + \Delta I_{CO}} \qquad (5.36)$$

Although Eq. 5.36 was developed for an *n-p-n* transistor, it is applicable to a *p-n-p* type because all the signs of the variables I_C, I_{CO}, and V_{BE} change. The use of Eq. 5.36 will be illustrated by carrying through a more precise design for the bias design of Example 5.5. The following values have already been determined or assumed.

EXAMPLE 5.6

$$
\begin{array}{lll}
T_{j_1} = 25°C & T_{j_2} = 71°C & \Delta T_j = 46°C \\
I_{C_1} = 1.9\ \text{mA} & I_{C_2} = 2.9\ \text{mA} & \Delta I_C = 1\ \text{mA} \\
I_{CO_1} = 4\ \mu\text{A} & I_{CO_2} = 97\ \mu\text{A} & \Delta I_{CO} = 93\ \mu\text{A} \\
V_{BE_1} = 0.2\ \text{V} & \beta_1 \simeq 95 & R_E = 1\ \text{k}\Omega
\end{array}
$$

We also assume that β rises to 150 at the maximum junction temperature and that $\Delta V_{BE}/°C = -2.5\ \text{mV}/°C$. Then $\Delta V_{BE} = -2.5 \times 10^{-3}(46) = -0.115$ V. Substituting these values into Eq. 5.36, $R_B = 8.5\ \text{k}\Omega$. Notice that R_b is less than the value calculated in Section 5.5 because changes in V_{BE} and β were taken into account.

The value of V_B can be calculated by using a known set of parameters at any given *q*-point. The values associated with I_{C_1} can be obtained most easily because I_{CO} is usually small compared with I_{C_1}/β_1 at this low temperature *q*-point. (See Eq. 5.32 for calculating I_B.) The value of V_B for this example was determined in Section 5.5 to be 2.3 V. The values of R_1 and R_2 for the circuit of Fig. 5.19*a*, with $V_{CC} = 25$ V, are $R_1 = 9.3$ k and $R_2 = 92\ \text{k}\Omega$.

When Eq. 5.36 yields values of R_b that are too low, and therefore seriously shunt the input currents, either a larger value of R_E is needed or a larger value of ΔI_C should be permitted. Perhaps replacement of a germanium transistor with a silicon transistor to reduce ΔI_{CO} may be the best solution. In some instances, negative values of R_b will be obtained from Eq. 5.36. The causes and possible solutions to this problem follow.

1. The numerator may become negative because ΔV_{BE} is negative and $(\Delta I_C - \Delta I_{CO})R_E$ is either (a) negative or (b) positive, but with smaller magnitude than ΔV_{BE}. In case (a), ΔI_{CO} is larger than ΔI_C and if ΔI_C cannot be increased considerably, a different transistor with lower ΔI_{CO} should be

selected. In case (b), the same solutions as listed for case (a) are applicable. In addition R_E may be increased.

2. The denominator may be negative because $I_{C_1}/\beta_1 - I_{C_2}/\beta_2$ is negative and has greater magnitude than ΔI_{CO}. This indicates that for the given temperature range ΔI_C has been chosen larger than can be obtained with any positive value of R_b. Decreasing ΔI_C will solve the problem. Also, if the value of ΔI_C gives positive but very large values of R_b so that V_B is greater than V_{CC}, R_1 will be negative. The solution to this problem is also to either reduce ΔI_C or arbitrarily reduce R_b to a practical value.

The stabilized bias circuit is also applicable to a mass-produced amplifier that must accept a specified variation in transistors due to manufacturing tolerances as well as temperature differences.

PROBLEM 5.12   Assume that the amplifier design, previously considered in Section 5.5 and continued in this Section, is to be mass produced using factory run transistors with $\beta_{min} = 50$ at 25°C and $\beta_{max} = 200$ at 75°C, $V_{BE}max = 0.25$ at 25°C and $V_{BE}min = 0.08$ at 75°C (at $I_E = 2$ mA). The value of $I_{CO} = 4.0\,\mu A$ at 25°C is the maximum for this type of transistor. Determine suitable values for R_b, V_B, R_1, and R_2 if the q-point shift, V_{CC}, and R_E remain the same as previously specified.
Answer: $R_b = 6.3$ kΩ, $V_B = 2.39$ V, $R_1 = 6.95$ kΩ and $R_2 = 66$ kΩ. ($R_1 = 6.8$ kΩ and $R_2 = 68$ kΩ would probably be used.)

Some signal shunting by R_b in the mass produced amplifier may not be objectionable because the higher gain transistors have higher input impedance (almost proportional to β) and therefore the shunting effect of R_b tends to make the variation (or spread) of amplifier gain much less than the spread of β.

A firm basis for determining the maximum permissible ΔI_C will be given in Chapter 8.

PROBLEM 5.13   Assume that you are hired by an electronics firm to design a transistor preamplifier that will amplify the signal from a phonograph pickup to at least a 1.0-V peak level for driving a main (power) amplifier. The phono-pickup generates a 10-mV peak signal, open circuit, and has an internal resistance of 2000 Ω. The preamp will be mass produced, and must operate satisfactorily over the ambient temperature range of 0°C to 50°C. The available supply voltage is $V_{CC} = 20$ V. An *n-p-n* silicon transistor with $I_{CO} = 0.01\,\mu A$ max at 25°C, $P_d max = 300$ mW at 25°C, and derating $= 2$ mW/°C above 25°C has been chosen. The max β spread (range) is 80 to 300 over the temperature range desired, and the recommended value of collector-circuit resistor R_C is 6.8 kΩ. The amplifier will operate properly if the q-point variation is limited to $I_{C_1} = 0.8$ mA and $I_{C_2} = 1.8$ mA. V_{BE}max $= 0.50$ V at $I_C = 0.8$ mA, $T = 25°$C. Choose a suitable value of R_E, calculate values for R_1 and R_2, and check the output voltage at both the minimum and maximum q-point values, assuming $h_{ie} \simeq (\beta + 1)r_e$, $h_{re} \simeq 0$, and h_{oe} is very small in comparison with G_L. The main amplifier has very high input resistance compared with R_C, so that R_C is the ac load resistance. Assume that the reactances of coupling and bypass capacitors are negligible. Is the voltage gain adequate?

Answer: (For $R_E = 1.5$ kΩ) $R_1 = 470$ kΩ, $R_2 = 1.3$ MΩ, yes (1200; 3300).

CHAPTER **6**

DEVICES WITH HIGH
INPUT IMPEDANCE

The input impedance of an amplifier frequently determines its suitability for a given application because maximum power is transferred and, hence, the maximum power gain is achieved when the load resistance matches the source resistance. Consequently, the common-emitter amplifier is usually preferred over the common-base configuration because the common emitter has reasonably good match between input and output impedance and, therefore, can be operated in cascade (output of one amplifier or stage provides the input signal for a following amplifier) without the use of transformers. Many signal sources, such as crystal or ceramic microphones and phonograph pickups require load impedances of the order of a megohm or higher, which is much higher than the input impedance of a common-emitter amplifier. Therefore, this chapter will be devoted to the common-collector configuration, or emitter follower, and the field-effect transistor, both of which are high input-impedance devices.

6.1 THE COMMON-COLLECTOR
AMPLIFIER

The common-collector amplifier, which is frequently known as an emitter follower, has the load in the emitter circuit (Fig. 6.1). The collector is normally connected directly to V_{CC}, which is at ac ground potential; thus, the name *common collector*.

120

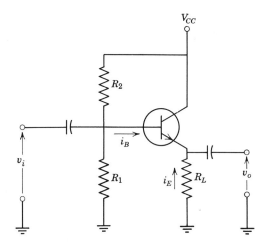

Fig. 6.1.   A common-collector amplifier or emitter follower.

Figure 6.1 shows that the input voltage v_i is equal to the output voltage v_o plus the ac voltage across the forward-biased emitter junction, provided the reactances of the coupling capacitors are small enough to be neglected. This voltage relationship can be expressed as follows:

$$v_i = v_{be} + v_o \tag{6.1}$$

Thus, the input voltage is larger than the output voltage and the voltage gain is less than 1. Usually the output voltage v_o is much larger than v_{be}, so the voltage gain is almost 1. Thus, the emitter voltage *follows* the base or input voltage very closely, as the name *emitter follower* indicates. Additional characteristics of the emitter follower may be deduced by observation of Fig. 6.1. Since the transistor input current is the base current and the current in the load is the emitter current, the current gain is approximately $(\beta + 1)$ for moderate values of load resistance. Therefore, Eq. 6.1 can be rewritten with $(\beta + 1)i_b R_L$ substituted for v_o and $i_b h_{ie}$ substituted for v_{be} to obtain

$$v_i \simeq i_b h_{ie} + (\beta + 1)i_b R_L \tag{6.2}$$

The input impedance R_i at the base of the transistor is v_i/i_b and, therefore, may be easily determined from Eq. 6.2

$$R_i \simeq h_{ie} + (\beta + 1)R_L \tag{6.3}$$

Thus, the input impedance of the emitter follower is increased over that of a common-emitter amplifier by approximately $(\beta+1)R_L$. Also, since the transistor output terminal is the emitter, the output impedance of the emitter follower should be the same order of magnitude as the input impedance of the common-base configuration, which is low.

The biasing circuit for the emitter follower may be determined in the same manner as described for the common-emitter amplifier in Chapter 5. However, since R_L is in the emitter circuit, the permissible values of base circuit resistance are usually quite large and the resistor R_1 may often be omitted to prevent the bias resistance from seriously reducing the input resistance.

EXAMPLE 6.1   Assume that the collector characteristics given in Fig. 6.2 are for the transistor in the amplifier of Fig. 6.1 and that $R_L = 2\,\text{k}\Omega$ is the total load resistance. The load line is drawn on the collector characteristics in Fig. 6.2 for $V_{CC} = 20\,\text{V}$ and the desired q-point is specified. Actually, the vertical axis of the collector characteristics should be i_E instead of i_C, but $i_E \simeq i_C$, so the available common-emitter curves are used. We assume the transistor is silicon with $v_{BE} = 0.5\,\text{V}$ at the q-point. From the figure we see that V_{CE} at the q-point is 10.5 V. The voltage across R_2 is then $V_{CE} - V_{BE} = 10\,\text{V}$ and with R_1 omitted, the value of R_2 is 10 V/ $50\,\mu\text{A} = 200\,\text{k}\Omega$. If h_{ie} is $1\,\text{k}\Omega$, the input resistance of the transistor is $1\,\text{k}\Omega + (101)\,2\,\text{k}\Omega$, which is approximately $200\,\text{k}\Omega$ and the input resistance to the amplifier is approximately $100\,\text{k}\Omega$, including the parallel biasing resistor.

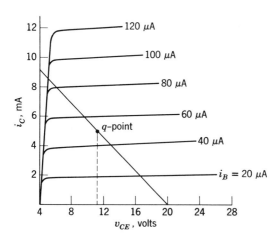

Fig. 6.2.   Collector characteristics and load line for the amplifier of Fig. 6.1.

PROBLEM 6.1   The load resistor R_L of the emitter follower of Figs. 6.1 and 6.2 ($V_{CC} \simeq 20$ V) is changed to 5 kΩ. Draw a load line, select a q-point and determine the approximate input resistance of the transistor and the circuit, using a single biasing resistor, R_2.

Answer: 500 kΩ, 250 kΩ if $V_{CB} = 10$ V.

6.2   EQUIVALENT CIRCUITS FOR THE COMMON-COLLECTOR CONFIGURATION

The common-collector equivalent-T circuit of Fig. 6.3a is the same as the common-emitter equivalent-T circuit of Fig. 5.7b except that the positions of the emitter and collector terminals have been exchanged. This circuit will be used to obtain the parameters for the h-parameter circuit given in Fig. 6.3b. The h-parameter circuit for the common-collector configuration is the same as the h-parameter circuits for the common-base and common-emitter configurations except for the values of the h-parameters.

The h-parameters for the common-collector configuration may be determined by observation of Fig. 6.3a while applying the definitions of these parameters. Since h_{ic} is the input impedance with the output shorted, the collector is shorted to the emitter for alternating current. But it makes no difference whether these terminals are exchanged, as compared with the common-emitter configuration, if they are shorted together. Therefore

$$h_{ic} = h_{ie} \tag{6.4}$$

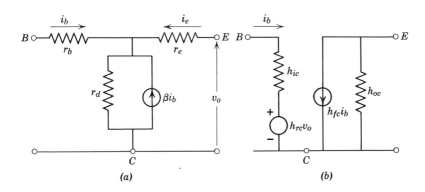

(a)   (b)

Fig. 6.3.   Equivalent circuits for the common-collector configuration. (a) Equivalent-T circuit and (b) h-parameter circuit.

Also, h_{fc} is the ratio of emitter current to base current with the output shorted. We are already familiar with this ratio in terms of β or h_{fe}. Notice, however, that i_e flows out of the transistor when i_b flows in, so

$$h_{fc} = -(\beta+1) = -(h_{fe}+1) \tag{6.5}$$

Since h_{oc} is the output admittance with the input open, this parameter can be obtained by letting $i_b = 0$ in Fig. 6.3a. Then

$$h_{oc} = \frac{1}{r_d + r_e} \tag{6.6}$$

But, as was shown in Fig. 5.7b, the common-emitter parameter h_{oe} has the same value, and, therefore,

$$h_{oc} = h_{oe} \tag{6.7}$$

Thus far, the common-collector h-parameters have proven to be either identical or almost equal to their common-emitter counterparts, so the impression might be developing that the common-collector configuration should behave very much like the common emitter. However, the final parameter, h_{rc}, is vastly different (Fig. 6.3a). If voltage v_2 is applied to the output terminals, nearly all of v_2 appears at the open circuit input terminals. The ratio of the voltage at the input terminals to v_2 is

$$h_{rc} = \frac{r_d}{r_d + r_e} \simeq 1 \tag{6.8}$$

The common-collector h-parameter circuit can now be redrawn as shown in Fig. 6.4 with the circuit components given in terms of the common

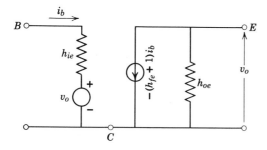

Fig. 6.4.   The common-collector h-parameter circuit labeled in terms of common-emitter h-parameters (except h_{rc}).

emitter h-parameters that are often furnished by the manufacturer or may be determined from the characteristic curves as discussed in Section 5.3.

As was mentioned previously, since all the h-parameter circuits have identical form, the equations that were developed in Section 5.3 for voltage gain, current gain, and input resistance may be used for any configuration providing the appropriate h-parameter values are substituted. Also, the circuit may be solved by observation of the equivalent circuit, assuming the input current i_b is known.

EXAMPLE 6.2   Let us find the current gain, and input resistance for the amplifier of Fig. 6.1, excluding the bias resistance. The load resistance for this amplifier is 2 kΩ and the common-emitter h-parameters at the specified q-point are $h_{fe} = 99$, $h_{ie} = 700\,\Omega$, $h_{oe} = 4 \times 10^{-5}$ mho, and $h_{re} = 2 \times 10^{-4}$. Then

$$v_o = -\frac{-(h_{fe}+1)i_b}{h_{oe}+G_L} = \frac{100i_b}{5.4 \times 10^{-4}} = 1.85 \times 10^5\, i_b$$

$$v_i = h_{ie}i_b + v_o = (700 + 1.85 \times 10^5)\, i_b$$

$$K_v = \frac{v_o}{v_i} = \frac{1.85 \times 10^5}{1.857 \times 10^5} = 0.997$$

$$i_L = \frac{v_o}{R_L} = \frac{1.85 \times 10^5}{2 \times 10^3}\, i_b = 97.5i_b$$

$$K_i = 97.5$$

$$R_i = \frac{v_i}{i_b} = \frac{i_b h_{ie} + v_o}{i_b} = \frac{1.857 \times 10^5}{i_b}\, i_b = 1.857 \times 10^5\,\Omega$$

The output resistance can be determined by the use of Fig. 6.5 if the source resistance R_s is known. Voltage v_2 is applied at the output terminals as shown and current i_e flows into the transistor. We can see that

$$i_e = h_{oe}v_2 - (h_{fe}+1)i_b \qquad (6.9)$$

But

$$i_b = -\frac{v_2}{h_{ie}+R_s} \qquad (6.10)$$

Then, substituting this value of i_b into Eq. 6.9,

$$i_e = h_{oe}v_2 + \frac{h_{fe}+1}{h_{ie}+R_s}v_2 \qquad (6.11)$$

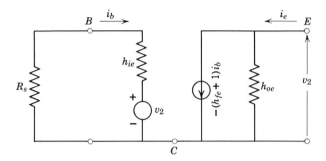

Fig. 6.5. Circuit used to determine the output resistance of the common-collector configuration.

and since the output admittance $Y_o = i_2/v_2$

$$Y_o = h_{oe} + \frac{h_{fe}+1}{h_{ie}+R_s} \qquad (6.12)$$

Thus, if the amplifier of Example 6.2 is driven by a resistive source with $R_s = 4{,}300\ \Omega$, by using the h-parameters given, the output admittance and impedance of the amplifier are $Y_o = 4 \times 10^{-5} + 100/(5 \times 10^3) = 2.004 \times 10^{-2}$ and $R_o = 50\ \Omega$.

You may have noticed from the preceding example that the input resistance is

$$R_i = h_{ie} + \frac{(h_{fe}+1)}{h_{oe}+G_L} \qquad (6.13)$$

But the second term on the right side of Eq. 6.13 is much larger than h_{ie} for a typical value of R_L. Also, G_L is usually much larger than h_{oe}, so $r_i \simeq (h_{fe}+1)R_L$. Equation 6.12 showed that the output admittance $Y_o = h_{oe} + (h_{fe}+1)/(h_{ie}+R_s)$, but the example indicated that h_{oe} was negligible in comparison with the second term. Therefore, $R_o = 1/Y_o \simeq (h_{ie}+R_s)/(h_{fe}+1)$. Thus, the emitter follower behaves like an impedance transformer with an impedance ratio equal to $(h_{fe}+1)$. However, unlike a conventional transformer the emitter follower has a power gain $K_p \simeq K_i \simeq (h_{fe}+1)$.

PROBLEM 6.2 A 2N2712 transistor (Appendix I) is used as an emitter follower with $R_L = 5\ \text{k}\Omega$. The driving source resistance $R_s = 3\ \text{k}\Omega$. Using the q-point for which the common-emitter h-parameters are given, determine the common-collector h-parameters. Draw an h-parameter equivalent circuit and use it to calculate the current gain, voltage gain, input impedance, and output impedance of the amplifier.

Answer: $K_i = 183$, $K_v = 0.975$, $R_i = 470\ \text{k}\Omega$, $R_o = 29\ \Omega$.

6.3   BOOTSTRAPPING TECHNIQUES

The input impedance of the common-collector amplifier is limited by the bias resistors, as was discussed in Section 6.1. Also, in the h-parameter circuit, the load resistance R_L is in parallel with h_{oe}. Therefore, the input resistance at the base of the transistor is $h_{ie} + (h_{fe} + 1)/(G_L + h_{oe})$ as given by Eq. 6.13. Thus, the maximum input resistance that is obtained when $R_L = \infty$ or $G_L \simeq 0$ is $h_{ie} + (h_{fe} + 1)/h_{oe}$. But $(h_{fe} + 1)/h_{oe} \simeq (h_{fe} + 1)r_d \simeq r_c$. Therefore, the maximum input impedance of the conventional emitter follower is approximately r_c, which may be a few megohms for a low-current transistor. Since the bias resistance is in parallel with this input resistance, the emitter follower may not have sufficiently high input impedance for many applications.

A technique that can be used to increase the input impedance of an emmiter follower is known as *bootstrapping*. This word comes from the phrase "lifting ones-self by his own bootstraps." The basic principle of bootstrapping is illustrated in the circuit of Fig. 6.6. In this circuit the capacitor C_2 is a dc blocking capacitor and has negligible reactance at the lowest signal frequency. Therefore, the output voltage v_o is applied at the junction of resistors R_1 and R_2. These resistors replace the single resistor that normally appears between base and ground in a stabilized bias circuit. But the ac voltage across R_2 is the difference between the input voltage v_i and the output voltage v_o, or $v_{R_2} = v_i - v_o$. Since v_o is nearly equal to v_i in an emitter follower, there is very little signal voltage across R_2 and, hence, very little signal current through R_2. This current can be determined by dividing the voltage by the resistance, or $i_{R_2} = (v_i - v_o)/R_2$. The effective

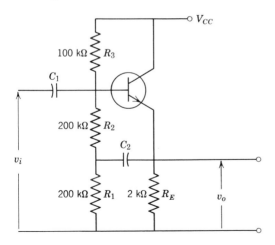

Fig. 6.6.   Bootstrapping used to increase biasing impedance.

resistance of the series combination of R_1 and R_2, as seen by the input voltage, is the ratio of the input voltage to the current i_{R_2}. Then

$$R_{\text{eff}} = \frac{v_i}{i_{R_2}} = \frac{v_i}{(v_i - v_d)/R_2} = \frac{R_2 v_i}{v_i - v_o} \tag{6.14}$$

If both the numerator and the denominator of Eq. 6.14 are divided by v_i, and it is recognized that v_o/v_i is the voltage gain K_v, Eq. 6.14 may be rewritten as

$$R_{\text{eff}} = \frac{R_2}{1 - K_v} \tag{6.15}$$

If the transistor in Fig. 6.6 is the same type as in Example 6.2, where K_v was found to be greater than 0.99 and $R_2 = 200$ kΩ, the effective resistance of R_1 and R_2 is at least $2 \times 10^5/0.01 = 2 \times 10^7$ Ω.

This high effective resistance is encouraging for the attainment of a high input resistance. The fact that the resistance of R_1 did not appear directly in the effective resistance formula may give the impression that the value of R_1 is unimportant. However, R_1 is in parallel with R_L, as far as the signal is concerned, and should be large in comparison with R_L. Otherwise, the voltage gain will be reduced, with a resulting reduction of effective resistance.

With the effective resistance of R_1 and R_2 increased to 20 MΩ, the bias resistor R_3 in Fig. 6.6 seriously limits the input resistance. This resistor may also be bootstrapped, as shown in Fig. 6.7. If the voltages and currents in the circuit of Fig. 6.7 are the same as their counterparts of Fig. 6.6, and

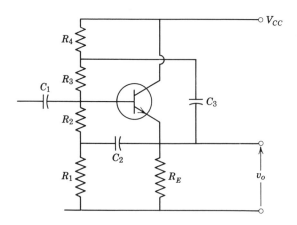

Fig. 6.7.   A circuit that bootstraps both R_2 and R_3.

the basic principles of bootstrapping are applied, the effective value of $R_3 + R_4$ becomes approximately equal to $100\,R_3\ \Omega$ and the resistance of R_4 is large in comparison with R_E. If R_3 is chosen to be $50\,\text{k}\Omega$, the effective resistance of the combination of R_3 and R_4 is $R_{\text{eff}} = 5\,\text{M}\Omega$.

The effective resistance of the bias circuit of Fig. 6.7 is probably higher than r_c, which is the limiting value of the input resistance of the transistor, as was previously discussed. The value of r_c is inversely proportional to the collector current in a given transistor, so that very high values of input impedance may be obtained if the q-point value of collector current is very small. However, the current gain or β should be high at this small value of collector current to maintain a high value of $(\beta + 1)R_L$, which in parallel with r_c, primarily determines the input impedance of the transistor.

PROBLEM 6.3   The amplifier of Fig. 6.7 has $h_{ie} = 5.0\,\text{k}\Omega$, $h_{oe} = 10^{-5}$ mho, $h_{fe} = 200$, $R_E = 10\,\text{k}\Omega$, $R_1 = 200\,\text{k}\Omega$, $R_2 = 50\,\text{k}\Omega$, $R_3 = 50\,\text{k}\Omega$, and $R_4 = 200\,\text{k}\Omega$. Determine the input impedance of the amplifier at frequencies for which the capacitors have negligible reactance.

Answer: 1.38 MΩ.

6.4   FIELD-EFFECT TRANSISTORS

The field-effect transistor (FET) is a voltage-controlled semi-conductor which has very high input impedance, particularly at low frequencies, such as audio frequencies. This transistor has only one p-n junction, as shown in Fig. 6.8. It is sometimes known as a unipolar field-effect transistor (UNIFET) because it has only one type of charge carrier. Most FET's are symmetrical, as indicated in Fig. 6.8, so that the source and drain terminals are interchangeable.

The schematic representation in Fig. 6.9 will be used to explain the

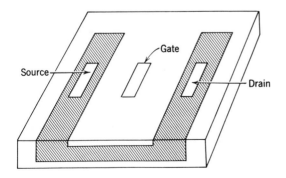

Fig. 6.8.   Typical field-effect transistor structure.

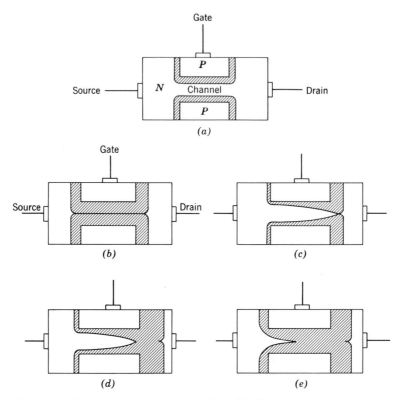

Fig. 6.9.   Schematic representation of the field-effect transistor. (a) $v_{GS} = 0$, $v_{DS} = 0$, (b) $v_{GS} = V_P$ (pinch-off), $v_{DS} = 0$, (c) $v_{GS} = 0$, $v_{DS} = V_P$, (d) $v_{GS} = 0$, $v_{DS} > V_P$, and (e) $v_{GS} = 0, v_{DS} > V_P$.

principles of operation of the field-effect transistor. Fig. 6.9a shows that a narrow semiconductor channel provides a conducting path between the *source* and the *drain*. This channel may be either an *n*- or *p*-type crystal. The *n*-type is used in this discussion. With no biases applied to the transistor, the channel conductance $G_c = \sigma(wt/l)$ where σ is the conductivity of the crystal and w, t, and l are the width, thickness, and length of the channel, respectively. For example, if $\sigma = 1$ mho/cm, $w = 0.1$ cm, $t = 0.01$ cm, and $l = 0.1$ cm, the channel $G_c = 0.01$ mhos and the channel resitance $R_c = 100\ \Omega$.

If reverse bias is applied between the gate and the source, the depletion region width is increased and the thickness of the channel is decreased; therefore, the conductivity of the channel is decreased. The gate bias required to just reduce the channel thickness to zero, as shown in Fig. 6.9b is called the *Pinch-off* voltage, V_p.

When the gate-source voltage v_{GS} is zero and the drain is made positive with respect to the source, electrons drift through the channel because of the electric field. The drain current i_D is equal to the drain-source voltage v_{GS} times the channel conductance G_c, providing v_{GS} is very small. However, the positive drain voltage reverse biases the p-n junction near the drain end of the channel, and when the drain voltage is increased to the pinch-off voltage the channel thickness is reduced to zero at a point near the drain end of the channel (Fig. 6.9c). The drain current does not stop when the drain voltage reaches pinch-off because a voltage equal to V_P still exists between the pinch-off point and the source, and the resulting electric field along the channel causes the free carriers in the channel to drift from the source to the drain.

As the drain voltage is increased above V_P, the depletion region thickness is increased between the drain and the gate (Fig. 6.9d). In fact, the additional drain voltage is absorbed by the increased field in the wider pinched-off region and the electric field between the original pinch-off point, and the source remains essentially unchanged. Therefore, the channel current and, hence, the drain current remains essentially unchanged. The carriers that arrive at the pinch-off point are swept through the depletion region in the same manner as carriers that are swept from the base into the collector region in a conventional transistor. Thus, whenever v_{DS} is higher than the pinch-off voltage V_P, the drain current is essentially independent of the drain voltage.

The field-effect transistor normally operates with the drain voltage v_{DS} beyond the pinch-off voltage V_P and reverse bias applied between the gate and the source. The electric field and, thus, the drain current in the channel is then controlled by the gate voltage v_{GS}. This action is similar to the control that the grid of a vacuum tube has on plate current by controlling the electric field between the grid and the cathode. Note that the *source* in the field-effect transistor compares with the emitter in a conventional transistor or vacuum tube. The effect of the gate voltage v_{GS} on the channel conductance is shown in Fig. 6.9e. The channel thickness is reduced as a result of the reverse gate bias. The drain current is essentially independent of the drain voltage whenever the sum of the drain voltage and the reverse-bias gate voltage exceeds the pinch-off voltage.

The drain characteristics of a typical n-channel FET are shown in Fig. 6.10. The drain current that flows when $v_{GS} = 0$ and $v_{DS} = V_P$ is known as I_{DSS}, or saturated drain current with input shorted. For the characteristics in Fig. 6.10, $I_{DSS} = 2.2$ mA. Observe that the output resistance $r_d = \Delta v_{DS}/\Delta i_D$ is very high in the normal operating range. Avalanche breakdown occurs at the junction whenever the drain-gate voltage exceeds a given value (about 35 V for the transistor of Fig. 6.10). Note that the drain

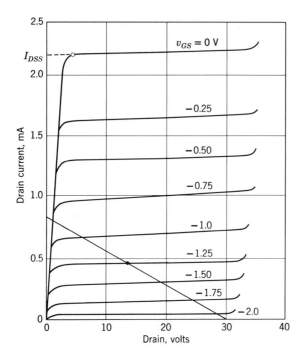

Fig. 6.10.   Drain characteristics of a typical *n*-channel FET.

characteristics of a FET closely resemble the plate characteristics of a pentode tube.

The input characteristics of a typical *n*-channel FET are shown in Fig. 6.11. The gate current is the minority current of the reverse-biased junction. This current is of the order of 10^{-9} A or 1 nanoampere (nA) for a low-power silicon FET at 25°C. The dynamic input conductance g_g is the slope of the input characteristics. Although the input resistance decreases rapidly as the junction becomes forward biased, Fig. 6.11*a* shows that the input resistance remains quite high (a megohm or more) as long as the forward bias does not exceed about 0.25 V in a silicon FET at 25°C.

EXAMPLE 6.3   Let us determine the output resistance r_d and the dynamic input conductance for the FET whose characteristics are given in Fig. 6.10 and Fig. 6.11. At the *q*-point (Fig. 6.10) where $v_{DS} = 20$ V and $v_{GS} = -0.75$ V, the value of r_d is $\Delta v_{DS}/\Delta i_D = 20$ V/0.05 mA $= 400$ kΩ. The dynamic or ac input resistance is found from Fig. 6.11 where $g_g = \Delta i_G/\Delta v_G = 10^{-9}$ A/4 $= 2.5 \times 10^{-10}$ mho. The input resistance, $r_g = 1/g_g = 1/(2.5 \times 10^{-10}) = 4 \times 10^{9}$ Ω for v_{GS} less than -1 V.

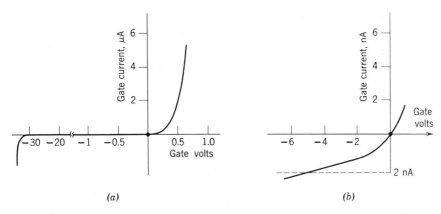

Fig. 6.11.   Input characteristics of a typical *n*-channel FET. (*a*) Expanded voltage scale and (*b*) expanded current scale.

The circuit diagram of an *n*-channel FET amplifier is given in Fig. 6.12. This circuit uses a bypassed resistor R_S in the source circuit to provide reverse bias for the transistor. If zero bias is desired this resistor may be omitted. The gate-circuit resistor R_G permits the saturation current that flows across the junction to flow to ground and the negative terminal of V_{DD}. Thus, the gate is maintained at approximately ground potential, for dc, which is negative with respect to the source. This type of bias is commonly used in vacuum tube circuits. Not more than about 0.1 V should be dropped across R_G.

PROBLEM 6.4   If the FET of Fig. 6.12 has the input characteristics of Fig. 6.11 and the *q*-point gate voltage is approximately −1 V, what should be the maximum resistance of R_G, assuming $T = 25°C$?   *Answer:* 100 MΩ.

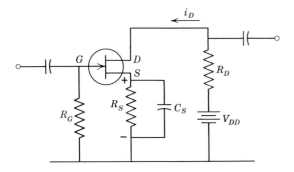

Fig. 6.12.   A circuit diagram for a FET amplifier.

The saturation current increases exponentially with temperature. Therefore, the value of R_G should be reduced for higher temperature operation. For example, if the saturation current doubles for each 10°C temperature increase and the maximum junction temperature is 55°C, the total gate current is $(2^3 + 1)$ nA $= 9$ nA and the value of R_G should not exceed approximately 0.01 V/9 nA $= 11$ MΩ.

The source-circuit resistor R_S (Fig. 6.12) tends to stabilize the drain current and, thus, provides stabilized bias because an increase of drain current increases the reverse bias between the gate and the source. This reverse bias tends to decrease the drain current.

There are two factors that cause the drain current to vary with temperature. One is the variation of the channel conductance. As the temperature increases, the carrier mobility decreases, as it does in practically all conductors, and the channel conductance decreases.[1] This effect tends to decrease the drain current as the temperature increases. The other effect is the variation of barrier voltage across the junction with temperature. As in a diode, an increase of temperature causes a reduction of barrier voltage, which has the same effect as a reduction of *reverse* bias. This tends to increase the drain current as the temperature increases. Since these two effects are subtractive, the temperature stability of a FET is quite good. In fact, it can be shown that for one specific value of gate source bias V_{GS} the temperature coefficient is zero. The theoretical value of this zero-drift bias for an n-channel transistor is

$$V_{GS} = V_P + 0.63 \text{ V} \qquad (6.16)$$

Notice that since the pinch-off voltage V_P is negative for an n-channel transistor, the zero-drift bias is 0.63 V above pinch-off. For example, the zero-drift bias for the transistor of Fig. 6.10 is approximately $-2.0 + 0.63 = -1.37$ V. All signs would be reversed for a p-channel transistor. The pinch-off voltage V_P can be most accurately determined by observing the value of gate-bias voltage v_{GS} (Fig. 6.10), which reduces the drain current to approximately zero. This value is -2.0 V for the FET with characteristics given in Fig. 6.10.

The voltage gain of the FET can be determined graphically by the technique used for a conventional transistor as shown by the following example.

---

[1]The charge-carrier density is essentially independent of temperature because the doping density is high in comparison with the thermally generated carrier density under normal conditions.

EXAMPLE 6.4 Let us assume that the drain characteristics of Fig. 6.10 are for the FET in the amplifier of Fig. 6.12. We then choose a q-point at $v_{DS} = 12$ V and $v_{GS} = -1.25$ V for approximately zero drift. Then let us choose a value of supply voltage $V_{DD} = 30$ V and draw a dc load line including the bias resistor R_S. The load line is drawn in Fig. 6.10. The value of this load resistance is 30 V/0.8 mA = 37.5 kΩ. The value of $R_S = V_{GS}/I_D = 1.25$ V/0.45 mA = 2.8 kΩ and the value of $R_D = 37.5$ k$\Omega - R_S = 34.7$ kΩ. When the input voltage varies 0.5 V peak-to-peak, the drain voltage v_{DS} varies 14 V peak-to-peak and the voltage gain $K_v = 14/0.5 = 28$.

PROBLEM 6.5 By using the FET characteristics of Figs. 6.10 and 6.11 for the amplifier of Fig. 6.12, and with the q-point $V_{DS} = 10$ V, $V_{GS} = -0.75$ V, determine suitable values for all the resistors in the circuit and the voltage gain.

Answer: $R_S = 1.3$ kΩ, $R_D = 18.7$ kΩ, $R_G = 10$ MΩ or so, $K_v = 25$.

6.5 AN EQUIVALENT CIRCUIT FOR THE FET

At low frequencies where the junction capacitance can be neglected, the equivalent circuit shown in Fig. 6.13 can be used. This circuit is the same type used for a pentode vacuum tube amplifier. The component values for the equivalent circuit can be determined from the FET characteristics.

EXAMPLE 6.5 The dynamic input resistance r_g is determined from the input characteristics. You may recall that the value found for the transistor of Fig. 6.11 was 5×10^9 Ω if reverse bias is used. The output conductance g_d is the slope of the drain characteristic at the q-point. The characteristics of Fig. 6.10 are given in Fig. 6.14a for convenience. The value of g_d at the q-point $v_{ds} = 12$ V, $v_{gs} = -1.25$ V is 0.05 mA/25 V = 2×10^{-6} mho, and $r_d = 1/g_d$ is 500 kΩ.

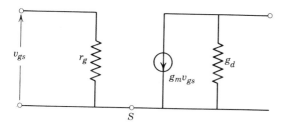

Fig. 6.13.   An equivalent circuit for a field-effect transistor.

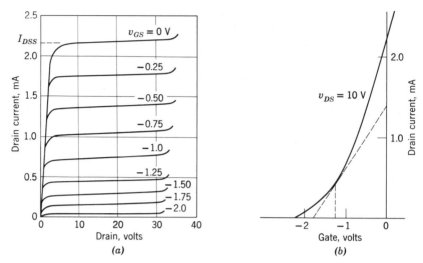

Fig. 6.14.   Drain and transfer characteristics of a typical n-channel FET.

The value of transconductance g_m can be most easily determined from the transfer characteristics given in Fig. 6.14b. This transfer curve can be obtained by plotting drain current as a function of gate voltage for a fixed value of drain voltage. In this example $v_{DS} = 10$ V. The transconductance is then the slope of this curve at a given point. For example, at the q-point $V_{GS} = -1.25$ V, $g_m = 1.4$ mA/1.7 V $= 8.25 \times 10^{-4}$ mho. The approximate value of g_m can be determined directly from the drain characteristics. For example, if the gate voltage v_{GS} is assumed to vary 0.5 V above and below the $V_{GS} = -1.25$ V q-point along a vertical line through the q-point, thus keeping the drain voltage constant, the change in drain current can be read from the scale on the drain-current axis, as indicated by the dashed lines on Fig. 6.14a. In this example, $\Delta i_D = (0.7 - 0.25) = 0.45$ mA and $g_m \simeq 9 \times 10^{-4}$ mho, approximately. If we assume the value of g_m previously determined from the transfer characteristic to be correct, the value determined from the drain characteristics is in error by 9 percent.

Figure 6.15 shows the equivalent circuit for the transistor amplifier with the parameters and load resistance previously determined for the zero-drift q-point. Also, a driving source with 100 kΩ internal resistance is shown. The value of g_m determined from the drain characteristics is used because this is an *average* value of g_m for a 0.5-V peak-to-peak input signal. The calculated results can then be realistically compared with the graphically determined results of Section 6.4.

The source-circuit resistor R_s does not appear in the equivalent circuit because it is assumed to be properly bypassed.

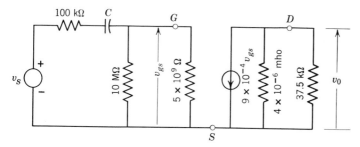

Fig. 6.15. The equivalent circuit for a FET amplifier with parameter values given.

The output voltage v_o (Fig. 6.15) in terms of v_{gs} is $v_o = 28\,v_{gs}$. The voltage gain $K_v = v_o/v_{gs} = 28$. The voltage gain from source to output is $K_v = v_o/v_s = 27.7$ ($R_G = 10$ MΩ). The output current $i_o = v_o/R_L = 28\,v_{gs}/37.7$ kΩ and the input current is approximately $i_i = v_{gs}/R_G = v_{gs}/10^7$. Therefore, the current gain $K_i = 7.4 \times 10^3$ and the power gain $K_P = 2.08 \times 10^5$.

The load conductance G_L is usually large compared with g_d, as you may have observed from the preceding example and problem. Therefore, sufficient accuracy is usually obtained from the approximate relationship

$$K_v \simeq g_m R_L \qquad (6.17)$$

The transconductance g_{mo} of the FET at $V_{GS} = 0$ and $V_{DS} = V_p$ where $I_D = I_{DSS}$ is given by the following relationship:[2]

$$g_{mo} = \frac{2I_{DSS}}{V_p} \qquad (6.18)$$

Also the transconductance at any given bias voltage V_{GS} can be readily calculated, providing g_{mo} is known, from the expression that follows:[3]

$$g_m = g_{mo}\left(1 - \frac{v_{GS}}{V_p}\right) \qquad (6.19)$$

The manufacturers usually give I_{DSS} and V_p so that the transconductance can be determined at any desired v_{GS} for a typical transistor of a given type without resorting to the characteristic curves. However, more accurate values can be obtained for a specific, in-hand transistor if a curve tracer is used to obtain the I_{DSS} and V_p for that transistor.

[2]Charles L. Alley and Kenneth W. Atwood, *Electronic Engineering*, Second Edition (New York: John Wiley and Sons, Inc., 1962) p. 226.
[3]Ibid.

PROBLEM 6.6   Determine g_m, g_d, and r_g for the FET of Figs. 6.14 and 6.11 at the point $V_{DS} = 10$ V, $V_{GS} = -0.75$ V. Use these parameters to determine the voltage gain of the amplifier with $R_L = 18.7$ kΩ. Compare this gain with the value obtained in Prob. 6.5.

Answer: $g_m = 1.3 \times 10^{-3}$ mho, $g_d = 4 \times 10^{-6}$ mho, $r_g = 5 \times 10^9$ Ω, $K_v = 23$.

PROBLEM 6.7   Use Eq. 6.18 to determine g_{mo} for the FET of Prob. 6.6; then use Eq. 6.19 to determine g_m at the specified q-point. Compare your results with the value of g_m obtained in Prob. 6.6.

6.6   INSULATED GATE FET'S

If extremely high input resistance is desired in a solid-state circuit, insulated gate FET's or IGFET's may be used. These devices are also known as metal-oxide semiconductor (MOS) field-effect transistors (FET) or metal-insulator semiconductor (MIS) field-effect transistors (FET). Usually, they are referred to simply as MOSFET's.

The MOSFET's are typically constructed as shown in Fig. 6.16. Notice that the gate is insulated from the semiconductor material (usually silicon) by a layer of oxide insulation (usually silicon oxide). Consequently, the input resistance is typically in the range of 10^{12} to 10^{14} Ω. Except for the insulated gate, the construction looks quite similar to a conventional *p-n-p* transistor. However, the electrical characteristics of the MOSFET are quite different than those of a conventional transistor.

Under typical operating conditions, the gate is maintained negative. The negative charge on the gate repels the electrons in the *n*-material which is known as the substrate, and attracts the holes. If the potential on the gate is sufficiently negative, the number of holes near the surface of the *n*-material will exceed the number of electrons. The surface of the *n*-

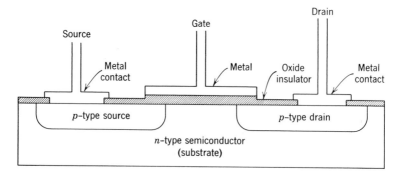

Fig. 6.16.   Construction of a *p*-channel MOSFET.

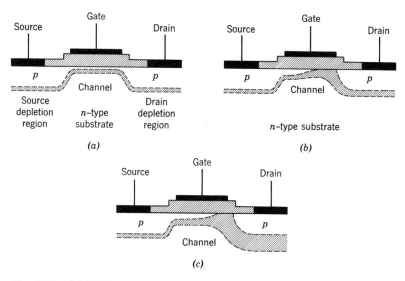

Fig. 6.17. A MOSFET with applied bias. (a) $V_G = -10$ V, $V_S = V_D = 0$ V, (b) $V_S = 0$, $V_G = -10$ V, $V_D = -5$ V, and (c) $V_S = 0$, $V_G = -10$ V, $V_D = -10$ V.

material is said to have *inverted* and now behaves as if it were *p*-material. Fig. 6.17*a* shows the configuration of a MOSFET with negative gate and zero potential on the source and drain. As the gate becomes more negative, the depth of the channel increases. Hence, the channel width is controlled by the gate potential, as in the regular unijunction FET. When a negative potential is applied to the drain as well as to the gate, the channel is distorted as shown in Figs. 6.17*b* and 6.17*c*, and current flows from the source to the drain.

When drain current flows along the channel, an *IR* drop is developed along the channel. This *IR* drop tends to cancel the field produced by the gate bias. When the cancellation is sufficient to *almost* prohibit the formation of the inversion layer, the channel *pinches off* and the drain current tends to *saturate* at a constant value independent of increased drain voltage. Of course, if the thickness of the inversion layer were actually zero, no current would flow to the drain. Thus the *pinch-off* or *threshold* voltage, V_{TH}, is the voltage from gate to channel necessary to just produce inversion in the channel.

The characteristics of a typical MOSFET are shown in Fig. 6.18*a*. Note that in contrast to the junction FET, the gate and drain potential have the same polarity. This means MOSFETS can be direct-coupled from the drain of one stage to the gate of the next stage with no isolating capacitors required.

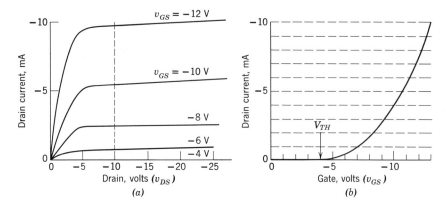

Fig. 6.18. Characteristics of a *p*-channel MOSFET. (*a*) Drain characteristics and (*b*) transfer curves.

If the drain voltage is maintained constant (for example, 10 V in Fig. 6.18*a*) and a plot of drain current versus gate voltage is obtained, this characteristic is known as a *transfer curve*. The transfer curve of the MOSFET in Fig. 6.18*a* is drawn in Fig. 6.18*b*. In this transistor, the threshold voltage, V_{TH}, is -4 V. The threshold voltage is readily apparent in either set of curves in Fig. 6.18. These curves illustrate another peculiarity of a MOSFET: *The drain current is proportional to the square of the input voltage in the saturation region.* This square-law effect may be desirable in some applications but can be a handicap in a linear amplifier.

The symbol for a *p*-channel MOSFET is given in Fig. 6.19. Note that a connection is made to the bulk or substrate (the *n*-material in Fig. 6.16) of the semiconductor as well as to the gate, to the source, and to the drain. In a *p*-channel MOSFET, the substrate is connected to the positive terminal of the power supply. This condition is necessary to insure the *p-n* junctions

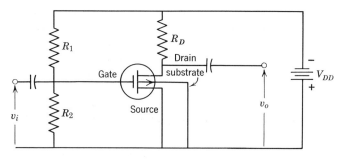

Fig. 6.19.   A MOSFET amplifier.

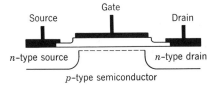

Fig. 6.20.   An *n*-channel MOSFET.

in the device do not become forward biased. (The substrate of an *n*-channel device is connected to the negative terminal of the power supply.) Of course, if the *p-n* junctions should become forward biased, the circuit will cease to behave as a MOSFET.

It is also possible to construct an *n*-channel MOSFET as shown in Fig. 6.20. However, the conditions created by the interfacing of the silicon and silicon oxide create contact potentials, which produce an *n*-channel when the gate potential is zero. Thus, drain current flows in an *n*-channel MOSFET when the gate potential is zero. Drain characteristics for a typical *n*-channel MOSFET is given in Fig. 6.21.

If the channel is cut off when zero gate bias is applied (Fig. 6.18), the MOSFET is said to be operating in the *enhancement mode* of operation. If channel current flows when zero gate bias is applied, as in Fig. 6.21, the MOSFET is operating in the *depletion mode*. A circuit for an *n*-channel depletion-mode transistor is given in Fig. 6.22. Since drain current flows with zero bias, the gate can be returned to the source as shown. Since there is no gate current, the value of R_G can be very large. Similarly, in Fig. 6.19 the two resistors R_1 and R_2 form a voltage dividing network to obtain the proper gate bias; but both R_1 and R_2 can be very large.

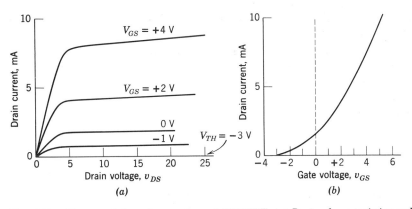

Fig. 6.21.   Characteristics of an *n*-channel MOSFET. (*a*) Drain characteristics and (*b*) transfer characteristics.

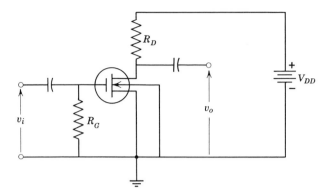

Fig. 6.22. An *n*-channel MOSFET amplifier.

In Figs. 6.19 and 6.22 the substrate lead is connected to the source. However, it is possible to change the characteristics of the device by applying a different bias to the substrate. To illustrate the effect of substrate bias, three sets of output characteristics for a *p*-channel MOSFET are given in Fig. 6.23. Note that if the gate voltage is maintained constant, the drain current decreases as the substrate (or bulk) voltage increases in the positive direction. The threshold voltage also changes with the substrate voltage. Since drain current control can be achieved by either the gate or the substrate, the gate is sometimes called the *front gate* and the substrate is referred to as the *back gate*. However, the input impedance of the substrate is much less than the input impedance of the gate. In fact, since the substrate to source or drain forms an *n-p* junction, the input impedance of the substrate is in the same order of magnitude as the input impedance of a conventional FET.

An equivalent circuit for a MOSFET with substrate maintained constant (as in Fig. 6.19 or Fig. 6.22) is given in Fig. 6.24. Since the impedance is so high, the effective input impedance is essentially an open circuit, as shown. (Actually, the MOSFET does have an input capacitance but this capacitor is usually in the order of fractions of a pico farad and so can be neglected except at very high frequencies.) The parameter g_m is defined by the equation

$$g_m = \frac{\Delta i_D}{\Delta v_{GS}}\bigg|_{v_{DS}=\text{constant}} \tag{6.20}$$

Of course, the value of g_m can be determined from the characteristic curves of the transistor. Typical values of g_m are in the range of 1000 to 2000 μmhos. From the curves in Figs. 6.18 or 6.21, we note that the value

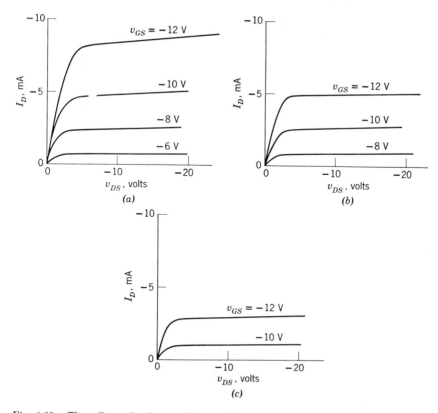

Fig. 6.23. The effect of substrate bias on the output characteristics of a MOSFET. V_{GS} = gate voltage, V_{DS} = drain voltage, V_{BS} = substrate voltage. (a) $V_{BS} = 0$ V, (b) $V_{BS} = +4$ V, and (c) $V_{BS} = +10$ V.

of g_m is a function of drain current. The value of g_m may be found from the following equation.

$$g_m = \frac{2i_D}{v_{GS} - V_{TH}} \qquad (6.21)$$

where i_D is the drain current in amperes, v_{GS} is the gate voltage in volts, and V_{TH} is the pinch-off voltage in volts.

The value of g_{ds} can be found from the characteristic curves by using the relationship

$$g_{ds} = \frac{\Delta i_D}{\Delta v_D}\bigg|_{v_{GS} = \text{constant}} \qquad (6.22)$$

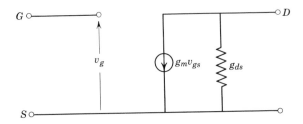

Fig. 6.24.   An equivalent circuit for a MOSFET.

PROBLEM 6.8   Determine g_{ds} and g_m for the MOSFET whose characteristics are given in Fig. 6.21. Use both Eqs. 6.20 and 6.21 to determine g_m, and compare the results. Assume the q-point is at $V_{GS} = 0$ V and $V_{DS} = 10$ V.   *Answer:* $g_{ds} \simeq 10\ \mu$mhos, $g_m = 1000$ to $1100\ \mu$mhos.

PROBLEM 6.9   Determine g_{ds} and g_m for the MOSFET whose characteristics are given in Fig. 6.18. Assume a q-point of $V_{GS} = -10$ V and $V_{DS} = -10$ V.

If both gate and substrate voltages are to be adjusted, the equivalent circuit of the device becomes more complicated. One approach is to use a circuit similar to that given in Fig. 6.24, but g_m is a function of the substrate voltage, v_{BS}. Thus, g_m in Fig. 6.24 for a p-channel MOSFET might be given as follows.

$$g_m = g_{m1} - D_{gm} v_{BS} \tag{6.23}$$

where g_{m1} is the value of g_m when $v_{BS} = 0$ and D_{gm} is the rate at which g_m changes with v_{BS}. An example will be given to clarify this concept.

EXAMPLE 6.6   Let us draw an equivalent circuit for the transistor whose characteristics are given in Fig. 6.23. We assume the q-point for this transistor is at $v_{GS} = -10$ V, and $v_{DS} = -10$ V. Notice that if $v_{BS} = 0$ V, the threshold voltage is $v_{TH} = -4$ V and the drain current (for $v_{GS} = -10$ V and $v_{DS} = -10$ V) is -5 mA. From Eq. 6.21, the value of g_{m1} is

$$g_{m1} = \frac{2(-5 \times 10^{-3})}{-10 - (-4)} = \frac{-10^{-2}}{-6} = 1667 \text{ mhos}$$

If v_{BS} is $+4$ V, the threshold voltage v_{TH} is -6 V and the drain current for the q-point is 2.6 mA. Then, Eq. 6.21 gives a value of g_m as follows.

$$g_m = \frac{2(-2.6 \times 10^{-3})}{-10 - (-6)} = \frac{-5.2 \times 10^{-3}}{-4} = 1300 \, \mu\text{mhos}$$

When v_{BS} is $+10$ V, the threshold voltage is $v_{TH} = -8$ V and the drain current (for $v_{GS} = -10$ V and $v_{DS} = -10$ V) is -0.8 mA. The value of g_m changes to the following value.

$$g_m = \frac{2(-0.8 \times 10^{-3})}{-10 - (-8)} = \frac{-1.6 \times 10^{-3}}{-2} = 800 \, \mu\text{mhos}$$

A plot of g_m versus v_{BS}, as determined from the foregoing information, is given in Fig. 6.25. Note that g_m decreases as v_{BS} increases, thus justifying the negative sign in Eq. 6.23. (An n-channel MOSFET would require a positive sign in this equation since v_{BS} is negative.) Observe from Fig. 6.25 that g_m decreases almost linearly as v_{BS} increases, thus verifying Eq. 6.21 (at least for the transistor in Fig. 6.23). Since D_{gm} is the slope $\Delta g_m / \Delta v_{BS}$, the average value of D_{gm} for this transistor is $(1667 - 800) \, \mu\text{mhos}/10 \text{ V} = 86.7 \, \mu\text{mho/V}$. Hence, Eq. 6.23 becomes

$$g_m = (1667 - 86.7 v_{BS}) \mu\text{mhos}$$

for the transistor of Fig. 6.23.

The equivalent circuit of the MOSFET in Fig. 6.23 (biased at the $v_{GS} = -10$ V, $v_{DS} = -10$ V q-point previously noted) would be as shown in Fig. 6.26. The substrate (or bulk) terminal is included as one of the input

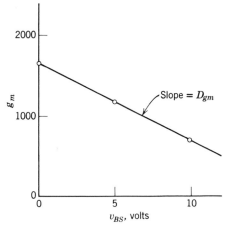

Fig. 6.25.   A plot of g_m versus v_{BS} for the transistor of Fig. 6.23.

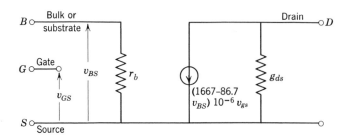

Fig. 6.26.   The equivalent circuit for the MOSFET of Fig. 6.23 when both gates are used for signal inputs.

terminals. Since this input terminal has a much lower input impedance than the insulated gate, an input resistor r_b is included. The output current is a function of both input circuits as noted.

PROBLEM 6.10   A MOSFET transistor is connected as shown in Fig. 6.27. The characteristics of this device are given in Fig. 6.23.

   (a) Determine v_o if $v_i = 0.1 \cos 1000\,t$ and $v_{BS} = 0$ V.
   (b) Determine v_o if $v_i = 0.1 \cos 1000\,t$ and $v_{BS} = 4$ V.

Answer: (a) $g_{ds} \simeq 10^{-5}$ mhos so $v_o = 0.64 \cos 1000\,t$. (b) $v_o = 0.5 \cos 1000\,t$.

PROBLEM 6.11   A circuit is connected as shown in Fig. 6.27. Find v_o if $v_i = 0.1 \cos (10^7 t)$ and $v_{BS} = 4 + 4 \cos (10^3 t)$.
Answer: $v_o = 0.5 \cos 10^7 t - 0.136 (\cos 10^3 t)(\cos 10^7 t)$. We have accomplished amplitude modulation!

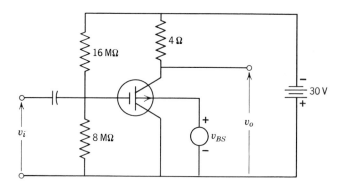

Fig. 6.27.   A MOSFET amplifier with adjustable substrate voltage.

Because of the extremely high input impedance of the MOSFET, one must be very careful when installing or handling them. To more fully understand this statement, consider the situation illustrated in Fig. 6.28. In this figure, a positive charge $(+q)$ is placed near the gate lead of a MOSFET. The positive charge attracts electrons to the end of the gate lead. The loss of these electrons creates a positive charge on the gate and a potential is developed across the insulation between the gate and substrate. In fact, the gate, insulation, and substrate is a capacitor. The voltage across the insulation in a capacitor is

$$V = \frac{Q}{C} \qquad (6.24)$$

As was mentioned previously, the input capacitance of the gate in a MOSFET is typically less than one picofarad. Thus, the voltage across the insulation is

$$V > 10^{12}Q \qquad (6.25)$$

Hence, a small charge can produce a large potential across the insulation.

The gate insulating material in a MOSFET is typically 10^{-5} to 2×10^{-5} cm thick. Consequently, gate to substrate voltages of about 50 V will cause breakdown of the insulation. Once breakdown has occurred, the insulating qualities of the insulator are destroyed and the MOSFET is ruined.

The charge shown in Fig. 6.28 can be a static charge on a person's finger or some tool he is using. Thus, *one can destroy a MOSFET without even touching it.* To prevent destruction, most MOSFET's are shipped with their leads twisted so that the gate is shorted to the source and/or substrate. Shorting leads should be clipped to the gate when soldering or installing MOSFET's.

Some MOSFET's are constructed with a reverse biased junction connected internally to the gate. This junction conducts (as a Zener diode) if the gate voltage exceeds the diode breakdown potential. The diode does protect the gate insulator but the total input impedance is now equal to the impedance of the reverse biased diode. Hence the input impedance is approximately equal to that of a conventional FET.

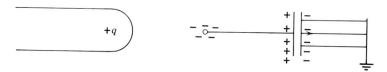

Fig. 6.28.   The effect of a static charge near the gate of a MOSFET.

PROBLEM 6.12   Design a two-stage direct-coupled MOSFET amplifier that uses a 20-V battery for a power supply. The q-point for both amplifiers is to be $V_{DS} = -10$ V and $V_{GS} = -10$ V.

PROBLEM 6.13   Design a preamplifier that will present a 20-MΩ input impedance to a condenser (capacitor) microphone but that will have low output impedance of about 100 Ω or less. The actual load is 1000 Ω resistive. Use an MPF 103 FET transistor, directly coupled to a 2N3903 transistor. Determine suitable component values and calculate the voltage gain of your amplifier. The ambient temperature is faily constant at about 25°C.

CHAPTER 7

RC COUPLED
AMPLIFIERS

Our attention has heretofore been focused on the principles of operation and the characteristics of semiconductor devices. Simple circuit applications have been given to illustrate the characteristics and usefulness of these devices and capacitors have been used to provide ac coupling and dc blocking between the particular device and its signal source and load. Also, capacitors have been used to bypass an ac signal around a resistor. However, little attention has been given to the values of these capacitors and no rules have been developed for determining their values. Beginning with this chapter, increased attention will be given to the circuit components associated with the semiconductor device and rules will be developed that will permit the calculation of component values in accordance with design specifications or requirements. In other words, the characteristics of an entire circuit will be considered, instead of restricting our attention to the semiconductor device. The circuits considered in this chapter will be typical of those used in sound amplifiers, audio amplifiers in communications equipment, video (or picture) amplifiers in television sets, and in many other applications.

7.1 THE LOW-FREQUENCY REGION

We have previously assumed that the reactance of a coupling capacitor or a bypass capacitor is negligible. This is true if the frequency is sufficiently high. However, there certainly will be frequencies for which the reactance of these capacitors cannot be neglected. By definition, this range of frequencies is known as the *low-frequency range* or *low-frequency region*.

A circuit diagram for a typical RC-coupled transistor amplifier is given in Fig. 7.1a. We initially assume that the emitter resistor R_E is perfectly bypassed by C_E, so that the effect of the coupling capacitor C can be considered alone. The h-parameter equivalent circuit has replaced the transistor in Fig. 7.1b.

The transistor input resistance is the only transistor parameter that affects the value of the coupling capacitor C. Therefore, the equivalent circuit of Fig. 7.1b can be simplified to that shown in Fig. 7.2a. As was previously observed, the transistor input resistance R_i is approximately equal to h_{ie}. This approximation will give adequate accuracy in the determination of the value of the coupling capacitor C.

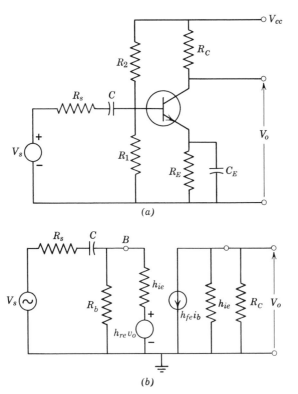

(a)

(b)

Fig. 7.1.   (a) A typical RC-coupled amplifier and (b) the equivalent circuit assuming R_E is perfectly bypassed.

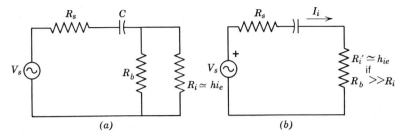

Fig. 7.2.   Simplified versions of the equivalent circuit of Fig. 7.1b.

The equivalent circuit of Fig. 7.2a may be further simplified by combining the parallel combination of bias resistance R_b and input resistance R_i into a single resistance R_i' (Fig. 7.2b). Usually, the bias resistance R_b is large compared to the input resistance R_i. This R_b is the total biasing resistance seen between the transistor base and either ground or V_{CC}, as was discussed in Chapter 5. If R_b is as much as five times R_i, adequate accuracy will be obtained in calculating the value of C if the assumption is made that $R_i' \simeq R_i \simeq h_{ie}$.

We assume that the source voltage V_s is sinusoidal. Figure 7.2b then shows that the transistor input current is

$$I_i = \frac{V_s}{R_s + R_i' - jX_C} \tag{7.1}$$

Equation 7.1 can be written in polar form, or in terms of magnitude and phase angle, as follows.

$$I_i = \frac{V_s}{\sqrt{(R_s + R_i')^2 + X_C^2}} \; \underline{/\tan^{-1} X_C/(R_s + R_i')} \tag{7.2}$$

The magnitude and phase of the input current are plotted as a function of frequency in Fig. 7.3a. Equation 7.2 and Fig. 7.3a show that the input current I_i is essentially proportional to the frequency when X_C is large in comparison with $R_s + R_i'$, but reaches a steady value of $I_s = V_s/(R_s + R_i')$ when X_C is small compared to $(R_s + R_i')$.

An interesting frequency is the frequency f_1. At this frequency $X_C = R_s + R_i'$, and Eq. 7.2 shows that the input current is

$$I_i = \frac{V_s}{\sqrt{2}(R_s + R_i')} \; \underline{/45°} \tag{7.3}$$

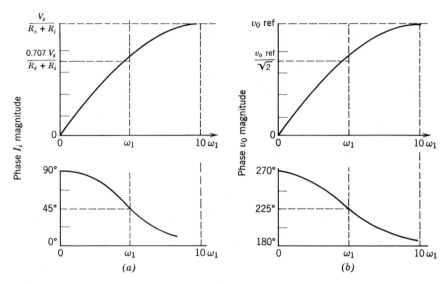

Fig. 7.3. The magnitude and phase of (*a*) the input current and (*b*) the output voltage as functions of frequency.

Since the power input to the amplifier is $|I_i^2|R_i'$, this input power is maximum and independent of frequency if X_C is negligibly small in comparison with $(R_s + R_i')$. The input power is then

$$P_i = \frac{V_s^2 R_i'}{(R_s + R_i')^2} \tag{7.4}$$

But, at the frequency ω_1 where the input current is reduced from its maximum value by the factor $\sqrt{2}$, Eq. 7.3 shows that the input power is

$$P_i = \frac{V_s^2 R_i'}{2(R_s + R_i')^2} \tag{7.5}$$

which is one-half the value obtained at much higher frequencies.

If we restrict the input signal current to a value small enough so the input resistance R_i is essentially constant, the input voltage is $I_i R_i'$ and the output voltage V_o is the product of the input voltage and the voltage gain G_v. But, a voltage polarity reversal occurs in a common-emitter amplifier, so that G_v for the amplifier, excluding the coupling capacitor, can be written $|G_v| \underline{/180°}$ where $|G_v|$ is the magnitude of the voltage gain and the 180° angle replaces the negative sign in the gain equation. Then, since the product is obtained by multiplying the magnitudes and adding the angles,

$$V_o = I_i R_i' G_v = |I_i R_i' G_v| \underline{/180° + \tan^{-1}\left(\frac{X_c}{R_s + R_i'}\right)} \tag{7.6}$$

Thus, the output voltage is proportional to the input current. Both the magnitude and phase of the output voltage as functions of frequency are sketched in Fig. 7.3b, where V_o ref is the voltage output at frequencies where X_C is negligible.

The frequency ω_1, at which the power output is one-half the value obtainable at higher frequencies, is commonly known as the *half-power frequency* or the *lower-cutoff frequency*. The latter name results from the commonly used definition of the *passband* of a circuit or amplifier, which has its edges set at the half-power frequencies. Since ω_1 is such an important frequency, it deserves further consideration. As was previously mentioned, at ω_1

$$X_C = \frac{1}{\omega_1 C} = R_s + R_i' \tag{7.7}$$

Then

$$\omega_1 = \frac{1}{(R_s + R_i')C} \tag{7.8}$$

Notice that the lower cutoff frequency, in radians per second, is the reciprocal of the time constant, which is the product of the capacitance and the total resistance in series with it. For example, if the values in the circuit of Fig. 7.2b are $R_s = 4\,\text{k}\Omega$, $R_i' = 1\,\text{k}\Omega$, and $C = 1\,\mu\text{F}$, the time constant $RC = 5 \times 10^{-3}$ s, $\omega_1 = 200$ radians/second (rad/s), and $f_1 = \omega_1/2\pi = 32$ Hz. The capacitance of C can be obtained in terms of ω_1 by rearranging Eq. 7.8.

$$C = \frac{1}{\omega_1(R_s + R_i')} \tag{7.9}$$

The low-frequency region is defined as the frequency range in which the reactance of the coupling and bypass capacitors cannot be neglected. This reactance can be neglected if it is less than one-tenth as large as the associated resistance. For example, if $X_C = 0.1\ (R_s + R_i')$, Eq. 7.2 shows that the magnitude of the input current I_i is reduced from the maximum value $V_s/(R_s + R_i')$ by the factor $\sqrt{1.01}$, which is a reduction of about 0.5 percent. Thus, the upper frequency limit of the low-frequency region is considered to be $10f_1$. At this frequency, the phase of the input current is 5.7° and the phase of the output voltage is 185.7° with respect to the source voltage V_s. For the example above, where $\omega_1 = 200$ rad/s or $f_1 = 32$ Hz, the upper limit of the low-frequency region is 2000 rad/s or 320 Hz.

The problem of coupling two transistors by a capacitor is the same as coupling a source to a transistor. The first transistor *is* the source for the second transistor. The source resistance is the output resistance of the first transistor (approximately $1/h_{oe}$) in parallel with the collector circuit resistor R_C, as shown in Fig. 7.4.

PROBLEM 7.1   The transistor T_1 in Fig. 7.4 has $h_{oe} = 5 \times 10^{-5}$ mho. Determine the source resistance R_s. Then, if the transistor T_2 has $h_{ie} = 400\ \Omega$, and the desired value of low-frequency cutoff $f_1 = 16$ Hz, determine the value of the coupling capacitor C.   *Answer: $R_s = 4$ kΩ, $C = 2.3\ \mu$F.*

7.2   THE EMITTER-BYPASS CAPACITOR

To this point, the emitter-bypass capacitor has been assumed to be very large so that its reactance is negligible for any frequency of interest. We now look more closely at this capacitor C_E (Fig. 7.4) and develop a formula for determining a suitable value for C_E. This time, the coupling capacitor C will be assumed to be very large so that the low-frequency cutoff ω_1 will be determined by C_E.

Your first impression may be that the emitter-bypass capacitor C_E need only have small reactance compared to R_E (for example 0.1 R_E) for effective bypassing. This premise follows the generally accepted rule for bypassing cathode or other resistors in vacuum-tube circuits. This would be a good rule if the impedance seen by looking into the emitter terminal of the transistor were not in parallel with R_E. But, this impedance is usually

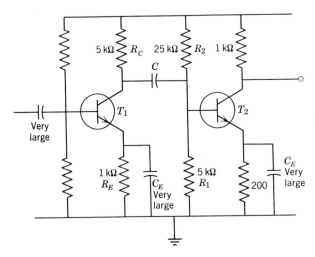

Fig. 7.4.   Two transistors with capacitive coupling.

much smaller than R_E, and therefore primarily determines the needed value of C_E. The impedance at the emitter terminal is essentially the same as the output resistance of an emitter follower, which was found to be approximately $(h_{ie} + R_s)/(\beta + 1)$ in Section 6.2.

Assume that the signal output power of transistor T_2 will be one-half its reference value when $X_{CE} = (R_s' + h_{ie})/(\beta + 1)$ where R_s' is R_s in parallel with R_b. Then, since $X_{CE} = 1/\omega_1 C_E$, this half-power frequency is

$$\omega_1 = \frac{\beta + 1}{(R_s' + h_{ie})C_E} \tag{7.10}$$

The value of C_E that will give a specified value of ω_1 is

$$C_E = \frac{\beta + 1}{\omega_1(R_s' + h_{ie})} \tag{7.11}$$

EXAMPLE 7.1   Consider the impedance of transistor T_2 (Fig. 7.4) at its emitter terminal, assuming $h_{ie} = 400\ \Omega$ and $\beta = 99$. First, observe that the source resistance R_s', as seen from the base of transistor T_2 toward the collector of transistor T_1, is the output resistance of the first stage in parallel with the biasing resistors of the second stage. The output resistance of the first stage ($1/h_{oe}$ of T_1, in parallel with R_C) was previously found (Prob. 7.1) to be about 4 kΩ and the biasing resistance R_B (5 kΩ in parallel with 25 kΩ) is also about 4 kΩ. Then $R_s' = 2$ kΩ and $(h_{ie} + R_s')/(\beta + 1) = 2.4$ kΩ/100 = 24 Ω. Since this value of resistance is so much smaller than the value of R_E, the latter may be neglected in calculating the value of C_E.

Thus, if the desired value of ω_1 is 100, the value of C_E required to bypass the emitter of transistor T_2 (Fig. 7.4) is $C_E = 100/(100 \times 2.4 \times 10^3) = 416\ \mu$F. Notice that this capacitance is greater than β times the coupling capacitance previously determined.

We will now show that the half-power frequency ω_1 does occur when $X_{CE} = (R_s' + h_{ie})/(\beta + 1)$. Look again at the impedance of the base circuit and the effect of X_{CE} on the input current. Figure 7.5a shows an equivalent circuit for the driving source of transistor T_2. This circuit is further simplified in Fig. 7.5b by replacing the circuit to the left of Points A and B with a Thevenins equivalent circuit, which has a source resistance R_s' (the parallel combination of R_s amd R_b). Our experience with the emitter follower can now be used to complete the simplified equivalent circuit of Fig. 7.5b. You will recall that the impedance Z_E in the emitter circuit causes an effective impedance $(\beta + 1)Z_E$ in the base circuit because $I_e/I_b = (\beta + 1)$. Therefore, the effective resistance is $(\beta + 1)R_E$ and, since

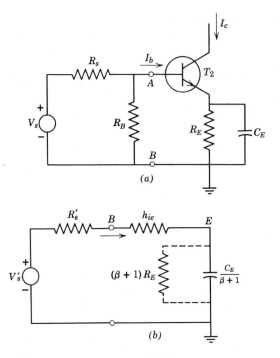

Fig. 7.5.   Equivalent circuits for the base circuit of a transistor with a bypassed emitter resistor.

X_C is inversely related to C, the effective capacitance is $C_E/(\beta+1)$. But we discovered in the preceding example that R_E is usually very large in comparison with X_{CE} at the frequency ω_1, and therefore, the effect of R_E may be neglected. Then, making this approximation,

$$I_b = \frac{V_s'}{R_s' + h_{ie} - j\dfrac{\beta+1}{\omega C_E}} \tag{7.12}$$

We have seen previously that I_b decreases by the factor $\sqrt{2}$ and the power drops to one-half when the magnitude of the reactive term in the denominator of Eq. 7.12 is equal to the total resistance. Since this relationship exists at the frequency ω_1 (or f_1),

$$\omega_1 = \frac{\beta+1}{(R_s' + h_{ie})C_E} \tag{7.10}$$

You may recall that this is the same value of ω_1 given by Eq. 7.10, which was obtained by assuming that the reactance of C_E was equal to the

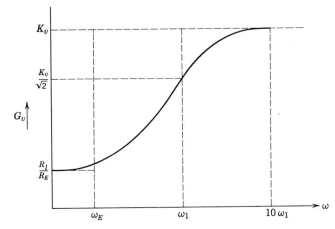

Fig. 7.6.   A sketch of voltage gain as a function of frequency, considering only the emitter bypass.

resistance looking into the emitter terminal. Then the value of C_E that is needed to provide a low frequency cutoff ω_1 is $C_E = (\beta+1)/(R_s' + h_{ie})\omega_1$, which was previously given as Eq. 7.11.

A sketch of voltage gain as a function of frequency is given in Fig. 7.6. The output is not zero at zero frequency when only the bypass capacitor is considered because the resistor R_E is the maximum limiting value of Z_E. Thus, there is base current at zero frequency. In fact, the voltage gain is easily determined when a resistance R_E is in the emitter circuit and is not bypassed. The analysis of the emitter follower showed that the signal voltage across the emitter resistor is essentially equal to the input voltage if $(\beta+1)R_E$ is large compared to h_{ie}. Then

$$V_i \simeq I_e R_E \tag{7.13}$$

$$V_o = I_c R_l \tag{7.14}$$

where R_l is the total load resistance, including R_C, R_B, and h_{ie} in parallel, if the transistor is used to drive a following transistor. Thus, when the emitter resistor is unbypassed,

$$K_v = \frac{V_o}{V_i} = \frac{I_c R_l}{I_e R_E} = \frac{\alpha R_l}{R_E} \simeq \frac{R_l}{R_E} \tag{7.15}$$

The main problem in calculating the voltage gain from Eq. 7.15 is in determining the load resistance R_l. To illustrate this point, let us consider an example.

EXAMPLE 7.2   An amplifier is connected as shown in Fig. 7.4. The transistor parameters are: $h_{ie} = 400$ Ω, $h_{fe} = 99$, $h_{oe} = 5 \times 10^{-5}$ mhos, $h_{re} \simeq 0$. Assume that the frequency is so low that neither emitter bypass is effective. In contrast, the coupling capacitor C is assumed to have negligible reactance.

The input resistance to transistor T_2 is $(h_{fe} + 1)R_E + h_{ie} = (99 + 1)200 + 400 = 20.4$ kΩ. This resistance is in parallel with the bias resistors, with R_C and $1/h_{oe}$. This combination (which is known as R_s') is equal to 2 kΩ. The total load on transistor T_1 is $R_l = (20.4$ k$\Omega \times 2$ k$\Omega)/22.4$ k$\Omega = 1.85$ kΩ. From Eq. 7.15, the voltage gain of transistor T_1 is $K_v = 1.85$ k$\Omega/1$ k$\Omega = 1.85$. The frequency at which $X_{CE} = R_E$ (shown as ω_E in Fig. 7.6) is the frequency below which the emitter-bypass capacitor C_E is ineffective and the emitter resistor R_E can be considered to be unbypassed.

We now know how to determine either the coupling capacitance C or the emitter-bypass capacitance C_E, separately, to obtain a given low-frequency cutoff ω_1. However, a stage of amplification (including one transistor and its associated components) usually includes both a coupling and a bypass capacitor. One solution is to let one capacitor determine ω_1 and make the reactance of the other capacitor negligible at ω_1. Since C_E must be so much larger than C, size and cost can be minimized by letting C_E determine ω_1, and using a value of C that will let $X_C = 0.1$ $(R_s' + h_{ie})$ at ω_1. The capacitance of C will then be ten times as large as the value that would be used if C were to determine ω_1. For example, in Example 7.1, we calculated $C_E = 416$ μF and from Eq. 7.9 the value of $C = 2.3$ μF if each individually determined $\omega_1 = 100$ rad/s. But, if both capacitors are to be used simultaneously and ω_1 is to remain at approximately 100 rad/s, an economical combination of values is $C_E = 416$ μF and $C = 23$ μF. The effect on ω_1 when several stages are included in one amplifier will be discussed in a later chapter.

PROBLEM 7.2   A single-transistor amplifier with $R_l = 2$ kΩ is capacitively coupled to a 6-kΩ driving source. If $h_{ie} = 700$ Ω, $\beta = 150$, $R_B = 10$ kΩ, and $R_E = 200$ Ω, determine values of coupling capacitance and emitter bypass capacitance that will provide a low-frequency cutoff, or half-power frequency, $f_1 = 16$ Hz, approximately.   *Answer:* $C = 16$ μF, $C_E = 340$ μF.

The coupling capacitor for a FET amplifier, or in fact any amplifier, can be calculated in precisely the same manner as for the bipolar transistor amplifier. The required coupling capacitance is much smaller, however, for the high-input impedance devices. Also, the capacitor leakage currents may become important when devices with extremely high-input impedance, such as MOSFET's, are capacitively coupled.

The FET source bypass capacitor must bypass the source-bias resistor R_S (not to be confused with the driving-source resistance) in parallel with the resistance looking into the FET source terminal. This source terminal resistance can be determined, as usual, by applying a signal voltage between the source terminal and ground and by noting the signal current that flows into the source terminal. But since there is no gate signal while this impedance test is being made, the gate potential remains at ground and the FET source-to-ground voltage is also applied between the source and gate. The signal current that flows is therefore approximately $g_m V_{gs}$, as was previously discussed, and the impedance seen at the FET source terminal is $1/g_m$. Therefore, the source bypass capacitor C_S must bypass the conductance $(g_m + G_S)$, where $G_S = 1/R_S$ or

$$C_S = \frac{G_S + g_m}{\omega_1} \tag{7.16}$$

PROBLEM 7.3   A given FET amplifier has $R_G = 5\ \text{M}\Omega$, $R_D = 10\ \text{k}\Omega$, and $R_S = 100\ \Omega$. If $g_m = 4 \times 10^{-3}$ mho and the driving source resistance is $100\ \text{k}\Omega$, determine suitable values for the input coupling capacitor and source-bypass capacitor to provide $f_1 = 16\ \text{Hz}$.

Answer: $C = 0.02\ \mu\text{F}$, $C_S = 140\ \mu\text{F}$ (or $150\ \mu\text{F}$).

7.3   TIME OR FLAT-TOP RESPONSE

In the preceding section, the input voltage was assumed to be sinusoidal and the output voltage was plotted as a function of frequency. This plot is usually known as a *frequency-response* curve. Frequently, however, the amplifier input is *not* sinusoidal and the output waveform, or the output voltage as a function of *time*, is of major interest. For example, consider the RC-coupled amplifier of Section 7.2 with a step voltage input

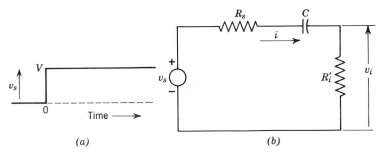

Fig. 7.7.   (a) A step voltage source and (b) the equivalent circuit of the source and the capacitively coupled transistor input.

instead of a sinusoidal input. The source voltage, which could be obtained with a battery and a switch, is shown in Fig. 7.7a and the equivalent circuit of the transistor input is repeated in Fig. 7.7b. The emitter resistor is assumed to be perfectly bypassed, and the coupling capacitor C is assumed to be initially uncharged.

As the step voltage V is applied at time $t = 0$, a current i flows through the circuit. The voltage across the capacitor is

$$v_c = \frac{Q}{C} \qquad (7.17)$$

But since the capacitor is initially uncharged ($Q = 0$), there is no voltage across the capacitor at the instant the voltage is applied. Thus, the entire voltage V appears across R_s and R_i', so that the current through the circuit at the instant the voltage V is applied is $i = V/(R_s + R_i')$ at $t = 0$. The current i causes a charge Q to build up on the capacitor and the resulting voltage (Eq. 7.17) opposes the applied voltage. Therefore, the voltage across the resistance, $V - v_c$, decreases as time increases. The charge, in coulombs, is equal to the product of the current i and the time t during which the current flows. Let us suppose that the current *could* continue at the initial value $V/(R_s + R_i')$. Then the charge on the capacitor would be the product of this current and time t, or $Q = Vt/(R_s + R_i')$ and

$$v_c = \frac{Q}{C} = \frac{Vt}{(R_s + R_i')C} \qquad (7.18)$$

The time τ required for the capacitor voltage v_c to build up to the supply voltage V can be found by letting $v_c = V$ and $t = \tau$ in Eq. 7.18. Then $V = V\tau/(R_s + R_i')C$ and

$$\tau = (R_s + R_i')C \qquad (7.19)$$

This τ is known as the *time constant* of the circuit and is the product of the capacitance and the total resistance from one capacitor terminal through the circuit to the other capacitor terminal. A sketch of the capacitor voltage v_c as a function of time for this hypothetical case is given as the dashed line in Fig. 7.8. The capacitor voltage does not follow this dashed line because the current decreases as the charge on the capacitor increases. Instead, the capacitor voltage increases exponentially toward V as shown by the solid line in Fig. 7.8, and actually charges to only 0.63 V in one time constant.

As the capacitor charges, the circuit current, or charging current,

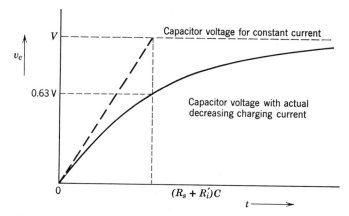

Fig. 7.8.   A sketch of the capacitor voltage v_c as a function of time.

actually decreases exponentially toward zero, as given by the equation

$$i = \frac{V}{R_s + R_i'} e^{-t/(R_s + R_i')C} \tag{7.20}$$

or

$$i = \frac{V}{R_s + R_i'} e^{-t/\tau} \tag{7.21}$$

A sketch of i as a function of time is given in Fig. 7.9. When $t = \tau$, or one time constant after the switch is opened, the current is reduced to e^{-1}, or 0.37, times its initial value. When $t = 2\tau$, the current is $e^{-2} = e^{-1} \times e^{-1} = 0.37 \times 0.37$ or 0.137 times its initial value. When $t = 3\tau$, the current is approximately 5 percent of its initial value and is 0.7 percent of its initial value when $t = 5\tau$. Thus, the current is within 1 percent of its final value in five time constants. The input voltage is iR_i' and therefore decreases exponentially in the same manner as the input current.

Figure 7.9 shows that the input current i decreases or *decays* exponentially when a step voltage is applied through a coupling capacitor to the input of a transistor. Since the output current and voltage are proportioned to the input current, the output voltage also decays exponentially and approaches zero after a time equal to about five time constants has elapsed. Therefore, the RC-coupled amplifier is not suitable as an amplifier for step-function inputs. Such an input is just a dc voltage, switched on at time $t = 0$.

However, if the input is switched *on* and then switched *off* during a period that is short compared with a time constant, the amplifier may satisfactorily preserve the input waveform as shown in Fig. 7.10. This

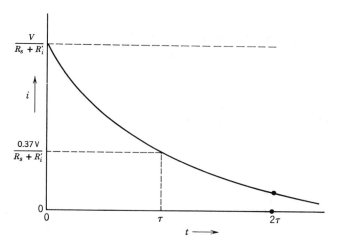

Fig. 7.9.   A sketch of current as a function of time in the circuit of Fig. 7.7.

type of input wave form is known as a *pulse*, which is usually understood to mean a rectangular pulse.

The top of the output pulse would ideally be *flat*, but the exponential decay caused by the charging of the coupling capacitor causes the output to *sag* (Fig. 7.10). The sag will be small if the pulse duration t_d is short compared to the time constant. A relationship between the sag and the time constant can be obtained easily if we observe that the output current

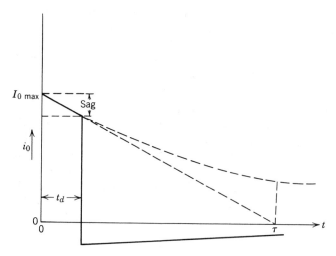

Fig. 7.10.   A sketch of the RC-amplifier output current as a function of time (the solid line) when the source voltage is a rectangular pulse of duration t_d.

would decay to zero in one time constant if it were to decay linearly at its initial rate. Then, if the pulse duration t_d is small in comparison with the time constant, we can see in Fig. 7.10 from similar triangles, that

$$\frac{\text{sag}}{t_d} = \frac{i_{o\,max}}{\tau} \qquad (7.22)$$

Solving for the sag,

$$\text{sag} = \frac{i_{o\,max}t_d}{\tau} \qquad (7.23)$$

Good amplifier design permits only small sag compared to $i_{o\,max}$. Thus, Eq. 7.23 is sufficiently accurate for calculating acceptable values of sag. The ratio of the sag to the initial current, or the fractional sag, is usually of greater significance than the actual sag. By using Eq. 7.23,

$$\text{Fractional sag} = \frac{\text{sag}}{i_{o\,max}} = \frac{t_d}{\tau} \qquad (7.24)$$

Since ω_1 is the reciprocal of the time constant τ, the fractional sag can also be written

$$\text{Fractional sag} = \omega_1 t_d \qquad (7.25)$$

For example, if an RC-coupled amplifier with $\omega_1 = 100$ rad/s has a 1 millisecond (ms) rectangular voltage pulse input, the fractional sag is 0.1.

PROBLEM 7.4 If the pulse duration in the foregoing RC-coupled amplifier is increased to 5 ms and the fractional sag is to be no greater than 0.1, what must be the low-frequency cutoff of the amplifier?
Answer: $\omega_1 = 20$ rad/s or $f_1 = 3.2$ Hz.

A series of rectangular pulses is usually applied to an amplifier. The output current or voltage then has the form shown in Fig. 7.11.

Since the average current through a capacitor must be zero, the crosshatched portion of the pulse (Fig. 7.11) must be equal to the crosshatched area below the I_q line and between the pulses, after a few pulses have passed through the amplifier. The fractional sag is determined by the relationship between the pulse duration t_d and the low-frequency cutoff of the amplifier. It makes little difference whether ω_1 is determined by the coupling capacitor, the emitter-bypass capacitor, or a combination of the two. The output pulse can be "squared up" by driving the transistor into cutoff and saturation. This type of nonlinear operation will be discussed in a later chapter.

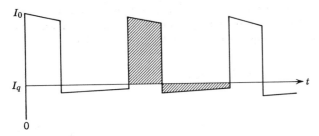

7.4 THE MIDFREQUENCY RANGE

The term midfrequencies is used to designate the frequency range in which all reactances are negligible. Thus, the coupling and bypass capacitors are assumed to have zero reactance in the midfrequency range.

An h-parameter circuit for a common-emitter amplifier is given in Fig. 7.12. This circuit can be simplified for the midfrequency range because the coupling capacitor C can be omitted. Also, since the amplifier load resistance is then the parallel combination of R_C, R_b, and R_i, the load resistance is usually quite small—not more than about 1000 Ω. Consequently, the voltage gain and output voltage v_o are small enough so that the voltage generator $h_{re}v_o$ can usually be omitted. Then $R_i \simeq h_{ie}$. A simplified midfrequency h-parameter circuit is given in Fig. 7.13.

Up to this point, the transistor load was assumed to be a single resistor in the collector or drain circuit. Thus, a single load line was drawn on the collector or drain characteristics for the purpose of graphical analysis. It now becomes apparent that the load resistance decreases from this value (R_C) at zero frequency to the parallel combination shown in Fig. 7.13 at midfrequencies. This combination is known as the ac load resistance. For the amplifier of Fig. 7.13, if $R_C = 5$ kΩ, $R_b = 10$ kΩ, and $h_{ie} = 1$ kΩ (for the second transistor), the ac load resistance $R_l = 770$ Ω. Notice that h_{oe} is

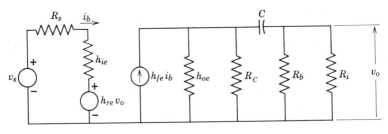

Fig. 7.12. A midfrequency h-parameter circuit.

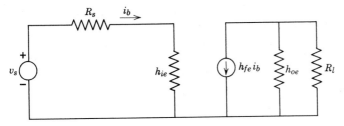

Fig. 7.13.   A simplified midfrequency h-parameter circuit.

neglected in this calculation because it is part of the transistor and not part of the load on the transistor.

Load lines can and should be drawn on the output characteristics for both dc and ac load resistances. The q-point must lie on both load lines and, therefore, must be at their intersection. One suitable technique is to select a q-point on the collector or drain characteristics in accordance with the current or voltage drive requirements of the following amplifier, then draw the dc and ac load lines through this point.

EXAMPLE 7.3   The transistor with collector characteristics given in Fig. 7.14 is to furnish a 1-mA peak-base signal current to the transistor that follows. The q-point must be chosen at a value of I_C greater than 1 mA because part of the signal collector current will be shunted through R_C and R_b. However, if R_C and R_b are both large in comparison with h_{ie}, most of the signal component of collector current will flow into the base of the following transistor. For this example, we choose the q-point at $I_C = 2\,\text{mA}$,

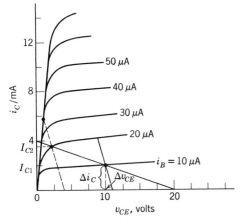

Fig. 7.14. q-point selections and load lines on the collector characteristics.

$v_{CE} = 10$ V with $V_{CC} = 20$ V. The dc load line is then drawn through this q-point and V_{CC}. This load line establishes the dc load resistance, which includes R_E, as was previously mentioned. For this example, $R_E + R_C = 20$ V/4 mA $= 5$ kΩ. Since R_C is usually much larger than R_E, we select $R_C = 4$ kΩ, $R_E = 1$ kΩ. Then, if h_{ie} of the following transistor is 500 Ω and $R_b = 10$ kΩ, the ac load resistance $R_l = 425$ Ω. The ac load line can be most easily drawn by using the relationship

$$\Delta v_{CE} = \Delta I_C R_l \qquad (7.26)$$

If the Δi_C is taken as the change from the q-point value of i_{C1} to $i_C = 0$ (2 mA) as shown in Fig. 7.14, the value of Δv_{CE} is $2 \times 10^{-3}(425) = 0.85$ V. The v_{CE}-axis intercept of the ac load line is thus located at $v_{CEq} + \Delta v_{CE} = 10.85$ V and the ac load line can be drawn through this intercept and the q-point.

The permissible excursion of the q-point along the dc load line can also be determined from the foregoing information. You may recall that the limits of collector current I_{C1} and I_{C2} must be established before the value of R_b in the stabilized bias circuit can be determined. We have already determined the minimum value of collector current I_{C1} as 2.0 mA, allowing a generous safety factor. The maximum permissible value of collector current can most easily be determined by recalling that the peak value of output signal voltage is $\Delta I_{c\,max} R_l = 0.85$ V. The minimum q-point value of v_{CE} must be at least this distance to the right of the transistor saturation region to prevent saturation. If the maximum saturation voltage is approximately 1.0 V at the maximum expected value of collector current (6 mA in this example) the minimum q-point value of v_{CE} is then 1.85 V. The value of 2.0 V used in Fig. 7.14 allows an additional margin of safety. The value of I_{C2} at this q-point can be determined from the slope of the dc load line and the desired value of $v_{CE\,min}$. Thus

$$I_{C2} = \frac{V_{CC} - v_{CE\,min}}{R_C + R_E} \text{ mA} \qquad (7.27)$$

For this example $I_{C2} = (20 - 2)$ V/5 kΩ $= 3.6$ mA.

The voltage and current gains at mid-frequencies, known as the reference gains K_v and K_i, can be determined from the h-parameter circuit of Fig. 7.13. Since $1/h_{oe}$ is very large in comparison with R_l,

$$v_o \approx -h_{fe} i_b R_l \qquad (7.28)$$

The signal input voltage of the first transistor is

$$v_{be} \approx h_{ie} i_b$$

Then
$$K_v = \left|\frac{v_o}{v_{be}}\right| = \frac{h_{fe}R_l}{h_{ie}} \tag{7.29}$$

The input current to the second transistor is, by using Eq. 7.26,

$$i_{b2} \approx \frac{v_o}{h_{ie2}} = \frac{-h_{fe}i_bR_l}{h_{ie2}} \tag{7.30}$$

Then
$$K_i = \left|\frac{i_{b2}}{i_b}\right| = \frac{h_{fe}R_l}{h_{ie2}} \tag{7.31}$$

PROBLEM 7.5 Determine K_v and K_i for the transistor amplifier of Example 7.3, where $h_{ie1} = 2.5\ \text{k}\Omega$, $h_{fe} = 200$, $h_{ie2} = 500\ \Omega$, and $R_l = 400\ \Omega$.
Answer: $K_v = 32, K_i = 160$.

EXAMPLE 7.4 We now consider the coupling of a field-effect transistor to a conventional transistor. Assume that the two-stage transistor amplifier of Fig. 7.12 is part of a sound amplifier that will be used with a crystal microphone. The crystal microphone requires a high load resistance (or amplifier input resistance) of the order of 1 MΩ; thus, the FET is a sensible choice for the first, or input, stage of the amplifier. A circuit diagram of the FET coupled to the conventional or bipolar transistor is given in Fig. 7.15a and the midfrequency equivalent of the FET stage is given in Fig. 7.15b.

We previously found that the transistor of Fig. 7.12 (Example 7.3) must provide a peak signal of 0.8 V and the h_{ie} of this transistor was given as 2.5 kΩ. Therefore, the input voltage and current is $v_{bc} = v_{ce}/K_v = 0.8/32 = 0.025$ V and $i_b = 0.025/2.5 \times 10^3 = 10^{-5}$ A. Also, we chose $R_E = 1$ kΩ, so that we can assume a current stability factor of 10 is suitable for this application; therefore, $R_b = 10$ kΩ will be used.

A typical output voltage for a crystal microphone is about 3-millivolt (mV) peak at a normal sound pressure input. Therefore, the voltage gain of the FET stage must be $25/3 = 8.33$.

Since the voltage gain of a FET is g_mR_l, and R_l is the parallel combination of $1/g_d$, R_D, R_b, and h_{ie}, we must choose a FET with a suitable combination of g_m and I_{DSS} to provide the required voltage gain. We assume that the drain resistance $r_d = 1/g_d$ is large compared to h_{ie}, but the drain circuit resistor R_D may not be high in comparison with h_{ie} and this resistor cannot be determined until the FET and its q-point are chosen. A basis for the choice of FET is given in the following relationship, which was Eq. 6.18 in Chapter 6.

$$g_{mo} = \frac{2I_{DSS}}{V_p} \tag{7.32}$$

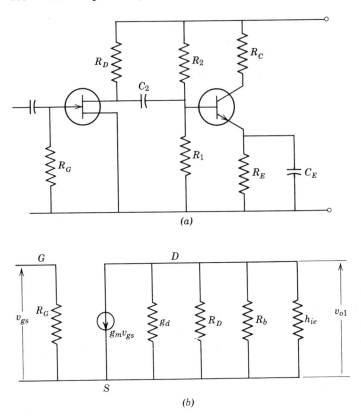

Fig. 7.15. (a) A FET coupled to a bipolar transistor and (b) a midfrequency equivalent circuit of the FET stage.

Where g_{mo} is the transconductance at $v_{GS} = 0$, $v_{DS} = V_p$, and I_{DSS} is the drain current at this point.

As a first approximation, we neglect R_D as well as r_d, and assume R_l is approximately equal to h_{ie} in parallel with R_b, or approximately 2.0 kΩ. The required value of g_{mo} is then greater than 8.33/2 kΩ $= 4.17 \times 10^{-3}$ mho. Thus, we must look for a FET with g_{mo} about 5×10^{-3} mho, since R_D has not been included in R_l. But to keep R_D large so that it will not seriously reduce R_l, I_{DSS} should be as small as possible. Equation 7.32 shows that a low pinch-off voltage V_p will give a high ratio of g_{mo} to I_{DSS}. The 2N3459 n-channel FET appears promising for this application. It has a typical $I_{DSS} = 2.0$ mA and typical $V_p = 0.8$ V. Then typical $g_{mo} = 5 \times 10^{-3} \mu$mhos. A family of typical drain characteristics for this transistor is given in Fig. 7.16. We must choose a q-point on the $v_{GS} = 0$ curve to obtain the value of $g_m = g_{mo}$. Also, the q-point should be well to right of the knee of the

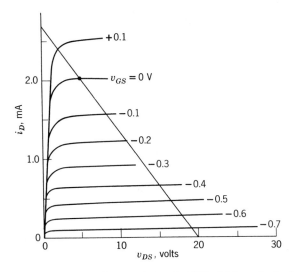

Fig. 7.16.   Drain characteristics of a FET with low pinch-off voltage.

characteristic curve so that g_d will be small. A value of $v_{DS} = 5.0$ V is indicated as a suitable value in Fig. 7.16. Using this q-point, and $V_{CC} = 20$ V, $R_D = 15$ V$/2.0$ mA $= 7.5$ kΩ.

PROBLEM 7.6   Determine R_l and K_v for the FET amplifier above.
Answer: $R_l = 1.67$ kΩ, $K_v = 8.35$.

7.5   HIGH-FREQUENCY CONSIDERATIONS

We learned in Section 7.1 that the low-frequency response of an RC-coupled amplifier is determined by the time constant of a series RC circuit. In this and the following sections we shall discover that the high-frequency response is determined by the time constant of the parallel or shunt capacitance and the associated resistance. The major problem with the shunt capacitance is that it does not appear on the circuit diagram and must be either calculated from the given transistor parameters or measured directly. Therefore, we shall first discuss the shunt capacitances in a transistor and then their effects on the circuit performance.

The junction capacitance associated with the depletion regions of a *p-n* junction has been previously discussed in considerable detail in connection with the junction diode. An additional capacitance, known as diffusion capacitance, which results from the stored charge caused by the injection of carriers across a junction, was discussed in connection with a junction diode in Chapter 3. This diffusion capacitance was also discussed in con-

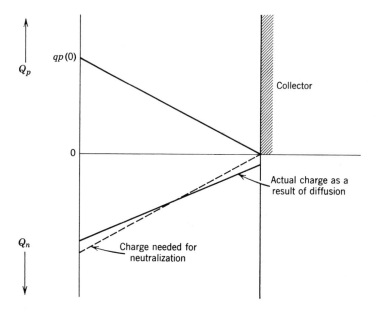

Fig. 7.17.   Stored charge in the base of a *p-n-p* transistor.

nection with the common-base amplifier in Chapter 4, but will be discussed in some detail here because it is a major element in the determination of the high-frequency characteristics of a transistor.

The diffusion capacitance in a transistor is almost entirely the result of the injection of carriers from the emitter into the base, assuming normal operation of the transistor. These carriers diffuse across the base region and create a stored charge in the base as illustrated in Fig. 7.17, which uses the *p-n-p* configuration as a model. But, as the injected carriers enter the base they produce an electric field that causes electrons to flow from the bias battery through the base terminal into the base in such a manner as to neutralize the charge. Notice that the current that flows as a result of this transfer of charge involves only the emitter and base currents and has no effect on the collector current. Therefore, each change of base current has two components. One component contributes to the stored charge in the base region as if a capacitor were connected from base to emitter. The second component causes the change of collector current desired in transistor operation. The shunting effect of the diffusion capacitance (the stored charge effect) on the base signal causes α and β to decrease as the frequency increases. The additional electrons that rush into the base to neutralize the injected charge do not appreciably increase the recombination rate in the base as long as their density is small compared to the doping density or concentration in the base. The electrons, like the holes, diffuse toward the collector, but the collector junction barrier potential prevents their entering the collector. Thus, a small net positive charge exists at the emitter junction as a result of the electron diffusion. This charge distribution creates an electric field that is just sufficient to offset the diffusion of electrons but *aids* the diffusion of holes and, thus, improves the performance of the transistor. This effect is enhanced in graded-base or drift-field transistors, which have fairly heavy base doping near the emitter junction, with the doping concentration decreasing to near zero at the collector junction. These transistors will be discussed later in this section.

An expression was developed in Chapter 4 for the diffusion capacitance due to the stored charge in the base of a transistor. This expression, which neglects the electric field in the base, was given as $C_D = w^2/2D_p r_e$ for a p-n-p transistor. As noted in Chapter 4, the emitter-junction capacitance C_{je} adds directly to C_D. The alpha-cutoff frequency, in radians per second was also shown to be the reciprocal of the time constant $r_e C_D$, providing C_D is large in comparison with C_{je}. Then

$$\omega_\alpha = \frac{1}{(w^2/2D_p r_e)r_e} = \frac{2D_p}{w^2} \qquad (7.33)$$

where w is the effective base width and D_p is the diffusion constant of the base material. Note that ω_α is independent of I_e as long as C_D is large compared to the emitter-junction capacitance C_{je}. Also observe from Eq. 7.33, as illustrated by the example in Chapter 4, that large values of ω_α, and perhaps good high-frequency performance, might be obtained by making the effective base width w very small and selecting the transistor material and doping arrangement so that the diffusion constant of the charge carriers in the base region is high. A gallium-arsenide n-p-n transistor holds great promise in this regard because of its high D_n.

The upper-cutoff frequency of a common-base amplifier might not be determined by the alpha-cutoff frequency of the transistor. As seen in Fig. 7.18, the collector-junction capacitance C_{jc} is in parallel with $(R_L + r_b)$ as well as r_c. At high frequencies, part of the current αi_e is shunted through C_{jc} and, therefore, does not pass through the load resistor R_l. Even if α is assumed to be independent of frequency, the load current will decrease to $0.707 i_e$, neglecting r_c, when $I_3 = I_c$. Since these currents are equal when $X_{Cjc} = R_l + r_b$, the cutoff frequency due to C_{jc} is

$$\omega_c = \frac{1}{(R_l + r_b)C_{jc}} \qquad (7.34)$$

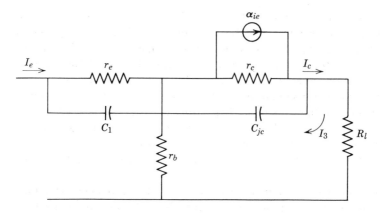

Fig. 7.18. An equivalent circuit showing the effect of the capacitor C_{jc} on the load current at high frequencies.

The upper-cutoff frequency f_2 of the amplifier is lower than either f_α or f_c. In fact, the effects of C_1 and C_{jc} in decreasing the collector current are additive, so that the upper-cutoff frequency ω_2 is the reciprocal of the sum of the time constants.

$$\omega_2 = \frac{1}{r_e C_1 + (R_l + r_b) C_{jc}} \tag{7.35}$$

EXAMPLE 7.5   A given transistor with $r_b = 200\ \Omega$, $C_1 = 200$ picofarads (pF), and $C_{jc} = 5$ pF at $I_E = 1$ mA and $V_{CB} = 10$ V is used in a common-base amplifier with $R_l = 5\ \text{k}\Omega$. We wish to determine the upper half-power frequency ω_2. First we must determine $r_e C_1 = 25 \times 200 \times 10^{-12} = 5 \times 10^{-9}$ and $(R_l + r_b) C_{jc} = 5.2 \times 10^3 \times 5 \times 10^{-12} = 2.6 \times 10^{-8}$. Then $\omega_2 = 1/(5 \times 10^{-9} + 2.6 \times 10^{-8}) = 3.2 \times 10^7$ and $f_2 = 5.1$ MHz.

PROBLEM 7.7   A given transistor has $r_b = 100\ \Omega$, $C_1 = 160$ pF, and $C_{jc} = 3$ pF at $I_E = 2$ mA and $V_{CB} = 10$ V. This transistor is used in a common-base amplifier at this q-point with $R_l = 6.8\ \text{k}\Omega$. Determine F_α and F_2.
                    Answer: $f_\alpha = 80$ MHz, $f_2 = 7$ MHz.

The development of the *graded-base* transistor caused a major advance in the high-frequency capabilities of the transistor. The graded base provides a built-in electric field in the base region, as was previously mentioned, and a small collector-junction capacitance C_{jc}. The electric field accelerates the carriers through the base and thus reduces the stored charge, or diffusion capacitance for a given collector current.

The term *graded base* means that the doping density in the base region

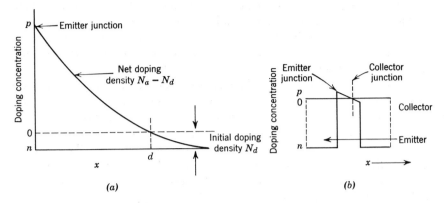

Fig. 7.19. Doping characteristics of a graded-base transistor. (a) Doping profile in base region and (b) doping profile in transistor.

is not uniform, but decreases as one traverses the region from the emitter junction to the collector junction. This doping distribution, or profile, which is shown in Fig. 7.19a. is obtained by diffusing the doping atoms into a heated semiconductor crystal. For example, assume that a graded-base n-p-n transistor is to be constructed. A bar of lightly doped n-type crystal is heated in an oven and the trivalent dopant is applied, in liquid or gaseous form, to one surface of the bar. Then the acceptor atoms diffuse into the semiconductor in a manner very similar to the diffusion of carriers from the emitter into the base when forward bias is applied to the emitter junction. However, the temperature must be high (near the melting point of the crystal) to permit the doping atoms to diffuse; the diffusion rate of these atoms is very slow compared to the diffusion of charge carriers. Since the effective doping density is the difference between the donor impurity concentration N_d and the acceptor impurity concentration N_a, the end of the bar exposed to the trivalent impurity soon becomes p-doped and the doping density decreases exponentially as the distance increases into the bar, as shown in Fig. 7.19a. At some distance d, the acceptor atom density is equal to the initial donor atom density and the crystal appears intrinsic. This point, which moves deeper into the material as the diffusion process continues, becomes the collector-base junction (when the diffusion is stopped by cooling the material) because the net doping changes from p-type to n-type at that point. The transistor is completed by replacing the trivalent doping material on the surface of the semiconductor with a heavily n-doped semiconductor, which serves as the emitter. A process known as *alloying* is usually used to attach the emitter. Alloying is similar to welding except that the continuous crystal structure is carefully maintained and the junction between the n-type and p-type materials is a plane surface. Also, the lightly n-doped collector material is usually replaced by heavily n-doped material (Fig. 7.19b) from a short distance to the right of the collector junction. This heavily doped material improves the conductivity of the collector region and decreases the saturation voltage for a given collector current. Either alloying or diffusion techniques can be used to increase the doping concentration in the collector. Of course, a p-n-p transistor can be produced by using the opposite type dopants. Finally, the leads are attached and the transistor is heated in an atmosphere of either oxygen or nitrogen to form a thin layer of silicon dioxide or silicon nitride (for a silicon semiconductor) on the surface of the semiconductor. This layer is a good insulator that protects the surface from contamination and reduces surface leakage. This process is known as *passivation*.

As was previously mentioned, the graded base produces a built-in electric field because the majority carriers in the base diffuse toward the collector but are not able to enter the collector because of the potential barrier. Therefore, they accumulate and an electric field builds up until the drift current, due to the electric field, is equal to the diffusion current. This electric field therefore aids the flow of minority carriers in the base that are injected from the emitter into the base. It therefore has the same effect as increasing the diffusion constant, which reduces the diffusion capacitance and increases f_α. The electric field also increases α and β because the minority carriers spend less time in the base and are less likely to recombine. The graded-base transistor also has better linearity because the electric field is proportional to the collector current in a uniform-base transistor, so that β strongly depends on collector current.

The graded collector junction in the graded-base transistor has very light doping on both sides of the junction in the vicinity of the junction. Therefore, the depletion region is wide on both sides of the junction, for a given collector voltage, as compared with the uniformly doped transistor. The collector voltage breakdown, therefore, is high and the junction capacitance is comparatively low — an important contribution to the high-frequency performance. The lightly doped layer on the collector side of the collector junction is known as an *epitaxial* layer.

The only disadvantage of the graded-base transistor is the low emitter-breakdown voltage that results from the comparatively heavy base doping at the emitter junction and the very heavy emitter doping required to give high emitter efficiency. This low breakdown voltage is a handicap only when the transistor is used for switching and the emitter junction needs to be reverse biased. However, in these applications a diode can be connected in series with the input lead and the diode will withstand the inverse voltage and protect the emitter-base junction in the transistor.

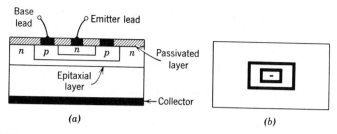

Fig. 7.20. Construction details of a typical transistor. (*a*) Cross-sectional view and (*b*) top view.

The physical layout of a typical graded-base, epitaxial, passivated transistor is shown in Fig. 7.20. The geometry of the transistor illustrates that it is designed for high β and good collector power dissipation when operated in the normal mode, in contrast with the *inverse mode* in which the collector is used as the emitter and vice versa.

7.6 THE HYBRID-π CIRCUIT AND BETA-CUTOFF FREQUENCY

Although the *h*-parameter equivalent circuit is very useful at low and middle frequencies, it is not easily adapted to the high-frequency range where diffusion and junction capacitances must be included. The chief problem is that the *h*-parameters vary with frequency and high-frequency values cannot be determined from the static characteristics. One equivalent circuit that gives good visualization of the high-frequency performance of a common-emitter transistor is the hybrid-π circuit shown in Fig. 7.21. This circuit is quite similar to the equivalent-T circuit of Fig. 5.7a. The following differences should be noted.

1. The diffusion and junction capacitances are included in the hybrid-π circuit.

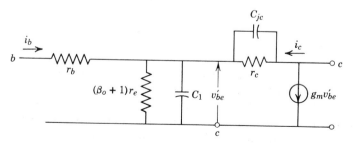

Fig. 7.21. A hybrid-π equivalent circuit for the common-emitter transistor configuration.

2. The current generator that appeared across the collector junction in the equivalent-T circuit appears between the collector and emitter terminals in the hybrid-π circuit. This current is expressed in terms of the voltage across the emitter junction instead of the input current.

3. The change in position of the dependent current generator eliminates the collector current through r_e. Therefore, only the input current i_b flows through the resistor $(\beta_0 + 1) r_e$. Therefore, the factor $(\beta_0 + 1)$ is required to give the proper voltage across the emitter junction and, hence, the proper input impedance for the transistor. β_0 is the low frequency value of β.

The current generator $g_m v'_{be}$ must provide the same current as the equivalent-T or h-parameter current generators. Thus,

$$g_m v'_{be} = \beta_0 i_b \tag{7.36}$$

But, at low frequencies where the reactance of C_1 and C_{jc} can be neglected (Fig. 7.21) $v'_{be} \simeq (\beta_0 + 1) r_e i_b$. Then

$$g_m = \frac{\beta_0 i_b}{v'_{be}} = \frac{\beta_0 i_b}{(\beta_0 + 1) r_e i_b} = \frac{\beta_0}{(\beta_0 + 1) r_e} \tag{7.37}$$

Since $\beta_0/(\beta_0 + 1) = \alpha_0$ and $g_e = 1/r_e = (q/kT) I_E$,

$$g_m = \frac{\alpha_0 q I_E}{kT} = \frac{q}{kT} I_C \tag{7.38}$$

At normal room temperatures, q/kT is approximately equal to 40, as was previously noted; thus, g_m can be easily determined as $40 I_C$, where I_C is the q-point value of collector current. The fictitious resistance $(h_{fe} + 1) r_e$ can also be expressed in terms of g_m, since $g_m = \alpha_0 g_e$,

$$(\beta_0 + 1) r_e = \frac{\alpha_0(\beta_0 + 1)}{g_m} = \frac{\beta_0}{g_m} \tag{7.39}$$

We designate this resistance β_0/g_m as r_π.

The hybrid-π circuit can be modified, as shown in Fig. 7.22. The output resistance r_o has replaced the resistance r_c, which was shown in Fig. 7.21. The resistance r_o is a function of the driving source resistance, but is approximately equal to $1/h_{oe}$. The resistance r_o is much smaller than r_c because, as seen from Fig. 7.21, a voltage applied at the output terminals causes a current to flow through r_c. This current causes a voltage v'_{be} which activates the current generator $g_m v'_{be}$. If a voltage v_o is applied to the

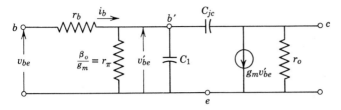

Fig. 7.22.   A modified hybrid-π circuit.

output terminals, the current which flows through r_c is approximately v_o/r_c. Then, if the input is open, the current flows through the resistance β_0/g_m and $v'_{be} = (v_o/r_c)(\beta_0/g_m)$. Then

$$g_m v'_{be} = \frac{\beta_0 v_o}{r_c} \tag{7.40}$$

Since the output resistance is the ratio of the voltage v_o to the current i_c, which flows as a result of v_o, and $i_c = v_o/r_c + \beta_0 v_o/r_c$

$$r_o = \frac{r_c}{\beta_0 + 1} \tag{7.41}$$

You may recall from Chapter 5 that $1/h_{oe} \simeq r_d = r_c/(\beta_0 + 1)$. The resistance r_c also has a slight loading effect on the input, but this can usually be neglected.

Figure 7.22 shows that the voltage v'_{be} decreases as the frequency increases, assuming i_b constant, because of the decreasing reactance of C_1. Also, observe that C_{jc} is in parallel with C_1 when the output is shorted (no load resistance). Then

$$v'_{be} = \frac{i_b}{(g_m/\beta_0) + j\omega(C_1 + C_{jc})} \tag{7.42}$$

Beta-cutoff frequency f_β is defined as the frequency at which the short-circuit current gain drops to $0.707 \beta_0$. This frequency occurs when the j term in the denominator of Eq. 7.42 is equal to the real term g_m/β_0, or $\omega_\beta(C_1 + C_{jc}) = g_m/\beta_0$. Then

$$\omega_\beta = \frac{g_m}{\beta_0(C_1 + C_{jc})} \tag{7.43}$$

PROBLEM 7.8   A given transistor has $\beta_0 = 100$, $C_1 = 190$ pF, and $C_{jc} = 10$ pF at $I_c = 5.0$ mA. Determine ω_β and f_β for this transistor.

Answer: $\omega_\beta = 10^7$ rad/s, $f_\beta = 1.6 \times 10^6$ Hz.

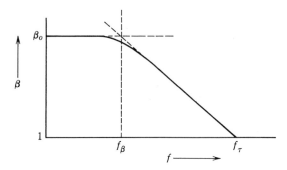

Fig. 7.23.   A sketch of β as a function of frequency, using log scales.

A sketch of β as a function of frequency is given in Fig. 7.23. The frequency at which $\beta = 1$ is defined as f_τ. Since the reactance of $(C_1 + C_{jc})$ is inversely proportional to frequency, β is also inversely proportional to frequency for frequencies well above f_β where this reactance is small in comparison with β_0/g_m. Therefore, the high-frequency part of the curve is a straight line if logarithmic scales are used. If this straight-line portion of the curve is extended upward until it intersects the β_0 line, intersection occurs at f_β. Therefore, while β decreases by the factor β_0, the frequency must increase by the factor β_0 and

$$f_\tau = \beta_0 f_\beta \qquad (7.44)$$

And from Eq. 7.43 we can see that

$$\omega_\tau = \frac{g_m}{C_1 + C_{jc}} \qquad (7.45)$$

Thus, the transistor with $\beta_0 = 100$, which was found to have $f_\beta = 1.6$ MHz, has a value of $f_\tau = 160$ MHz.

The magnitudes of f_τ and the alpha-cutoff frequency f_α may be compared if we recall that $\omega_\alpha = 1/r_e C_1 = g_e/C_1$. Since $g_m = \alpha_0 g_e$ and C_{jc} is usually small compared to C_1, $\omega_\tau \simeq \omega_\alpha$. Some manufacturers specify f_τ for their transistors and others specify f_α. A sufficiently accurate value of f_β may be obtained by dividing either f_α or f_τ by β_0.

PROBLEM 7.9   A given transistor has $\beta_0 = 100$, $C_{jc} = 3.6$ pF, and $f_\tau = 200$ MHz at $i_C = 2$ mA and $v_{CE} = 10$ V. Find f_β and C_1.

Answer: $f_\beta = 2$ MHz, $C_1 = 60$ pF.

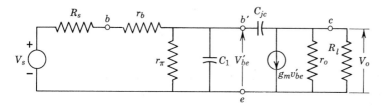

Fig. 7.24. A hybrid-π circuit including source and load.

7.7 THE UPPER-CUTOFF FREQUENCY OF A COMMON-EMITTER AMPLIFIER

The beta-cutoff frequency f_β is the upper-cutoff frequency of the common-emitter amplifier when the load resistance is very near zero and the driving source resistance is infinite because these are the conditions assumed when the output is shorted and the input current is constant, or independent of frequency. We shall now investigate the effects of the load resistance and the source resistance on the upper-cutoff frequency f_2 or ω_2.

The hybrid-π equivalent circuit with load and driving source included is shown in Fig. 7.24. The collector-junction capacitance C_{jc} is usually small in comparison with C_1, but the effect of C_{jc} may be surprisingly large because of the comparatively high signal voltage across it. This voltage is the difference between the voltage v'_{be} and the output voltage v_0. Since v_0 is usually large compared to v'_{be}, the signal current through C_{jc} may be comparable with, or even larger than the current through C_1. The current through C_{jc} is

$$I_{jc} = (V'_{be} - V_o) j\omega C_{jc} \tag{7.46}$$

But, as seen in Fig. 7.24, $V_o = - g_m V'_{be} R'_l$, where R'_l is the parallel combination of R_l and r_o. The negative sign occurs because of the polarity reversal of V_o as compared with V'_{be}. Then

$$I_{jc} = V'_{be} (1 + g_m R'_l) j\omega C_{jc} \tag{7.47}$$

The susceptance of C_{jc} as viewed from the $b'e$ terminals is

$$B_{jc} = \frac{I_{jc}}{V'_{be}} = j\omega (1 + g_m R'_l) C_{jc} \tag{7.48}$$

The effective capacitance of C_{jc}, as viewed from the terminals $b'e$ is, therefore,

$$C_{\text{eff}} = \frac{B_{jc}}{j\omega} = (1 + g_m R'_l)C_{jc} \qquad (7.49)$$

This capacitance magnification that results when a capacitance exists between the input and the output of an amplifier is often known as the *Miller effect*. The development of the pentode tube was a result of the effort to reduce this type of capacitance in a triode.

The equivalent circuit of Fig. 7.25 shows the effect of the junction capacitance C_{jc}. The capacitance $(1 + g_m R'_l)C_{jc}$ has been added to the input circuit to account for the effect of C_{jc} at the terminals $b'e$. Transistor manufacturers usually give a capacitance C_{ob} (output capacitance with the base grounded), which differs only from C_{jc} by the header capacitance. Therefore, this C_{ob} can be used as an approximate value for C_{jc}.

EXAMPLE 7.6   Let us consider the example of the transistor that had $C_1 = 190\ \text{pF}$, $C_{jc} = 10\ \text{pF}$, $\beta_0 = 100$, and $r_\pi = \beta_0/g_m = 500\ \Omega$ at $I_C = 5\ \text{mA}$. If this transistor has $R'_l \simeq R_l = 1\ \text{k}\Omega$, then $C_{\text{eff}} = 2010\ \text{pF}$ and the total capacitance in parallel with r_π is 2200 pF. We now assume that the source resistance R_s is so large that the input current i_b does not vary with frequency even though the input impedance decreases as the frequency increases. The upper-cutoff frequency occurs when the reactance of the total shunt capacitance is equal to r_π. In this example, $\omega_2 = 10^{12}/(500 \times 2200) = 9 \times 10^5$ rad/s or $f_2 = 1.44 \times 10^5$ Hz. Observe that this upper-cutoff frequency is less than one-tenth as high as the 1.6 MHz beta-cutoff frequency that was previously calculated for this transistor.

Now let us consider a transistor amplifier that is driven by a voltage source with moderate resistance R_s (Fig. 7.25). The effect of this source resistance can be most easily seen if the input circuit to the left of r_π in Fig. 7.25 is converted to a Norton's or current-source equivalent circuit

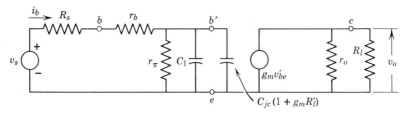

Fig. 7.25.   An equivalent circuit, which shows the effect of the collector junction capacitance.

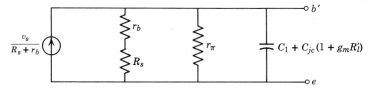

Fig. 7.26.   An equivalent circuit, which illustrates the effect of the source resistance R_s on the upper-cutoff frequency.

as shown in Fig. 7.26. This circuit shows that the series combination of r_b and R_s is actually in parallel with r_π. Thus the voltage at the terminals $b'e$, and, hence, the output current, drops to 0.707 times the midfrequency value when the reactance of the total effective capacitance is equal to the total shunt resistance, consisting of β_0/g_m in parallel with the series combination of r_b and R_s. We shall indicate this combination as $(R_s + r_b) \| r_\pi$, where the two vertical lines mean "in parallel with." The upper-cutoff frequency of the total amplifier can now be written

$$\omega_2 = \frac{1}{(R_s + r_b) \| r_\pi [C_1 + (1 + g_m R_l')C_{jc}]} \qquad (7.50)$$

EXAMPLE 7.7   Assume that the transistor of the preceding Example (7.6), which had $r_\pi = 500\ \Omega$ and a total effective capacitance of 2200 pF, has $r_b = 100\ \Omega$ and $R_s = 1\ \text{k}\Omega$. Then the parallel combination of r_π and $(r_b + R_s)$ is $500 \times 1.1\ \text{k}\Omega / (500 + 1.1\ \text{k}\Omega) = 344\ \Omega$ and $\omega_2 = 1/(344 \times 2.2 \times 10^{-9}) = 1.32 \times 10^6$ rad/s. Thus $f_2 = 210$ kHz, which is about 1.5 times as high as the f_2 obtained with the current source drive, or very high R_s.

The preceding examples and Eq. 7.50 show that the maximum upper-cutoff frequency occurs when both load and source resistances are small (actually zero). But small values of load resistance result in small values of voltage gain. Therefore, since the bandwidth of the amplifier is essentially equal to the upper-cutoff frequency, the bandwidth decreases as the voltage gain increases.

The frequency and phase response of the entire frequency range of the RC-coupled amplifier is given in Fig. 7.27. The phase is less than 180° at high frequencies, as seen from Fig. 7.28, and the relationships that follow.

$$V_{be}' = \frac{I_s}{G + j\omega C} = \frac{I_s}{\sqrt{G^2 + (\omega C)^2} \big/ \tan^{-1} \dfrac{\omega C}{G}} \qquad (7.51)$$

$$V_o = \frac{g_m I_s R_l \big/ 180°}{\sqrt{G^2 + (\omega C)^2} \big/ \tan^{-1} \dfrac{\omega C}{G}} \qquad (7.52)$$

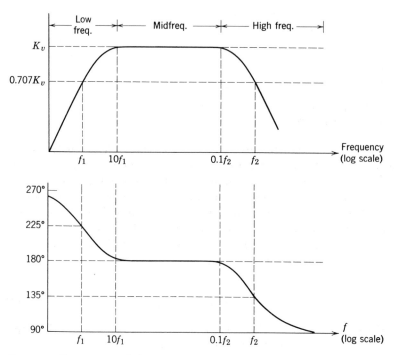

Fig. 7.27.   Frequency and phase response of an RC-coupled amplifier.

The 180° was used in Eq. 7.52 to replace the negative sign which results from the polarity reversal. Therefore,

$$\frac{V_o}{I_s} = \frac{g_m R_l}{\sqrt{G^2 + (\omega C)^2}} \left/\underline{180° - \tan^{-1} \dfrac{\omega C}{G}}\right. \tag{7.53}$$

Since $I_s = V_s G_s$

$$\frac{V_o}{V_s} = \frac{g_m G_s R_l}{\sqrt{G^2 + (\omega C)^2}} \left/\underline{180° - \tan^{-1} \dfrac{\omega C}{G}}\right. \tag{7.54}$$

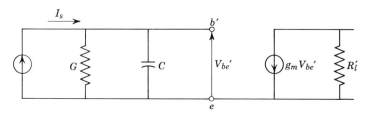

Fig. 7.28.   Equivalent circuit, which illustrates the phase relationships at high frequencies.

PROBLEM 7.10 A given transistor has $\beta_0 = 80$, $r_b = 100\,\Omega$, $C_{ob} = 3\,\text{pF}$, $f_\tau = 300\,\text{MHz}$, and $r_o = 20\,\text{k}\Omega$ at $I_C = 4\,\text{mA}$ and $V_{CE} = 10\,\text{V}$. This transistor is used with $R_l = 1\,\text{k}\Omega$ and driving source resistance $R_s = 2\,\text{k}\Omega$. Determine the upper-cutoff frequency f_2.   *Answer:* $f_2 = 712\,\text{kHz}$.

7.8 FREQUENCY RESPONSE OF THE FET AND THE CASCODE AMPLIFIER

The high-frequency characteristics of the field-effect transistor depend primarily on the junction capacitance, the source resistance, and the voltage gain of the amplifier. The relationship between these factors will be developed with the aid of the equivalent circuit of the FET amplifier shown in Fig. 7.29. Since drain current flows because of an electric field and *not* by diffusion, there is no diffusion capacitance in the FET. However, junction capacitance does exist and this capacitance is separated into two parts: capacitance between the gate and the source C_{gs}, and capacitance between the gate and the drain C_{gd}.

Observe from Fig. 7.29 that the two parts of the FET junction capacitance appear in the same relative positions as the two junction capacitances in the bipolar or conventional transistor. Therefore, the gate-drain capacitance C_{gd}, which appears between the output terminal and the input terminal, must be multiplied by the factor $(1 + g_m R_l')$ to obtain its effect on the input circuit. The total effective input capacitance is therefore $C_{\text{eff}} = C_{gs} + (1 + g_m R_l')C_{gd}$. Since r_d is usually large in comparison with the ac load resistance R_l, we may usually neglect r_d in calculating the voltage gain.

The upper-cutoff frequency of the amplifier can be easily visualized and obtained if the input portion of the circuit of Fig. 7.29 is simplified, as shown in Fig. 7.30. Observe that the Thevenins' equivalent voltage source was transformed to a Norton's equivalent current source so that the resistances R_s and R_G could be combined to produce the modified source resistance R_s'. Since R_G is usually much larger than R_s, R_G can often be neglected. The output current and voltage are proportional to V_{gs}. There-

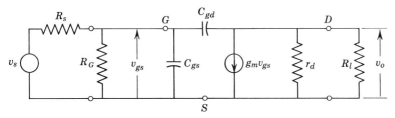

Fig. 7.29.   A FET amplifier equivalent circuit suitable for medium and high frequencies.

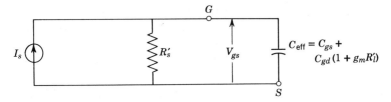

Fig. 7.30.   A simplified equivalent circuit representing the FET input and driving source.

fore, the upper-cutoff frequency occurs when R'_s is equal to $1/\omega_2 C_{eff}$ and

$$\omega_2 = \frac{1}{R'_s C_{eff}} \qquad (7.55)$$

At first glance, the FET may appear to have better high-frequency characteristics than the bipolar transistor because its effective capacitance is less due to the elimination of diffusion capacitance and because of the likelihood of smaller voltage gain. However, the comparatively large values of driving source resistance R_s commonly used in conjunction with a FET may seriously limit the high-frequency response. For example, let us consider a typical, small signal FET that has $C_{gs} = 5$ pF, $C_{gd} = 4$ pF, $g_m = 10^{-3}$ mho, $g_d = 6$ μmho, and $R_l = 10$ kΩ at a given q-point. The driving source resistance is 10^6 Ω. Neglecting r_d and R_G, which may be 10^7 Ω or more, $C_{eff} = 49$ pF, $\omega_2 = 1.98 \times 10^4$ rad/s and $f_2 = 3 \times 10^3$ Hz, which is too low for even a good audio-frequency amplifier. Of course, a MOSFET with lower capacitances would be useful in this case.

The frequency response of the FET amplifier can be improved, of course, by decreasing either the source resistance or the voltage gain of the amplifier. However, the high source resistance was probably the basis for the choice of the FET in the first place. One novel way of improving the high-frequency performance is to use two FET's in a series arrangement, as shown in Fig. 7.31. This arrangement is commonly known as a *cascode* amplifier.

Figure 7.31 shows that the impedance looking into the source of transistor T_2 is the load impedance of transistor T_1 in the cascode configuration. Also, if the capacitor C is large enough to maintain the gate of transistor T_2 at ground potential for all signal frequencies of interest, the impedance presented by the source of transistor T_2 is resistive and is equal to $\Delta V_{gs2}/\Delta I_{d2} = 1/g_{m2}$, as was previously discussed. The voltage gain of transistor T_1 is then $K_v = g_{m1}/g_{m2}$. Therefore, if the two transistors are similar so that $g_{m1} \simeq g_{m2}$, the voltage gain of T_1 is *one* and the effective input capacitance of the amplifier is $(C_{gs} + 2C_{gd})$. In the preceding FET amplifier where $R_s = 10^6$ Ω, $C_{gs} = 5$ pF and $C_{gd} = 4$ pF, the upper-cutoff

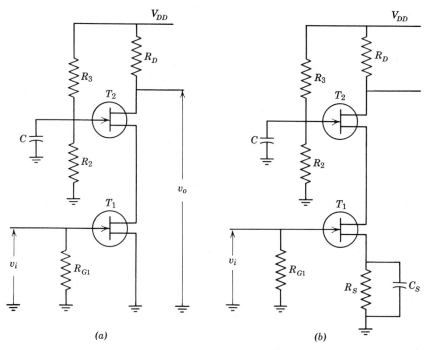

(a)   (b)

Fig. 7.31.   A FET cascode amplifier.

frequency was 3 kHz. However, if the same FETs are used in a cascode amplifier, $f_2 = 12.2$ kHz. Note that moderate values of source resistance would provide high-cutoff frequencies. Since the input resistance of transistor T_2 is low ($1/g_m$) the input capacitance of this transistor (C_{gs}) does not influence the upper-cutoff frequency of the amplifier. The gate grounding capacitor C must be very large compared to C_{gs} or C_{gd} to maintain the gate at signal ground. The resistors R_2 and R_3 establish the q-point drain voltages V_{DS} across both transistors. In the amplifier shown in Fig. 7.31a, transistor T_1 has zero gate bias; therefore, transistor T_2 must also have zero gate bias, if the transistors are identical, since their drain currents are the same. This equality of gate biases also requires that both transistors have drain voltages V_{DS} higher than the pinch-off voltage V_p. In fact, this requirement provides the basis for determining R_2 and R_3. With zero gate bias the voltage drop across R_2 is equal to the q-point drain voltage V_{DS} of transistor T_1. This voltage should be at least one or two volts greater than the pinch-off voltage. Therefore, noting that R_2 and R_3 form a voltage divider,

$$V_{DS1} = \frac{V_{DD}R_2}{R_2 + R_3} \qquad (7.56)$$

and
$$R_2 = \frac{V_{DS1}R_3}{V_{DD} - V_{DS1}} \tag{7.57}$$

EXAMPLE 7.9 Let us assume that the pinch-off voltage of each transistor in Fig. 7.31 is 2 V and the drain supply voltage V_{DD} is 30 V. We then choose $V_{DS1} = 5$ V and $R_3 = 220$ kΩ, which will not load the power supply. Then $R_2 = 5 \times 220$ kΩ/25 = 44 kΩ. The drain supply voltage for transistor T_2 is $30 - 5 = 25$ V. The design of the cascode amplifier can now proceed as though it were a single FET with $V_{DD} = 25$ V. The design depends, of course, on the input requirements of the amplifier that follows.

Reverse bias will sometimes be desired, depending on the temperature stability requirements, the magnitude of input signal or the impedance of the load. A bias resistor and bypass capacitor must then be included in the source circuit of transistor T_1, as shown in Fig. 7.31b, to provide the proper bias. Transistor T_2 must have the same drain current and will automatically adjust to the same bias if both transistors have V_{DS} greater than V_p. However, for the n-channel transistors shown, the gate potential of transistor T_2 is more negative than the drain potential of transistor T_1 by the amount of the bias voltage V_{GS}. But this same amount of voltage is dropped across the bias resistor of transistor T_1, so that the voltage drop across R_2 is still equal to the drain-to-source voltage V_{DS} of transistor T_1. Therefore, the design of the biased-cascode amplifier can proceed in the same manner as for a single FET with source bias but the effective drain supply voltage is V_{DD} minus the V_{GS} of transistor T_1.

The absence of coupling between the output and the input circuits of the cascode amplifier is a great advantage in some applications, particularly in tuned radio-frequency amplifiers. These amplifiers will be discussed in Chapter 10. Also, we should note that stray wiring capacitance should be kept to an absolute minimum since it adds directly to the FET capacitance, which is very small in the cascode configuration.

PROBLEM 7.11 A given FET has $g_m = 2 \times 10^{-3}$ mho, $C_{gd} = 2$ pF, and $C_{gs} = 3$ pF at the q-point $I_D = 1.5$ mA, $V_{DS} = 5$ V, $V_{GS} = -0.1$ V. If $g_m = 3 \times 10^{-3}$ mho, $R_l = 6$ kΩ, and the driving source resistance is 50 kΩ, determine the upper-cutoff frequency f_2 for the amplifier.

Answer: $f_2 = 77.7$ kHz.

PROBLEM 7.12 Calculate values of R_S and R_2 (Fig. 7.31b) for a cascode amplifier to replace the amplifier of Prob. 7.11 using two of the FET's of Prob. 7.11 at the q-points given, with $V_{DD} = 25$ V and $R_3 = 100$ kΩ. Determine the upper-cutoff frequency f_2.   *Answer:* $R_S = 67$ Ω, $f_2 = 455$ kHz.

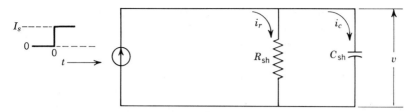

Fig. 7.32.   A simplified circuit representing the input of an amplifier driven by a current source.

7.9   RISE TIME

We shall now investigate the effect of the shunt capacitance on the output waveform of an amplifier when the input signal is a rectangular pulse. The simplified equivalent input circuit of such an amplifier is shown in Fig. 7.32. In this circuit C_{sh} represents the effective shunt capacitance of the amplifier as seen from the input terminals, plus the stray wiring capacitance and the capacitance seen looking back into the driving source. Also R_{sh} represents the entire shunt resistance, including the driving source resistance.

As the current I_s is suddenly applied at time $t = 0$, the capacitor, which is initially uncharged, begins to charge. As C_{sh} charges, the voltage across C_{sh} increases, but the voltage across R_{sh} also increases and the current through R_{sh} increases until finally all of the source current I_s passes through R_{sh} and C_{sh} is completely charged at the voltage $V = I_s R_{sh}$. As shown before, the capacitor would charge to voltage V in one time constant if the initial current I_s continued to flow into the capacitor until the voltage reached $V = I_s R_{sh}$. But since the capacitor current decreases as the voltage increases, the voltage increases exponentially, as shown in Fig. 7.33. This relationship is expressed by the following equation, which can be obtained by the use of calculus.

$$v = I_s R_{sh}(1 - e^{-t/\tau}) \tag{7.58}$$

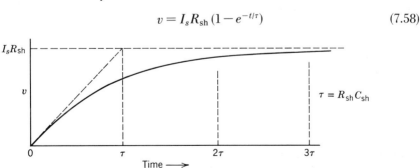

Fig. 7.33.   Exponential voltage rise as a function of time.

where $\tau = R_{sh}C_{sh}$ is the time constant. Thus, at time $t = \tau$, $v = I_s R_{sh}$ $(1 - 1/e) = 0.63 \; I_s R_{sh}$ and at time 2τ (two time constants) $v = I_s R_{sh}$ $(1 - 1/e^2) = 0.88 \; I_s R_{sh}$. Note that the voltage v never quite reaches the value $I_s R_{sh}$, although the difference is negligible after a time equal to about five time constants has elapsed. In fact, the voltage v, and hence the output current, is considered to have risen to an acceptable level when $v = 0.9 I_s R_{sh}$. The voltage rises to this 90 percent level in approximately 2.3 time constants. However, the *rise time*, t_r, is *defined* as the time required for the voltage to rise from 10 percent to 90 percent of its final value. Since the rise from zero to 10 percent requires about 0.1 time constant, the rise time is

$$t_r = 2.2\tau \tag{7.59}$$

But we have found that the upper-cutoff frequency is the reciprocal of the time constant of the effective shunt capacitance and the shunt resistance. Therefore, the rise time can be expressed in terms of the upper-cutoff frequency ω_2.

$$t_r = \frac{2.2}{\omega_2} \tag{7.60}$$

For example, an amplifier that has an upper-cutoff frequency of one MHz will have a rise time $t_r = 3.5 \times 10^{-7}$ s when the input is a rectangular pulse. Notice that short rise times are associated with amplifiers having high-cutoff frequencies.

After the output rises essentially to the full expected output, it then begins to sag because of the coupling capacitors, as was previously discussed. Good design would permit only a small percent sag by the end of the input pulse. The output then *falls*, as shown in Fig. 7.34. The sudden termination of the input current would cause the output voltage also to drop instantly except for the fact that the shunt capacitance C_{sh} must discharge through the shunt resistance R_{sh} before the voltage can drop to zero. Therefore, the *fall time* t_f depends on the time required for C_{sh} to discharge through R_{sh}. Since the same time constant is involved in both the fall time and the rise time, one would expect that the fall time is equal to the rise time, or

$$t_f = \frac{2.2}{\omega_2} \tag{7.61}$$

This equality holds only if the amplifier is linear, that is, not driven into either cutoff or saturation.

The rise and fall times are so short compared with the sag time in a well-designed amplifier that they cannot be both viewed at once on an

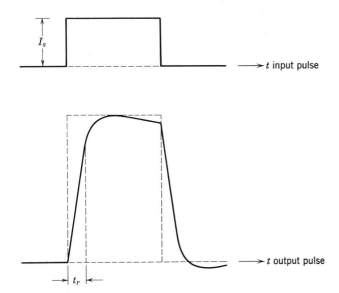

Fig. 7.34.   Comparison of the output pulse with the input pulse of an amplifier.

oscilloscope as Fig. 7.34 might infer. In fact, the sweep speed of an oscilloscope may need to be increased by a factor of 100 to 1000 to view the rise time, as compared with the sag. For example, an amplifier that has a low-frequency cutoff $f_1 = 20$ Hz and a high-frequency cutoff $f_2 = 2$ MHz, would require an input pulse of 8×10^{-4} s duration to produce a 10 percent sag, while the rise or fall time would be only 1.75×10^{-7}. Thus, the ratio of pulse duration to rise time for this example is 4.57×10^3.

PROBLEM 7.13   The video amplifier in a television set must have a rise time no greater than about 0.08 microsecond (μs) in order not to appreciably deteriorate the available picture quality. What must be the minimum upper-cutoff frequency of the video amplifier?   *Answer: $f_2 = 4.44$ MHz.*

PROBLEM 7.14   A transistor with $\beta_0 = 80$, $r_b = 100\ \Omega$, $C_{ob} = 1.5$ pF, and $f_\tau = 600$ MHz at $I_C = 10$ mA, $V_{CE} = 10$ V is chosen for the television video amplifier of Prob. 7.13. The driving source resistance is 1.0 kΩ. Determine the load resistance R_l needed to give the required upper-cutoff frequency f_2 and calculate the midfrequency voltage gain v_o/v_s for the amplifier.
     Answer: $R_l = 180\ \Omega$, $K_v = 11$.

PROBLEM 7.15   A given FET has $g_m = 4 \times 10^{-3}$ mho, $C_{gd} = 2$ pF, and $C_{gs} = 3$ pF at $I_D = 3.0$ mA, $V_{DS} = 5$ V, $V_{GS} = -0.2$ V. What voltage gain

v_o/v_s could be obtained for the television video amplifier of Prob. 7.14 ($f_2 = 4.44$ MHz, $R_s = 1$ kΩ) if this FET is used at the given q-point? What must be the value of load resistance?   *Answer: $K_v = 15.5, R_l = 387$ Ω.*

PROBLEM 7.16   What values of gain and bandwidth could be obtained for a cascode video amplifier using two of the FET's of Prob. 7.15 at the q-points given if $V_{DD} = 30$ V and the picture tube input impedance is very high so that $R_l \simeq R_D$? ($R_s = 1$ kΩ.)
   Answer: $f_2 = 22.7$ MHz, $K_v = 26.7$, (neglecting stray wiring capacitance).

PROBLEM 7.17   Assume that the low-frequency cutoff of the video amplifier of the preceding problems is 8 Hz. Completely design the video amplifier, using any one of the preceding three approaches. Draw a circuit diagram and calculate or otherwise select all the component values.

PROBLEM 7.18   Choose a bipolar transistor from current catalogs with the goal of obtaining maximum voltage gain for the video amplifier of Prob. 7.13. What is the proper load resistance and what is the mid-frequency voltage gain of your amplifier?

CHAPTER 8

TRANSFORMER-COUPLED AMPLIFIERS

The transformer has some inherent advantages and also some disadvantages as a coupling device, as compared with RC coupling. For example, the ohmic or dc resistance of the windings may be small in comparison with the ac impedance. As a result, the efficiency may be much higher than in an RC-coupled amplifier. In addition, the turns ratio may be chosen so that impedance matching may be utilized to obtain maximum power gain and, hence, maximum voltage and current gains. On the other hand, the cost, size, and weight of a transformer may not compare favorably with the resistors and capacitors of the RC-coupled amplifier in many applications.

8.1 TRANSFORMER CHARACTERISTICS

Some fundamental characteristics of the transformer will be reviewed in this section. The iron-core tightly coupled transformer illustrated in Fig. 8.1 will be the only type discussed at this time. Loosely coupled transformers will be discussed in connection with tuned amplifiers.

The transformer shown in Fig. 8.1 consists of two windings wound on a laminated iron core. In actual practice, both the primary winding with n_1 turns the secondary winding with n_2 turns are wound on the same leg of the transformer to increase the coefficient of coupling k, which is the ratio of the magnetic flux that cuts or links with the secondary to the flux produced by the primary. In a well-designed transformer k may be well above 0.99.

Let us first assume that there is no load on the secondary of the transformer, so no secondary current flows. Then, when an ac voltage V_1 is applied to the primary, a back electromotive force (EMF) is generated in the primary winding. This EMF must be essentially equal to the applied voltage if the ohmic resistance of the primary is negligible in comparison with its inductive reactance. In other words, the primary winding acts simply as an inductive reactance with some unavoidable but hopefully small series resistance. If the primary inductance is L_1 and the input

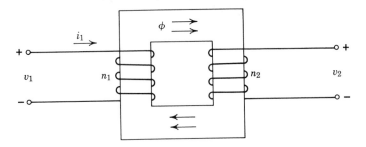

Fig. 8.1.   Representation of an iron-core transformer.

voltage is sinusoidal with rms voltage V_1 and frequency f, the primary current is approximately $I = V_1/j2\pi f L_1$. This no-load primary current is known as the magnetizing current I_m.

You will recall that the voltage induced in a coil by a changing magnetic flux[1] $d\phi/dt$ is $v = n d\phi/dt$. Then the primary voltage v_1 can be expressed in terms of the primary turns n_1 and the rate of change of flux $d\phi/dt$.

$$v_1 = n_1 \frac{d\phi}{dt} \tag{8.1}$$

But essentially all the primary flux links with the secondary turns n_2. Then the voltage induced in the secondary winding is

$$v_2 = n_2 \frac{d\phi}{dt} \tag{8.2}$$

---

[1]The symbol $d\phi/dt$ is a term used in calculus. It simply means the rate at which flux changes with time. It can also be expressed as $\Delta\phi/\Delta t$ where Δ means a small change. As Δ becomes very small (approaching zero) the Δ is replaced by d.

Thus, the ratio of the secondary voltage to the primary voltage is

$$\frac{v_2}{v_1} = \frac{n_2}{n_1} \tag{8.3}$$

We shall now connect a load resistance R_L to the secondary terminals as shown in Fig. 8.2. Then the secondary voltage v_2 causes a current i_2 to flow through the load resistor. But this secondary current produces a magnetic flux ϕ_2 in the core that opposes the primary flux (Lenz' law) and, thus, tends to reduce the total flux in the core. But this total flux must be maintained at the initial level established by the magnetizing current to

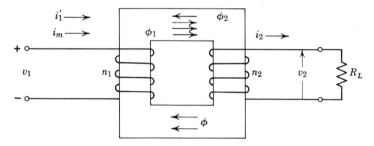

Fig. 8.2.   The effect of secondary current on primary current.

produce a back EMF in the primary winding equal to the applied voltage. Thus, the tendency of the secondary current to reduce the flux is balanced by an increased primary current i_1' (Fig. 8.2) and the total flux remains about the same. Therefore, the increased magnetomotive force (MMF) of the primary due to the extra current i_1' must be equal to the opposing MMF of the secondary due to the secondary current i_2. Since MMF is the product of amperes and turns, ni

$$n_1 i_1' = n_2 i_2 \tag{8.4}$$

Thus, the ratio of additional primary current i_1' to secondary current i_2 is, by using Eq. 8.4,

$$\frac{i_1'}{i_2} = \frac{n_2}{n_1} \tag{8.5}$$

The total primary current is, of course, $i_1 = i_m + i_1'$. The extra primary current i_1' must be in phase with the secondary current in order to cancel its flux. Thus, the phase relationship of the secondary current and voltage determines the phase relationship of the extra primary current i_1' and the primary voltage v_1. For example, if the load impedance is a pure resistance, the current i_1' is in phase with v_1 while the magnetizing current i_m lags v_1 by 90°. When the transformer is carrying full load, the extra current i_1' is large compared to the magnetizing current, so that the total primary current is almost equal to i_1' and the current ratio is said to be inversely proportional to the turns ratio.

The impedance presented by the transformer to the driving source is $V_1/I_1 = V_1/(I_1' + I_m)$. We first assume $I_1 \simeq I_1'$. Then, since $R_L = V_2/I_2$

$$Z_i \simeq \frac{V_1}{I_1'} = \frac{V_1}{(n_2/n_1)I_2} = \frac{(n_1/n_2)V_2}{(n_2/n_1)I_2} = \left(\frac{n_1}{n_2}\right)^2 R_L \qquad (8.6)$$

The total input impedance of the transformer can be represented by the equivalent circuit shown in Fig. 8.3. This circuit is not adequate at high frequencies because the distributed capacitance of the windings has been neglected. Also, the ohmic resistance and leakage inductance have been neglected. These parameters will be discussed in connection with the high-frequency characteristics in Section 8.3. As was mentioned previously, the magnetizing current is usually neglected, or in other words, the inductive reactance of the primary is assumed to be large in comparison with $(n_1/n_2)^2 R_L$.

Until now we have assumed that one transformer winding is designated as the primary and the other the secondary. However, either winding may be used as the primary. In fact, a transformer may have several windings that may be connected in any desired arrangement.

The polarity of the secondary voltage as compared to the primary depends on the relative winding directions of the two windings. The relative winding polarities are often indicated by black dots as shown in Fig. 8.4. The two dotted winding ends have the same polarities. That is, when one dotted end is positive, the other dotted end is also positive. An example may help to clarify these concepts.

EXAMPLE 8.1   Let us assume that we have an iron-core transformer with $n_2 = 10\, n_1$ and $L_1 = 1$ henry (H). What will be the primary current and

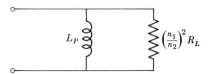

Fig. 8.3.   An equivalent input circuit for the transformer.

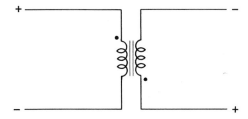

Fig. 8.4. Winding polarity marks.

primary impedance if 20 V at 50 Hz is applied to the n_1 side and a 2 kΩ resistor is connected across the n_2 terminals? The voltage across the 2 kΩ resistor is 10×20 V $= 200$ V which causes 200 V$/2$ k$\Omega = 0.1$ A to flow through this resistor. The primary current caused by this secondary, or load, current is 10×0.1 A $= 1.0$ A. The magnetizing current is 20 V$/\omega L =$ 20 V$/314\,\Omega = 0.064$ A. Since this current is less than one-tenth of the primary current due to the load, and the two must be added vectorally, the primary current is approximately 1.0 A and the primary impedance is $V_p/I_p = 20\,\Omega = (n_1/n_2)^2 R_L$.

PROBLEM 8.1   Determine the required turns ratio of a transformer that has a $Z_i = 300\,\Omega$ when $Z_L = 2700\,\Omega$. Find I_1, I_2, V_2, P_1, and P_2 for this transformer if $V_1 = 30$ V.
Answer: $n_2 = 3n_1$, $I_1 = 0.1$ A, $I_2 = 0.033\text{-}1/3$ A, $V_2 = 90$ V, $P_1 = P_2 = 3$ W.

PROBLEM 8.2   A student accidentally connected the transformer in Example 8.1 up backwards so that $n_1 = 10n_2$. Repeat Example 8.1 for this connection.

PROBLEM 8.3   A given transformer has 3000 primary turns and 100 secondary turns. The primary inductance is 70 H. If 50 V at 100 Hz is applied to the primary and a 5-Ω resistor is connected across the secondary, find V_2, I_2, I_1, and Z_i.
Answer: $V_2 = 1.67$ V, $I_2 = 0.33$ A, $I_1 \simeq 11$ mA, $Z_i \simeq 4.5$ kΩ.

8.2 TRANSFORMER COUPLING

Since the transformer transfers time varying signals but not direct currents, it is a useful device for coupling amplifier stages. Figure 8.5 shows two common-emitter stages coupled by a transformer. Notice that the bias resistors R_1 and R_2 do not rob any signal current because they are connected to the end of the transformer that is maintained at signal ground potential by capacitor C. This capacitor should have low reactance

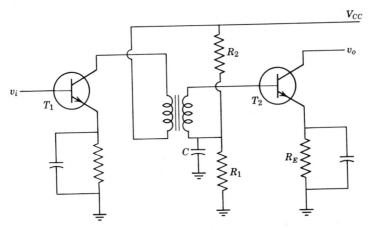

Fig. 8.5.   Two common-emitter stages transformer coupled.

in comparison with h_{ie} of transistor T_2, so that nearly all of the signal voltage will be applied between the base and emitter of transistor T_2.

An h-parameter equivalent circuit for the amplifier of Fig. 8.5 is given in Fig. 8.6. We assume that the input resistance of transistor T_2 is h_{ie2} and that the output resistance of transistor T_1 is $1/h_{oe1}$. These are not very good approximations when transformer coupling is used because of the high voltage gains encountered. However, designs based on these approximations yield power gains that are very nearly equal to the maximum values attainable. Thus, the simplified equivalent circuit of Fig. 8.7 is drawn, using

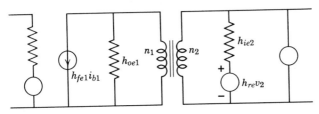

Fig. 8.6.   An h-parameter equivalent circuit for the amplifier of Fig. 8.5.

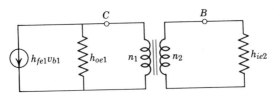

Fig. 8.7.   A simplified equivalent circuit of Fig. 8.6.

these approximations. Note that the load resistance of transistor T_1 is, neglecting the transformer magnetizing current

$$R_{L1} = \left(\frac{n_1}{n_2}\right)^2 h_{ie2} \tag{8.7}$$

The maximum power transfer theorem states that maximum power is transferred from a source to a load when the load resistance is equal to the source resistance. Thus, maximum power gain is achieved when this impedance match is achieved. Since the turns ratio of the transformer can be selected to provide any desired load resistance, an impedance match can be easily obtained. Figure 8.7 shows that an impedance match is obtained, approximately, when $R_{L1} = 1/h_{oe}$. Then, by using Eq. 8.7

$$1/h_{oe1} = \left(\frac{n_1}{n_2}\right)^2 h_{ie2} \tag{8.8}$$

Thus, the turns ratio that will provide an approximate impedance match can be determined from Eq. 8.7

$$\frac{n_2}{n_1} = (h_{oe1} h_{ie2})^{1/2} \tag{8.9}$$

The current gain from the base of transistor T_1 to the base of transistor T_2 can be easily determined for the matched amplifier. Since the impedance presented by the transformer primary is $Z_i = 1/h_{oe1}$, the signal current into the primary is equal to the current through h_{oe} and from Fig. 8.7, $I_1 = h_{fe1} I_{b1}/2$. Then the base current of transistor T_2, which is the transformer secondary current, can be expressed in terms of the primary current and the turns ratio.

$$I_{b2} = \frac{h_{fe1} I_{b1}}{2} \frac{n_1}{n_2} \tag{8.10}$$

The reference current gain for the stage is

$$K_i = \frac{I_{b2}}{I_{b1}} = \frac{h_{fe1} n_1}{2 n_2} \tag{8.11}$$

Let us again consider a numerical example to illustrate these concepts.

EXAMPLE 8.2 Determine the proper turns ratio and the current gain if a transformer is used to couple a transistor with $h_{fe1} = 100$ and $h_{oe1} = 4 \times 10^{-5}$ mhos to a transistor with $h_{ie2} = 1000 \ \Omega$.

The turns ratio for maximum power gain is (from Eq. 8.9) $n_2/n_1 = (4 \times 10^{-5} \times 10^3)^{1/2} = (4 \times 10^{-2})^{1/2} = 0.2$ or $n_1/n_2 = 5$. Then, from Eq. 8.11, the current gain $K_i = 100 \times 5/2 = 250$.

PROBLEM 8.4   Find the voltage gain V_{be2}/V_{be1} of the transformer-coupled stage in Example 8.2 if $h_{ie1} = 1000\ \Omega$ and $h_{re1} = 0$. Note that $V_{be1} = I_{b1} h_{ie1}$.

Answer: $K_v = 250$.

Note that the current gain can be much greater than h_{fe} when transformer coupling is used. Also, even though a step-down transformer is used, the voltage gain is increased because the input voltage of transistor T_2 is proportional to the input current, assuming small signal, linear operation. We should not forget that the preceding relationships are based on some rather loose but useful approximations.

The transformer-coupled amplifier can be designed without the aid of characteristic curves, providing the essential transistor parameters are known. However, we shall draw the dc and ac load lines on a set of collector characteristics, as shown in Fig. 8.8, to improve our understanding of the transformer-coupled amplifier. Notice that the dc load line is steep because the dc resistance of the transformer primary is small. In fact, the dc load resistance may be essentially the bias stabilizing resistance

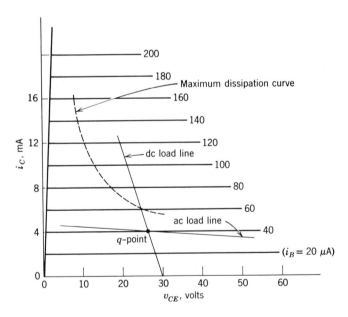

Fig. 8.8.   Load lines for a transformer-coupled transistor.

in the emitter circuit. Bias stabilization is very important in the transformer-coupled amplifier because the steep dc load line could cross the maximum dissipation line, as shown, and cause destruction of the transistor unless adequate q-point stability is provided. One sure way to protect the transistor is to use a sufficiently large value of emitter-circuit resistance plus transformer primary ohmic resistance to keep the dc load line below the maximum dissipation curve. This technique may not be desirable in some applications because of the low resultant efficiency.

Observe that the ac load line is much less steep than the dc load line. In fact, if an impedance match is achieved, the ac load line has about the same slope magnitude as the collector-characteristic curve but has opposite sign or slopes the opposite direction. Note also, that the signal voltage at the transformer primary alternately adds to and subtracts from the dc q-point voltage. Therefore, the maximum applicable voltage rating of the transistor must be at least twice the q-point voltage and the q-point voltage may be nearly equal to V_{CC}.

The stabilized bias circuit, which was discussed in connection with the RC-coupled amplifier, is suitable for use with the transformer-coupled amplifier. The same design procedure may be used, but you should be aware that the resistance of the secondary winding, as shown in Fig. 8.5, is part of the base-circuit resistance R_b. This resistance is often small enough to be neglected.

PROBLEM 8.5   A transistor with $h_{oe} = 100$ μmho is coupled to a transistor with $h_{ie} = 500\ \Omega$. The h_{fe} of the driving transistor, or input stage, is 120. Determine the approximate transformer turns ratio for maximum power gain, and calculate the reference current gain of the first stage.

$$Answer:\ n_1/n_2 = 4.5, K_i = 270.$$

8.3 LOW-FREQUENCY RESPONSE AND SAG OF TRANSFORMER COUPLED AMPLIFIERS

The low-frequency characteristics of the transformer-coupled amplifier can be obtained from the equivalent circuits given in Fig. 8.9. The circuit of Fig. 8.9a shows a Norton's equivalent driving source, which represents a transistor or any other device, transformer-coupled to a load resistance R_L. In the simplified circuit of Fig. 8.9b, the resistance R is the parallel combination of R_s and $(n_1/n_2)^2 R_L$. This figure shows that the signal voltage V_p across the transformer primary is the product of the source current I_s and the parallel combination of R and $j\omega L_p$, assuming the source current to be sinusoidal. Expressing this relationship in an equation

$$V_p = I_s \frac{j\omega L_p R}{R + j\omega L_p} \tag{8.12}$$

Both the numerator and the denominator of Eq. 8.12 can be divided by $j\omega L_p$ to obtain

$$V_p = \frac{I_s R}{1 - j(R/\omega L_p)} \tag{8.13}$$

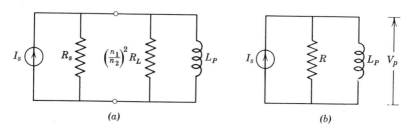

(a)   (b)

Fig. 8.9. Low-frequency equivalent circuits for a transformer-coupled amplifier. (a) Equivalent circuit and (b) simplified equivalent circuit.

Observe that the numerator $I_s R$ is the primary voltage at frequencies that are high enough so that ωL_p is large in comparison with R. We call these *midfrequencies*. The low-frequency cutoff or half-power frequency, therefore, is the frequency at which the denominator of Eq. 8.13 has a magnitude equal to $\sqrt{2}$. This frequency occurs when $R/\omega_1 L_p = 1$, or

$$\omega_1 = \frac{R}{L_p} \tag{8.14}$$

The primary voltage is sketched as a function of frequency in Fig. 8.10. The secondary voltage varies in similar fashion since $V_s = (n_2/n_1)V_p$. Also, the secondary inductance L_s is $(n_2/n_1)^2 L_p$ for tightly coupled transformers.

Usually, the total shunt resistance R and the desired low-frequency cutoff ω_1 are known or easily calculated, so that the required primary inductance is the dependent variable. The required primary inductance may be obtained by rearranging Eq. 8.14

$$L_p = \frac{R}{\omega_1} \tag{8.15}$$

Figure 8.10 shows that the low-frequency response characteristics of a transformer-coupled amplifier are very similar to those of an RC-coupled amplifier. The only difference is in the method of determining ω_1. We have previously learned that cutoff frequencies can be related to time constants. Thus, we would suspect that ω_1 is the reciprocal of the time constant L_p/R of the inductive circuit shown in Fig. 8.9b. This is indeed the case. Therefore, the fractional sag of a transformer-coupled amplifier for an input pulse of duration t_d can be expressed in terms of the cutoff frequency ω_1. As given by Eq. 7.25, fractional sag $= \omega_1 t_d$ or in terms of the circuit

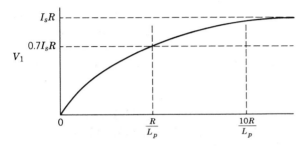

Fig. 8.10.   A sketch of primary voltage as a function of frequency.

time constant,

$$\text{fractional sag} = \frac{R\,t_d}{L_p} \tag{8.16}$$

Transformers are sometimes used in amplifiers that are designed to amplify short-duration pulses. The primary inductances of these transformers may be very small if the pulse duration is very short.

A numerical example will be used to illustrate the impressive value of primary inductance that is required for good low-frequency response if the source resistance is high and maximum gain is desired.

EXAMPLE 8.3   Determine the required value of L_p and L_s for an amplifier with an output resistance of $10\,\text{k}\Omega$ which is transformer coupled to a transistor with an input resistance of $1\,\text{k}\Omega$. The desired low-frequency cutoff is 32 Hz and maximum power gain is desired.

For optimum power gain, the input impedance of the transformer should be equal to the output resistance of the device ($10\,\text{k}\Omega$ in this example). Thus, R is the parallel combination of two 10-kΩ resistors or

5 kΩ. Then, from Eq. 8.15, $L_p = 5 \times 10^3/(2\pi \times 32) = 5 \times 10^3/200 = 25$ H and $L_s = 25/10$ or 2.5 H.

Let us determine the duration of a rectangular pulse which, when applied to the input of this amplifier will have 10 percent sag in the output pulse. From Eq. 8.16, $t_d = 0.1L_p/R = 2.5/5 \times 10^3 = 0.5$ ms. In contrast, if the foregoing amplifier is to be used to amplify 1 μs pulses (or shorter) and a sag of 10 percent is permitted, then the primary inductance may be reduced to (Eq. 8.16) $L_p = 5 \times 10^{-3}/0.1 = 0.005$ H. Thus, short-period-pulse transformers may be quite small physically compared to long-period-pulse transformers.

PROBLEM 8.6   The average value of h_{oe} for a 2N3903 is 20 μmhos. The desired load on this transistor is 8 Ω. If transformer coupling is desired to obtain a maximum power transfer, find the value of L_p and the turns ratio. The desired low-cutoff frequency is 32 Hz. Determine the sag for 0.5-ms pulses.

8.4   HIGH-FREQUENCY RESPONSE AND RISE TIME OF TRANSFORMER-COUPLED AMPLIFIERS

The high-frequency response and rise time of a transformer-coupled amplifier are not as easily determined as the low-frequency response and sag. One reason for this is the lack of availability of data concerning the leakage inductance and distributed capacitance of the transformer windings. The effect of these parameters on the high frequency performance of the amplifier is shown by the equivalent circuit of Fig. 8.11. The leakage inductance L_{1p} is the inductance that results from the lack of complete coupling in the transformer. This inductance, which is equal to $(1 - k)L_p$, is the only inductance remaining in the primary when

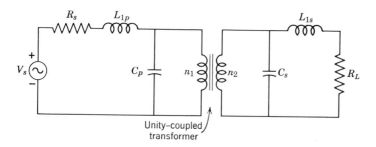

Fig. 8.11.   A high-frequency equivalent circuit for a transformer-coupled amplifier.

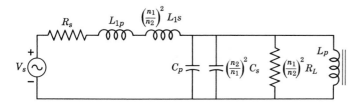

Fig. 8.12.   A transformer-coupled amplifier equivalent circuit with all elements referred
to the primary side of the transformer.

the secondary is short circuited. The capacitance C_p is the distributed
capacitance between turns in the primary winding. The leakage inductance
L_{1s} and the distributed capacitance C_s are similarly defined for the secondary winding.

The elements in the secondary circuit of Fig. 8.11 can be placed in the
primary circuit of the transformer, as shown in Fig. 8.12, if the secondary
impedances are multiplied by $(n_1/n_2)^2$, as was previously discussed. This
equivalent circuit is not very accurate for two reasons. First, the leakage
inductance and distributed capacitance are both distributed along the
winding and cannot be accurately represented by lumped elements.
Second, the output capacitance of the source and the input capacitance of
the load should be added to the circuit. However, the complexity of the
circuit is increased considerably with these additions so they will be
omitted in the interest of gaining a simple but only qualitative understanding of the circuit response. In pursuit of this purpose, the elements of
Fig. 8.12 are combined where possible to form the simplified equivalent
circuit of Fig. 8.13. The primary inductance L_p has been omitted from this
equivalent circuit because ωL_p is very large compared to $(n_1/n_2)^2 R_L$ at
high frequencies.

Figure 8.13 shows that the transformer behaves like a low Q series-tuned
circuit at high frequencies. The Q is low because the source resistance R_s

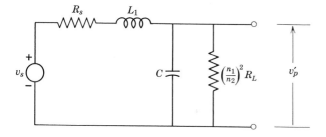

Fig. 8.13.   A simplified high-frequency transformer-coupled amplifier circuit.

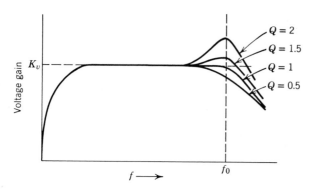

Fig. 8.14.   Typical frequency response of a transformer-coupled amplifier.

is in series with the leakage inductance L_p, and the reflected load resistance $(n_1/n_2)^2 R_L$ is in parallel with the distributed shunt capacitance C. The upper-cutoff frequency of the amplifier is somewhat above the resonant frequency f_0 of the circuit and the shape of the frequency response curve in the high-frequency region depends on the circuit Q as shown in Fig. 8.14. The frequency f_0 is defined here as the frequency at which $X_L = X_C$. To illustrate these ideas, let us consider another example.

EXAMPLE 8.4   Determine the frequency response of a transformer-coupled amplifier that has $R_s = 10\,\text{k}\Omega$ and $(n_1/n_2)^2 R_L$ also equal to $10\,\text{k}\Omega$. The transformer has a primary inductance $L_p = 25\,\text{H}$, a coupling coefficient $k = 0.996$, and effective shunt capacitance $C = 200\,\text{pF}$.

The value of L_1 is $(1 - k)L_p = (1 - 0.996)25 = 0.1\,\text{H}$ and $f_0 = 1/2\pi(L_1 C)^{1/2}$ $= 1/2\pi(2 \times 10^{-11})^{1/2} = 36\,\text{kHz}$. The Q of the source resistance (R_s) and leakage inductance (L_1) combination is $Q_L = \omega_0 L_1/R_s = 2\pi \times 3.6 \times 10^4 \times 0.1/10^4 = 2.26$. The Q of the shunt capacitance (C) and the reflected resistance $(n_1{}^2 R_L/n_2{}^2)$ combination is $Q_C = R_p/\omega_0 C = 10^4/6.28 \times 3.6 \times 10^4 \times 2 \times 10^{-10} = 2.2$. These values of Q can be combined in the same manner as parallel resistors to obtain the total circuit Q_T. Thus, $Q_T = Q_L Q_C/(Q_L + Q_C) = 2.26 \times 2.2/(2.26 + 2.2) \simeq 1.1$. Notice from Fig. 8.14 that this value of Q will give a fairly uniform response up to 36 kHz.

PROBLEM 8.7   Determine the frequency response of the circuit in Example 8.4 if the coefficient of coupling k is reduced to 0.990.

The time response of a transformer-coupled amplifier also depends on the resonant frequency and the circuit Q, as should be expected. The rise-time response to a square-wave input is shown in Fig. 8.15a for several values of Q with f_0 constant. Figure 8.15b shows the effect of the value of f_0

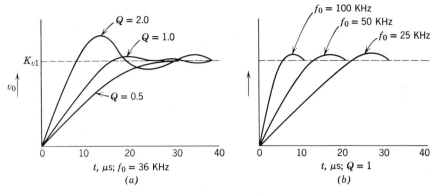

Fig. 8.15.   Time response as a function of (a) Q and (b) f_0.

on rise time with Q held constant. Observe that fast rise times are obtained when the resonant frequency is high and the Q is high. However, high values of Q produce large overshoot and ringing, which result from the oscillatory nature of an LC circuit. A value of Q that provides a good compromise between rise time and overshoot is $Q = 1.0$.

The winding resistance has been neglected in the previous discussion. This resistance is effectively in series with the source resistance and tends to lower the circuit Q. Good quality transformers have high-resonant frequencies with values of Q about one or less. Very tight coupling is required to achieve these characteristics. Bifilar winding is sometimes used to produce broad-band transformers. In this type of winding the primary and secondary wires are first formed into a pair and then wound together on one leg of a magnetic core. The impedance ratios are somewhat limited with this type of winding.

Recommended impedance levels are usually specified for a transformer. When a transformer is used at other impedance levels the frequency characteristics and efficiency are usually degraded.

PROBLEM 8.8   A field-effect transistor with $r_d = 100 \text{ k}\Omega$ is transformer coupled to a bipolar transistor with $h_{ie} = 4 \text{ k}\Omega$.
 (a) Determine the required primary inductance if $f_1 = 32$ Hz and impedance matching is achieved.   *Answer:* 250 H.
 (b) What is the voltage gain of the matched FET stage (to the base of the bipolar transistor) if $g_m = 4 \times 10^{-3}$ mho for the FET?
                                                              Answer: $K_v = 40$.

PROBLEM 8.9   A transformer-coupled amplifier is required to amplify a rectangular pulse with 100 μs duration. The output resistance of the amplifier is $20\,k\Omega$ and the transformer achieves an impedance match. What must be the primary inductance of the transformer if the sag is limited to 5 percent and what must be the minimum resonant frequency of the transformer, assuming $Q = 1$, if the maximum permissible rise time is $5\,\mu s$?   *Answer:* $L_p = 20\,H, f_0 = 100\,kHz.$

PROBLEM 8.10   A transformer is used to couple two silicon transistors, as shown in Fig. 8.5. Transistor T_1 has $h_{oe} = 40$ μmhos and P_d max $= 200\,mW$. Transistor T_2 has I_C (q-point) $= 10\,mA$, $h_{ie} = 150\,\Omega$, and $h_{fe} = 100$. Assume $V_{BE} = 0.6\,V$, $V_{CC} = 25\,V$. The transformer has ohmic resistance $R_{pri} = 300\,\Omega$ and $R_{sec} = 10\,\Omega$.

(a) Determine the transformer-turns ratio for maximum power gain.

Answer: 9.1.

(b) Determine the minimum value for the emitter-circuit resistance of T_1 that will insure that the dissipation rating of T_1 will not be exceeded.   *Answer:* $R = 750\,\Omega.$

(c) Determine values for R_1, R_2, R_E, and C (Fig. 8.5) that will provide a 2.0 V drop across R_E, a current stability factor $\simeq$ 10, and $\omega_1 = 200$.

Answer: $R_E = 200, R_1 = 22\,k\Omega, R_2 = 18\,k\Omega, C = 330\,\mu F.$

CHAPTER 9

SMALL-SIGNAL-TUNED AMPLIFIERS

The need frequently arises for an amplifier that will amplify only those frequencies that lie within a given frequency range or band. This type of amplifier is known as a tuned amplifier or band-pass amplifier. Radio and television receivers, for example, use tuned amplifiers to select one radio signal from the many that are being broadcast. Several types of tuned amplifiers are discussed in this chapter. The gain and bandwidth of each type will be of interest. It will be assumed, in the discussion of tuned amplifiers, that the input signal is a modulated signal with a basic frequency ω_0 and the amplifier is tuned to this frequency. The required bandwidth will depend on the characteristics of the modulation. For amplitude modulation, the required bandwidth is two times the highest modulating frequency of interest.[1] Pulse modulated signals require a bandwidth, in hertz, approximately twice the reciprocal of the pulse duration, or $B = 2/t_d$.

9.1 SINGLE-TUNED, CAPACITIVELY COUPLED AMPLIFIERS

A tuned FET amplifier is shown in Fig. 9.1. Observe that the basic difference between this amplifier and the RC-coupled amplifier is the

---

[1]Bandwidth requirements for am and fm signals are developed in *Electronic Engineering*, C. L. Alley and K. W. Atwood, John Wiley & Sons, Inc., New York.

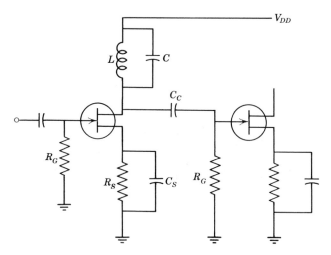

Fig. 9.1.   A tuned FET amplifier.

tuned circuit that replaces the resistor in the drain circuit. To determine
the voltage gain and the bandwidth of the amplifier, the equivalent circuit
of Fig. 9.2 is given. The capacitor C includes the shunt capacitance of the
amplifiers and the distributed wiring capacitance. This absorption of the
shunt capacitance into the tuned circuit makes possible the amplification
of very high frequencies.

The equivalent circuit of Fig. 9.2 would have only parallel elements and,
hence, the analysis would be comparatively easy except for the series
resistance of the coil R_{ser}. Therefore, we shall determine a parallel combination of resistance and inductance that will have the same impedance
over the pass band as the series combination. This will be accomplished by
finding the admittance of the series combination.

$$Y = \frac{1}{Z} = \frac{1}{R_{ser} + j\omega L} \tag{9.1}$$

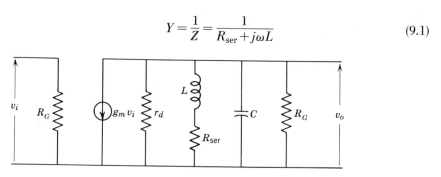

Fig. 9.2.   An equivalent circuit for the tuned amplifier of Fig. 9.1.

Rationalizing,

$$Y = \frac{R_{\text{ser}} - j\omega L}{(R_{\text{ser}} + j\omega L)(R_{\text{ser}} - j\omega L)} = \frac{R_{\text{ser}} - j\omega L}{R_{\text{ser}}^2 + (\omega L)^2} \tag{9.2}$$

But the ratio of ωL to R_{ser} is the Q of the coil, known as Q_0. Therefore, if Q_0 is ten or higher, which is usually the case, $(\omega L)^2$ is at least 100 times R_{ser}^2; so this latter term can be neglected in the denominator of Eq. 9.2. Using this simplification,

$$Y \simeq \frac{R_{\text{ser}} - j\omega L}{(\omega L)^2} = \frac{R_{\text{ser}}}{(\omega L)^2} - j\frac{1}{\omega L} \tag{9.3}$$

This admittance is of the form $G + jB$ and represents a conductance $G = R_{\text{ser}}/(\omega L)^2$ in parallel with an inductive susceptance $B = 1/(\omega L)$, as shown in the dashed enclosure in the equivalent circuit of Fig. 9.3. Since the Q of a series circuit can be written as $Q_0 = \omega L/R_{\text{ser}}$,

$$G_p = \frac{R_{\text{ser}}}{(\omega L)^2} = \frac{1}{Q_0 \omega L} \tag{9.4}$$

This conductance may also be expressed as a resistance

$$R_{\text{par}} = \frac{1}{G_p} = Q_0 \omega L \tag{9.5}$$

Observe that this equivalent parallel resistance is a function of frequency. However, the tuned amplifier that uses a high Q circuit $(Q \geqslant 10)$ amplifies only a narrow band of frequencies near the resonant frequency ω_0. Therefore, the effective parallel resistance of the coil is assumed to be constant

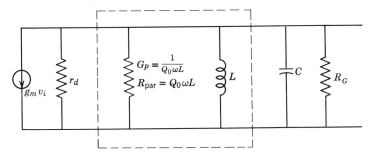

Fig. 9.3.   An equivalent circuit containing only parallel elements.

over the pass band with the value

$$R_{par} = Q_0 \omega_0 L \tag{9.6}$$

Notice that the Q_0 of the coil is the ratio $R_{par}/\omega_0 L$ and that small values of series resistance R_{ser}, which provide high Q_0 give large values of effective parallel resistance R_{par} because R_{par} is proportional to the Q_0 of the coil. In fact, rearranging Eq. 9.6 yields $Q_0 = R_p/\omega_0 L$. To emphasize this behavior, let us consider a numerical example.

EXAMPLE 9.1 A given coil has an inductance of 0.1 H and a series resistance of 20 Ω. This coil is connected in parallel with a 0.1 μF capacitor. Determine the parameters of an equivalent parallel circuit.

We assume that we have a high Q circuit ($Q = 10$ or more). Then, $\omega_0 = 1/(LC)^{1/2} = 1/(10^{-1} \times 10^{-7})^{1/2} = 10^4 \text{ rad/s}$. The value of $\omega_0 L$ is $10^4 \times 10^{-1} = 1000$ Ω. The Q_0 of the coil is $Q_0 = \omega_0 L/R_{ser} = 1000/20 = 50$. From Eq. 9.3, the size of the inductor in the parallel circuit is the same as the inductor in the series circuit. However, from Eq. 9.6, $R_{par} = Q_0 \omega_0 L = 50 \times 1000 = 50,000$ Ω. Thus, the equivalent parallel circuit is a 0.1 μF capacitor in parallel with a 0.1 H coil, which is also in parallel with a 50,000-Ω resistor.

Note that if the series resistance of the coil is reduced to 5 Ω, the Q_0 becomes 200 and the equivalent parallel resistance becomes 200,000 Ω.

PROBLEM 9.1 A 1-mH coil with $Q_0 = 100$ is connected in parallel with a 1000-pF capacitor. Determine ω_0 and R_{par}.

$$\textit{Answer: } \omega_0 = 10^6 \text{ rad/s}, R_{par} = 10^5 \text{ Ω}.$$

The effective parallel resistance of the coil can be combined with the other parallel resistance elements in the equivalent circuit of Fig. 9.3 to produce the simplified equivalent circuit of Fig. 9.4 where R represents the parallel combination of r_d, R_p, and R_G of Fig. 9.3. In calculating Q_0 (the coil Q) we have used $R_p/\omega_0 L$ where R_p accounts for the energy loss in

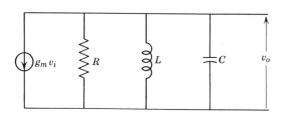

Fig. 9.4. A simplified equivalent circuit for a tuned amplifier.

the coil. However, the characteristics of a tuned circuit depend on the energy loss of the *entire* circuit. Thus, a *circuit Q* will be defined as

$$Q = \frac{R}{\omega_0 L} \tag{9.7}$$

where R is the effective total shunt resistance in parallel with L and C as shown in Fig. 9.4.

The symbol Q, given without subscripts, will always represent the *circuit Q* in this discussion. The circuit Q is a very important parameter in a tuned circuit because it determines the bandwidth and affects the amplifier gain. Since the impedance of a lossless parallel tuned circuit is infinite at the resonant frequency, Fig. 9.4 shows that the total load impedance is R and the output voltage v_o of the FET amplifier at the resonant frequency is

$$v_o = -g_m v_i R \tag{9.8}$$

Then, the voltage gain at resonance is

$$K_v = \frac{v_o}{v_i} = -g_m R \tag{9.9}$$

Observe that the voltage gain is proportional to R, which is equal to $Q\omega_0 L$ at the resonant frequency. Thus, the gain is proportional to both the circuit Q and the inductive reactance of the coil. Let us consider an example.

EXAMPLE 9.2   A circuit is connected as shown in Fig. 9.1. The inductance L has a value of 1 mH and a Q_0 of 100. The capacitance is 1000 pF and $R_G = 1$ MΩ. The FET has $r_d = 2.5 \times 10^5$ Ω and $g_m = 2 \times 10^{-3}$ mhos. Determine the voltage gain of this circuit.

As was noted in Prob. 9.1, the value of ω_0 is 10^6 rad/s and $R_{par} = 100{,}000$ Ω. The parallel combination of R_{par} (100,000 Ω), r_d(2.5 $\times 10^5$ Ω), and R_G(10^6 Ω) is $R = 67$ kΩ. Then, the circuit Q is $Q = R/\omega_0 L = 6.7 \times 10^4/10^6 \times 10^{-3} = 67$. Finally, the voltage gain at resonance (Eq. 9.9) is $K_v = -g_m R = 2 \times 10^{-3} \times 6.7 \times 10^4 = 134$.

PROBLEM 9.2   In the tuned FET amplifier of Fig. 9.1, $L = 1$ mH, $Q_0 = 100$, $C = 500$ pF, and $R_G = 1$ MΩ. The FET has $g_m = 2 \times 10^{-3}$ and $r_d = 2.5 \times 10^5$. Determine the resonant frequency and the voltage gain.
Answer: $f_0 = 225$ kHz, $K_v = 167$.

The desired circuit Q is determined by the bandwidth requirement of the amplifier, however, and not by the desired voltage gain. The relationship between the bandwidth and the circuit Q can be obtained by writing a general expression for the output voltage V_o. From Fig. 9.4,

$$V_o = g_m V_i Z = \frac{g_m V_i}{G + j\omega C + (1/j\omega L)} \qquad (9.10)$$

where $G = 1/R$. If we now multiply both numerator and denominator of the right hand side of Eq. 9.10 by R and then divide both sides of the equation by V_i,

$$G_v = \frac{g_m R}{1 + j\omega CR + (R/j\omega L)} \qquad (9.11)$$

But $R/\omega_0 L = Q$ and similarly $R\omega_0 C = Q$, since $\omega_0 L = 1/\omega_0 C$ at the resonant frequency ω_0. Making these substitutions into Eq. 9.11,

$$G_v = \frac{g_m R}{1 + jQ\left(\dfrac{\omega}{\omega_0} - \dfrac{\omega_0}{\omega}\right)} \qquad (9.12)$$

Since the numerator in Eq. 9.12 is equal to the voltage gain K_v at resonance, when $\omega = \omega_0$, the gain decreases to the half-power value when the magnitude of the j-part of the denominator is equal to unity. But there are two frequencies at which half-power gain occurs, as shown in Fig. 9.5. These frequencies are designated as ω_L and ω_H. Thus, when ω is equal to ω_L (or ω_H) the magnitude of the j-part of the denominator of Eq. 9.12 is equal

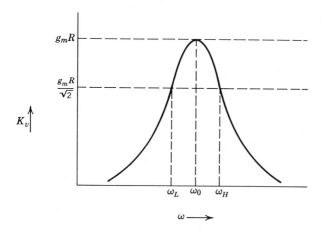

Fig. 9.5.   Frequency response of the single-tuned amplifier.

to one.

$$\left| Q \left(\frac{\omega_H}{\omega_0} - \frac{\omega_0}{\omega_H} \right) \right| = 1 = \left| Q \left(\frac{\omega_L}{\omega_0} - \frac{\omega_0}{\omega_L} \right) \right| \qquad (9.13)$$

and

$$\frac{\omega_H}{\omega_0} - \frac{\omega_0}{\omega_H} = \frac{\omega_0}{\omega_L} - \frac{\omega_L}{\omega_0} \qquad (9.14)$$

Rearranging terms, we have

$$\frac{\omega_H}{\omega_0} + \frac{\omega_L}{\omega_0} = \frac{\omega_0}{\omega_L} + \frac{\omega_0}{\omega_H} \qquad (9.15)$$

$$\frac{\omega_H + \omega_L}{\omega_0} = \omega_0 \frac{\omega_H + \omega_L}{\omega_H \omega_L} \qquad (9.16)$$

Therefore,

$$\omega_0^2 = \omega_H \omega_L \qquad (9.17)$$

Again, by using Eq. 9.13, we find

$$Q \left(\frac{\omega_H^2 - \omega_0^2}{\omega_0 \omega_H} \right) = 1 \qquad (9.18)$$

and substituting the value of ω_0^2 in Eq. 9.17,

$$Q \left(\frac{\omega_H^2 - \omega_L \omega_H}{\omega_0 \omega_H} \right) = Q \left(\frac{\omega_H - \omega_L}{\omega_0} \right) = 1 \qquad (9.19)$$

But $\omega_H - \omega_L$ is the bandwidth B, in radians per second, as seen from Fig. 9.5. Then

$$B = \frac{\omega_0}{Q} \qquad (9.20)$$

Notice that both sides of Eq. 9.20 may be divided by 2π to obtain the bandwidth in Hertz.

$$B \text{ (Hertz)} = \frac{f_0}{Q} \qquad (9.21)$$

The circuit designer usually knows the desired bandwidth and his assignment is to produce an amplifier to meet the specification. Therefore, we shall use this philosophy in the example to follow.

EXAMPLE 9.3 Let us assume that we need to build an intermediate frequency amplifier with a resonant, or center, frequency $f_0 = 455$ kHz. This amplifier is for use in a standard broadcast receiver so that the desired bandwidth is 10 kHz. Therefore, the required circuit Q is 45.5. Field effect transistors with $r_d = 200$ kΩ at the desired q-point are chosen for the amplifier. We select $R_G = 10^6$ Ω and the shunt resistance, exclusive of the coil loss, is then 167 kΩ. The effective parallel resistance of the coil is not yet known and, in fact, cannot be determined from the product $Q_0\omega_0 L$ because the required inductance is not yet known, but needs to be determined. However, we may know from experience or from Q-meter measurements that a typical value of coil Q_0 at this frequency is 100. We also know that $R_{par} = Q_0\omega_0 L$ and $R = Q\omega_0 L$. Dividing the first equation by the second, and solving for R_{par},

$$R_{par} = \frac{RQ_0}{Q} \tag{9.22}$$

We do not yet know R, which is the parallel combination of R_{par}, and the known shunt resistance, which excludes R_{par}. We call this known shunt resistance R_k. Then $R = R_{par}R_k/(R_{par} + R_k)$. Substituting this value of R into Eq. 9.22 and solving for R_{par},

$$R_{par} = R_k \frac{(Q_0 - Q)}{Q} \tag{9.23}$$

For our design with $Q_0 = 100$, $Q = 45.5$ and $R_k = 167$ kΩ, $R_{par} = 167$ kΩ $(54.5/45.5) = 200$ kΩ and $R = 91$ kΩ. Then $L = R/Q\omega_0 = 7 \times 10^{-4}$ H and $C = 1/\omega_0^2 L = 1.75 \times 10^{-10}$ F. Note that this tuning capacitance includes the output capacitance of the driving transistor and the effective input capacitance of the following transistor. This input capacitance may be a major part of the required 175 pF tuning capacitance and, therefore, the tuning would depend strongly on the voltage gain of the following amplifier unless the cascode connection is used. Either the inductance may be variable by use of a slug-tuned core or the capacitance may be variable in order to tune the circuit to 455 kHz.

The voltage gain of the amplifier may be easily determined if the transconductance of the FET is known at the desired q-point. Assume that $g_m = 10^{-3}$ mhos. Then the gain of the 455-kHz amplifier is $K_v = -g_m R = -91$ at the resonant frequency.

The design of a single-tuned capacitively coupled amplifier is summarized below:

1. The desired resonant frequency and bandwidth are assumed known

or have been determined from the signal frequency and modulation characteristics.

2. The circuit Q is determined from the relationship $Q = f_0/B$.

3. The resistance R_k is determined as the parallel combination of the output resistance of the driving transistor and the input resistance of the following stage.

4. The effective parallel resistance required from the tuned circuit is $R_{par} = R_k(Q_0 - Q)/Q$ (Eq. 9.23). The coil Q_0 is measured, given, or assumed.

5. The total shunt resistance is $R = R_{par}R_k/(R_{par} + R_k)$.

6. The required tuning inductance $L = R/Q\omega_0 = R_{par}/Q_0\omega_0$ (Eq. 9.6 or 9.7).

7. The required tuning capacitance, including the output and input capacitances, is $C = 1/\omega_0^2 L$, or $C = 1/RB$ where B is the bandwidth in radians per second.

8. The voltage gain $K_v = -g_m R$ for a FET (or vacuum tube).

The preceding type of circuit is not used with bipolar transistors. The coupling circuits discussed in the following paragraphs will be more generally applicable to the bipolar types. Instability problems are being ignored here, but will be discussed later.

PROBLEM 9.3 An RF amplifier is needed for a standard broadcast receiver. It is to have a 20-kHz bandwidth at 1.0 MHz center frequency. Design a single-tuned capacitively coupled amplifier for this frequency and bandwidth, using a FET with $g_m = 2 \times 10^{-3}$ mho, $R_o = 250$ kΩ, R_g (of the following transistor) $= 1.0$ MΩ, $C_o = 5$ pF, and $C_i = 25$ pF for the transistors used at the recommended q-points. Determine the values for L, C, and the voltage gain, if the Q_0 of the coil is assumed to be 150.

Answer: $L = 477 \, \mu\text{H}, C = 330 \, \text{pF}, K_v = -300.$

9.2 INDUCTIVELY COUPLED TUNED AMPLIFIERS

The capacitively coupled tuned circuit of Section 9.1 did not permit impedance transformation. Therefore, field-effect transistors were used because they have high input impedance as well as high output impedance and do not need an impedance-transform type of coupling circuit. However, bipolar transistors can give considerably higher gain if the coupling circuit can provide impedance transformation, as was discussed in Chapter 8. One such type of coupling circuit is the inductively coupled circuit of Fig. 9.6. Observe that this amplifier is almost identical to the transformer-coupled amplifier discussed in Chapter 8. In fact, it is a transformer-coupled ampli-

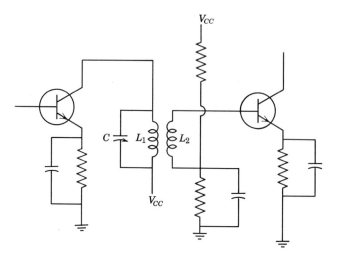

Fig. 9.6.   An inductively coupled tuned-transistor amplifier.

fier in which the primary of the transformer is tuned by capacitor C and the secondary is not so tightly coupled because of the different type core that is perhaps air or powdered iron.

A simplified h-parameter circuit, in which h_{re} is neglected, is given in Fig. 9.7. The resonant frequency f_0 is assumed to be much lower than f_β so that h_{ie} is resistive. Since the transformer does not meet the tight-coupling, large-inductance requirements of the untuned transformer discussed in Chapter 8, we must use a different approach to determine the effective impedance of the transformer primary. The mutual impedance of the transformer $j\omega M$, where M is the mutual inductance, is defined as the ratio of the voltage induced in one winding to the current flowing in the other winding. Therefore, by referring to Fig. 9.7,

$$V_1 = (j\omega L_1 + R_1)I_1 - j\omega M I_2 \qquad (9.24)$$

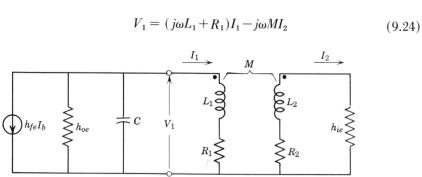

Fig. 9.7.   A simplified h-parameter equivalent circuit.

where the currents and voltages are assumed to be sinusoidal. Since the voltage induced in the secondary is $j\omega M I_1$,

$$I_2 = \frac{j\omega M I_1}{j\omega L_2 + h_{ie} + R_2} \tag{9.25}$$

Substituting this value of I_2 into Eq. 9.24,

$$V_1 = (j\omega L_1 + R_1)I_1 + \frac{(\omega M)^2 I_1}{j\omega L_2 + h_{ie} + R_2} \tag{9.26}$$

Therefore, the impedance seen at the primary terminals is

$$Z_1 = \frac{V_1}{I_1} = j\omega L_1 + R_1 + \frac{(\omega M)^2}{j\omega L_2 + h_{ie} + R_2} \tag{9.27}$$

The third term on the right-hand side of Eq. 9.27 is the impedance coupled into the primary (in series) as a result of the secondary current I_2. Notice that this term is complex and, thus, reactance as well as resistance will be coupled into the primary. This reactance, which will affect the tuning of the primary, is a function of h_{ie}. Thus, the coupling circuit will be detuned by any change that affects h_{ie}. This undesirable detuning effect can be essentially eliminated if h_{ie} is large in comparison with $R_2 + j\omega L_2$ and the impedance seen at the primary terminals is

$$Z_1 \simeq j\omega L_1 + R_1 + \frac{(\omega M)^2}{h_{ie}} \tag{9.28}$$

This transformer primary impedance can replace the transformer in the equivalent circuit of Fig. 9.7 to produce the simplified primary circuit shown in Fig. 9.8.

We now need a basis for choosing the primary inductance L_1. One sensible basis might be the obtaining of maximum power transfer and thus

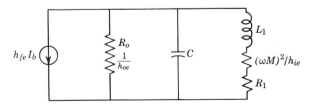

Fig. 9.8.   A simplified equivalent circuit with $h_{ie} \gg \omega L_2$.

maximum gain, as was discussed in Chapter 8. However, maximum power into the primary may not yield maximum power in the load because part of the primary power is lost in the primary resistance R_1. This problem can be eliminated if we transform this loss resistance to its effective parallel value and combine it with R_o to obtain a modified output resistance R'_o (Fig. 9.9). Using the relationship $R_{par} = Q_0 \omega_0 L_1$ at the resonant frequency (Eq. 9.6) and $R'_o = R_{par} R_o / (R_{par} + R_o)$,

$$R'_o = \frac{R_o Q_0 \omega_0 L_1}{R_o + Q_0 \omega_0 L_1} \tag{9.29}$$

If we now match the impedance of the tuned primary, Fig. 9.9, to R'_o, maximum power will be transferred to the load, since the power loss in the transformer has been removed. Therefore, we shall transform the series resistance $(\omega M)^2 / h_{ie}$ to its equivalent parallel value and equate it to R'_o.

$$R'_o = \frac{(\omega L_1)^2}{(\omega M)^2 / h_{ie}} = \frac{L_1{}^2}{M^2} h_{ie} \tag{9.30}$$

But from basic coupled-circuit theory we know that the mutual inductance is

$$M = k \sqrt{L_1 L_2} \tag{9.31}$$

where k is the coefficient of coupling. Substituting this value for M in Eq. 9.30, we have

$$R'_o = \frac{L_1{}^2 h_{ie}}{k^2 L_1 L_2} = \frac{L_1 h_{ie}}{k^2 L_2} \tag{9.32}$$

We can now determine L_1 from the bandwidth requirement. Since the circuit Q is f_0/B, which is known, and the total shunt resistance R is $R'_o/2$ in

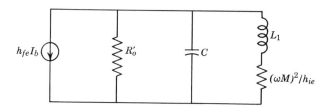

Fig. 9.9. An equivalent circuit showing the modified output resistance R'_o, which includes the primary winding loss.

the matched amplifier, we can write

$$\frac{R_o'}{2} = Q\omega_0 L_1 \tag{9.33}$$

and

$$R_o' = 2Q\omega_0 L_1 \tag{9.34}$$

Substituting the value of R_o' given in Eq. 9.29 into Eq. 9.34, we have

$$\frac{R_o Q_0 \omega_0 L_1}{R_o + Q_0 \omega_0 L_1} = 2Q\omega_0 L_1 \tag{9.35}$$

and

$$\frac{R_o Q_0}{R_o + Q_0 \omega_0 L_1} = 2Q \tag{9.36}$$

Then

$$R_o Q_0 = 2QR_o + 2QQ_0 \omega_0 L_1 \tag{9.37}$$

Solving for L_1, Eq. 9.37 becomes

$$L_1 = \frac{R_o(Q_0 - 2Q)}{2Q\,Q_0 \omega_0} = \frac{R_o}{\omega_0}\left(\frac{1}{2Q} - \frac{1}{Q_0}\right) \tag{9.38}$$

Notice that an impedance match cannot be obtained unless the primary Q_0 is greater than two times the required circuit Q.

Let us now consider the design of a transistor amplifier using these ideas.

EXAMPLE 9.4   Design a transistor amplifier with $f_0 = 455$ kHz and bandwidth $B = 15$ kHz using transistors with $h_{oe} = 5 \times 10^{-5}$ mho and $h_{ie} = 1$ kΩ. We assume $f_\beta > f_0$ so that R_i is real. The Q_0 of the primary of the coupling transformer is assumed to be 100. The circuit $Q = 455/15 = 30.3$ and $R_o \simeq 1/h_{oe} = 20$ kΩ. Then, by using Eq. 9.38, the primary inductance $L_1 = 4.55 \times 10^{-5}$ H; $C = 1/\omega_0^2 L = 2.68 \times 10^{-9}$ F, and $R_o' = 2Q\omega_0 L_1 = 7.9$ kΩ. The secondary inductance L_2 and the coefficient of coupling k remain to be determined. Both of these parameters appear in Eq. 9.32. We have the requirement that $\omega_0 L_2$ be small in comparison with h_{ie} and Eq. 9.32 shows that minimum L_2 will occur when k is maximum. Therefore, the secondary should be tightly coupled to the primary. Values of $k \simeq 0.7$ can be obtained in air core coils if the secondary is wound on top of the primary or vice versa. Somewhat higher values of k can be obtained with ferrite cores. We assume $k \simeq 0.7$. Then L_2 can be determined from Eq. 9.32.

$$L_2 = \frac{L_1 h_{ie}}{k^2 R_o'} \tag{9.39}$$

The required secondary inductance is $L_2 = 11.8$ μH. We now need to verify the assumption that $\omega_0 L_2$ is much smaller than h_{ie}. Since $\omega_0 L_2 = 33.7$ Ω its magnitude is only 3.4 percent of h_{ie}, thus justifying the initial assumption. The technician sometimes finds it necessary to construct the coils for a given application. The number of turns of wire for a cylindrical coil configuration can be determined from the empirical formula

$$L = \frac{n^2 r^2}{9r + 10l} \times 10^{-6} \tag{9.40}$$

where r is the mean radius of the coil in inches, l is the length of the coil in inches, and L is the inductance in henries. If the coil is slug tuned, the maximum value of inductance will be about twice the value obtained from the formula.

We shall select a coil form with a 1/2-in. diameter and a 3/4-in. winding space for the coupling coil of this example. Then the number of primary turns $n_1 = 84$ and the number of secondary turns $n_2 = 43$. A wire table can be used to determine the size wire that will fill the winding space. More than one layer may be wound for each winding, but the Q_0 will not be as high for a multilayer winding because of the increased distributed capacitance. Therefore, the winding should be confined to a single layer, if practical. If multilayer windings must be used, they can be wound with a coil winding machine to achieve high Q.

The procedure for designing an inductively coupled tuned circuit is summarized below:

1. The center frequency f_0 or ω_0 and the bandwidth requirement is either specified by the user or determined from the known signal characteristics.

2. The circuit Q is determined from the relationship $Q = f_0/B$.

3. The primary inductance of the primary coil is determined from the relationship $L_1 = (1/2Q - 1/Q_0)R_o/\omega_0$.

4. The tuning capacitance is $C = 1/\omega_0^2 L_1$.

5. The required secondary inductance is $L_2 = L_1 R_i/k^2 R_o'$.

6. If you are winding your own coils, transpose Eq. 9.40 to obtain $n = L(9r + 10l)^{1/2}/r$, where L is the inductance in μH.

The voltage gain from the base to the collector of the tuned amplifier is the current gain times the ratio of the collector-load impedance to the transistor-input impedance, as discussed in previous chapters. The magnitude of the voltage gain at resonance with the load matched for maximum power transfer is, therefore,

$$K_v = \frac{\beta R'_o}{2h_{ie}} \tag{9.41}$$

The values of β and h_{ie} must be determined at the resonant frequency, f_0. Low frequency values may be used if f_0 is well below f_β (beta-cutoff frequency). However, if f_0 is *above* f_β, β can be determined from the relationship $\beta = f_r/f_0$, since β is inversely proportional to frequency when the frequency is well above f_β, and β decreases to 1 at f_r. The relationship $h_{ie} = r_b + (\beta + 1)r_e$ was developed in Chapter 5 for low frequencies. This relationship can also be used at high frequencies, but best accuracy will be obtained if it is recognized that β has a phase angle of 45° at f_β and increases to approximately 90° at frequencies above 10 f_β. You may recall that $r_e \simeq 25/I_E$ (mA).

The ratio of the secondary voltage to the primary voltage, across the coupling transformer, is kn_2/n_1. The current ratio must be the reciprocal of the voltage ratio because the power on both sides of the transformer must be the same. (Remember that we extracted the effective transformer loss resistance R_p and lumped it with R_o to obtain R'_o.) Thus the voltage gain for the entire stage (base-to-base) for the matched amplifier at resonance is

$$K_v = \frac{\beta R'_o k n_2}{2h_{ie}n_1} \tag{9.42}$$

The current gain is

$$K_i = \frac{\beta n_1}{2kn_2} \tag{9.43}$$

The power gain is the product of the voltage gain, the current gain, and the cosine of the phase angle between them. But this phase angle θ is the phase angle of h_{ie}, and $h_{ie} \cos \theta$ is the resistive part of h_{ie}, or R_i. Therefore, the power gain for the matched amplifier is

$$K_p = \frac{\beta^2 R'_o}{4R_i} \tag{9.44}$$

This power gain neglects the effect of feedback through the transistor. This problem will be discussed later.

PROBLEM 9.4   An amplifier is needed to amplify a 10-MHz signal. The required bandwidth is 300 kHz. The transistor has $\beta_0 = 100$, $R_o = 10\,k\Omega$, $h_{ie} \simeq R_i = 500\,\Omega$, and $f_r = 250\,MHz$ at $I_C = 2\,mA$ and $V_{CE} = 10\,V$. The load on the amplifier is another transistor with $R_i = 500\,\Omega$. Design an inductively coupled tuned circuit to couple the amplifiers assuming $Q_0 = 120$ and $k = 0.7$. Determine the power gain of the amplifier.

Answer: $L_1 = 1.1\,\mu H, C_1 = 230\,pF, L_2 = 0.25\,\mu H, K_p = 2.6 \times 10^3$.

9.3   TAPPED-TUNED CIRCUITS

A tuned circuit will provide impedance transformation if a tap is provided on a single coil as shown in Fig. 9.10. The coil acts as an auto-transformer and has an impedance ratio approximately equal to the square of the turns ratio $(n/n_2)^2$, the same as a tightly coupled autotransformer. The proof of this statement is beyond the scope of this discussion; however, a reasonable argument for its validity is given with the aid of Fig. 9.11.

The impedance across the entire tuned circuit, as shown in Fig. 9.11a is $(\omega_0 L)^2/R'_s$ at resonance, as discussed in Section 9.1, where R'_s is the effective series resistance which will give the proper circuit Q. However, the impedance between the coil tap and V_{CC} (or ground) is only $(\omega_0 L_2)^2/R'_s$ at resonance, where L_2 is the effective inductance of that part of the coil which has turns n_2. Note that L_2 is tuned by the effective capacitance C_{eff}, which represents the difference between the capacitive reactance of the tuning capacitor C and the inductive reactance of the part of the coil with turns n_1. Therefore, the effective $Q_2 = \omega_0 L_2/R'_s$, as seen at the tap, is

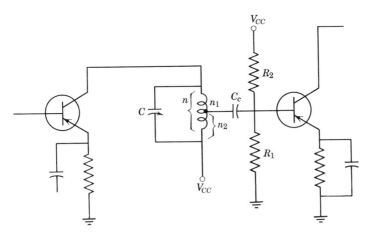

Fig. 9.10.   A tuned amplifier that uses a tapped coil for impedance transformation.

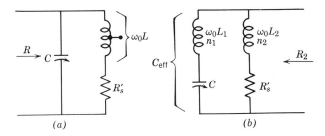

Fig. 9.11. The tapped-tuned circuit as seen (*a*) across the entire coil and (*b*) between the tap and ground.

proportional to L_2 and the impedance transformation ratio is

$$\frac{R}{R_2} = \frac{(\omega_0 L)^2}{(\omega_0 L_2)^2} = \left(\frac{L}{L_2}\right)^2 \tag{9.45}$$

Regardless of the magnetic coupling between turns, the inductance ratio L/L_2 is very nearly equal to the turns ratio n/n_2. Then

$$\frac{R}{R_2} \simeq \left(\frac{n}{n_2}\right)^2 \tag{9.46}$$

The tapped-tuned circuit has an advantage over the inductively coupled circuit because of the simplicity of the single coil and the elimination of the need to estimate the coefficient of coupling. However, part of the signal is shunted through the bias resistors R_1 and R_2. An example may help clarify the tapped-coil approach.

EXAMPLE 9.5 Use a tapped-coil impedance transformation to design the 455-kHz intermediate frequency (IF) amplifier with a 15-kHz bandwidth considered in Example 9.4. Then, $R_0' = 7.9\,\text{k}\Omega$ and $h_{ie} = 1\,\text{k}\Omega$. If impedance matching is desired, the turns ratio $n/n_2 = \sqrt{7.9} = 2.81$. The turns required for the primary winding of the inductively coupled circuit were previously calculated in Example 9.4 to be 84. If this coil is tapped instead of inductively coupled, the number of turns between V_{CC} and the tap should be $n_2 = 84/2.81 = 30$ turns.

Impedance transformation can also be accomplished by tapping the capacitor (or more correctly by using two capacitors) in a tuned circuit, as shown in Fig. 9.12. The impedance transformation occurs in the same manner as described for the tapped coil except an $\omega_0 L$ for the coil is

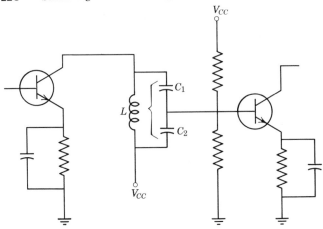

Fig. 9.12.   A tapped-tuned circuit that uses two capacitors.

replaced by a $1/\omega_0 C$ for the corresponding capacitor. Therefore, using Eq. 9.45, we can write

$$\frac{R}{R_2} = \frac{(1/\omega_0 C)^2}{(1/\omega_0 C_2)^2} = \left(\frac{C_2}{C}\right)^2 \tag{9.47}$$

Let us use two capacitors instead of tapping the coil for the next example.

EXAMPLE 9.6   Design an amplifier similar to that shown in Fig. 9.12 with $f_0 = 455$ kHz and $B = 15$ kHz. Assume $R'_o = 7.9$ kΩ, $h_{ie} = 1$ kΩ, and $L = 4.55 \times 10^{-5}$, as found in Example 9.4.

In this amplifier, $R/R_2 = 7.9$ kΩ/1 kΩ = 7.9. In Example 9.4 we found $C = 2.69 \times 10^{-9}$ F. Then, from Eq. 9.45, $C_2 = \sqrt{R/R_2}\, C = (7.9)^{1/2} \times 2.69 \times 10^{-9} = 7.8 \times 10^{-9}$ F. Since C is the total tuning capacitance, which is essentially the series combination of C_1 and C_2, the value of C_1 can be found from the relationship $C = C_1 C_2/(C_1 + C_2)$ or

$$C_1 = \frac{C_2 C}{C_2 - C} \tag{9.48}$$

In this example, the value of C_1 is $C_1 = (7.8 \times 10^{-9})(2.68 \times 10^{-9})/(7.8 - 2.68)10^{-9} = 4.1 \times 10^{-9}$ F.

Sometimes a combination of the impedance transforming techniques are used in a single coupling circuit, as illustrated in Fig. 9.13. Let us refer back to Example 9.4, in which we used inductive coupling for a 455-kHz

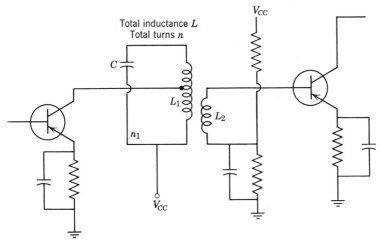

Fig. 9.13. An inductively coupled circuit with tapped primary.

amplifier, and consider the need for the tapped primary coil. In that example we found that proper bandwidth was obtained with $L_1 = 45.5 \, \mu\text{H}$. We also assumed that the coil Q_0 was 100. However, a little experience in measuring the Q_0 of coils will show that each coil has a maximum Q_0 over a certain frequency range and that large coils having high inductance have maximum Q_0 at low frequencies as compared with small coils, which have maximum Q_0 at high frequencies. Therefore, we may not be able to find or wind a coil that has $45.5 \, \mu\text{H}$ with $Q_0 = 100$ at 455 kHz, since this is a low frequency for such a small coil. We almost certainly could increase the coil Q_0 at this frequency by using a higher inductance. The total primary inductance can be almost any desired value if the primary is tapped, as shown in Fig. 9.13. To illustrate this concept, let us consider another example.

EXAMPLE 9.7 Assume that in order to obtain a Q_0 of 100 at 455 kHz, the total primary inductance L must be at least $200 \, \mu\text{H}$. Then, we wish to design the 455-kHz amplifier with $B = 10$ kHz, as specified in Example 9.4.

In Example 9.4, we found the required inductance from collector to ground (V_{CC} is at ground potential for ac signals) is $45.5 \, \mu\text{H}$. To obtain this same value of inductance from collector to ground, the value of L_1 must now be $45.5 \, \mu\text{H}$. Then, if we assume the inductance is proportional to the square of the number of turns, we have

$$\frac{n}{n_2} = \left(\frac{L}{L_1}\right)^{1/2} \tag{9.49}$$

In Example 9.4, we found 84 turns produced an inductance of 45.5 μH. Then, to obtain the 200 μH inductance for L, we have $n = n_1(L/L_1)^{1/2} = 84(200/45.5)^{1/2} = 176$ turns. Thus, our inductance is a 176-turn coil with the tap 84 turns above the V_{CC} terminal. Of course, the tuning capacitance must be decreased by the same ratio by which L is increased in order to maintain the same resonant frequency. Thus, $C = 2.68 \times 10^{-9}(45.5/200) = 6.1 \times 10^{-10}$ F.

PROBLEM 9.5 The amplifier of Prob. 9.4, which had $R_o = 10$ kΩ, $R_i = 500$ Ω, $L_1 = 1.1$ μH, $f_0 = 10$ MHz, $Q_0 = 120$, and $C = 230$ pF, is to be coupled to its load ($R_i = 500$ Ω) by tapping the primary coil, rather than by inductive coupling. Determine the location of the tap point n_2/n_1.
Answer: $n_2/n_1 = 0.33$.

PROBLEM 9.6 If two capacitors are used to obtain the impedance transformation for the amplifier of Prob. 9.5 instead of tapping the coil, determine the values for the capacitors.
Answer: $C_2 = 690$ pF, $C_1 = 345$ pF.

PROBLEM 9.7 Assume that the coupling circuit of Prob. 9.4 with $L_1 = 1.1\mu$H and $C = 230$ pF has 6 primary turns and that we wish to increase the primary inductance to 12 μH by using a tapped primary. How many total turns should the primary have and what tuning capacitance is needed?
Answer: $n = 20$, $C = 70$ pF.

9.4 DOUBLE-TUNED CIRCUITS

Increased selectivity and improved bandpass characteristics can be obtained if two tuned circuits are used in the amplifier coupling circuit, as shown in Fig. 9.14. The inductances and capacitances of the two tuned circuits are usually chosen to be equal and the tap on the secondary coil is normally chosen so that the two circuit Q's are equal, or in other words, so that the effective parallel resistance is the same for both tuned circuits.

Two equivalent circuits for the double-tuned inductively coupled amplifier are given in Fig. 9.15. The circuit in Fig. 9.15a is the type we have used consistently heretofore in this chapter. However, the equivalent series circuit of Fig. 9.15c will show more easily the required coefficient of coupling and the frequency response characteristics. The equivalent series circuit is obtained by first transforming the parallel resistance R_o' to the equivalent series resistance R_{ser} as shown in Fig. 9.15b by using the relationship developed in Section 9.1. Specifically, using Eq. 9.4,

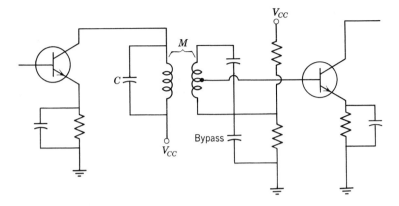

Fig. 9.14. A double-tuned transistor amplifier.

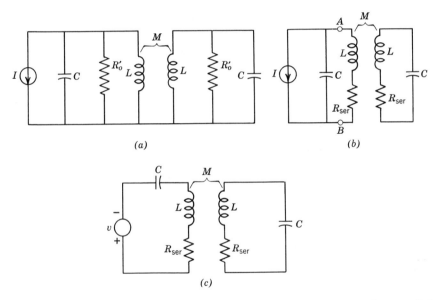

Fig. 9.15. Equivalent circuits for the double-tuned amplifier. (*a*) Equivalent parallel circuit, (*b*) parallel resistance transformed to series resistance, and (*c*) equivalent series circuit.

$$R_{ser} = \frac{(\omega_0 L)^2}{R'_o} \tag{9.50}$$

The current source to the left of points AB (Fig. 9.15*b*) is then transformed to an equivalent voltage source to produce the series circuit of Fig. 9.15*c*.

Referring to Fig. 9.15c, maximum power will be transferred to the secondary circuit when both circuits are resonant and when the resistance coupled into the primary is equal to the effective source resistance R_{ser}. You may recall that an expression for the impedance coupled into the primary was developed in Section 9.2, Eq. 9.28. When the inductance and the tuning capacitance are included in the secondary this equation becomes

$$Z = \frac{(\omega M)^2}{R_{ser} + j\left(\omega L - \dfrac{1}{\omega C}\right)} \tag{9.51}$$

As was mentioned, maximum power is transferred at resonance, with $\omega_0 L = 1/\omega_0 C$, when $Z = R_{ser}$. Then, making these substitutions into Eq. 9.51 $R_{ser} = (\omega_0 M)^2/R_{ser}$. Therefore,

$$R_{ser} = \omega_0 M \tag{9.52}$$

We now substitute the value of $M = k\sqrt{L_1 L_2}$ into Eq. 9.52. But, $L_1 = L_2 = L$; so with this simplification,

$$R_{ser} = \omega_0 k L \tag{9.53}$$

The coefficient of coupling k in Eq. 9.53 is the value that provides maximum power transfer at resonance and is known as critical coupling k_c (sometimes called transitional coupling). This value of coupling can be expressed explicitly by using Eq. 9.53.

$$k_c = \frac{R_{ser}}{\omega_0 L} = \frac{1}{Q} \tag{9.54}$$

Observe from Eq. 9.54 that the optimum coefficient of coupling for a double-tuned circuit is much smaller than that for a single-tuned inductively coupled circuit. For example, a circuit with $Q = 50$ for both primary and secondary circuit would require a value of $k_c = 0.02$.

The frequency response characteristic of a typical double-tuned circuit with $k = k_c$ is shown in Fig. 9.16. This response is flatter on top and steeper on the sides than the response curve for a single-tuned circuit, thus the bandpass and selectivity characteristics are improved.

Since the total shunt resistance is $R_0'/2$ when the impedance is matched, the voltage or power gain of a double-tuned amplifier to the transformer primary at resonance is the same as the gain of a single-tuned amplifier with matched impedance, providing the other parameters are equal. The voltage across the total secondary is essentially equal to the primary

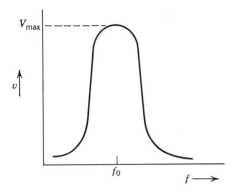

Fig. 9.16.   Frequency response characteristic of a typical double-tuned circuit.

voltage as seen from the following relationships. The primary current is

$$I_p = \frac{V_p}{j\omega_0 L} \qquad (9.55)$$

The induced voltage in the secondary winding $V_i = j\omega_0 M I_p$, but with critical coupling $M = k_c L = L/Q$. Then

$$V_i = \frac{j\omega_0 L I_p}{Q} = \frac{j\omega_0 L V_p}{j\omega_0 L Q} = \frac{V_p}{Q} \qquad (9.56)$$

The voltage across either L or C in the tuned secondary is Q times the induced voltage. Then $V_s = Q V_i = V_p$.

The bandwidth and frequency response characteristics of a double-tuned circuit can be varied considerably by the adjustment of the coefficient of coupling k. We shall see from the following relationships that the double-tuned circuit has two additional resonant frequencies on either side of ω_0 when the coefficient of coupling is greater than the critical value k_c. We shall investigate this phenomenon by looking again at the input impedance of the coupling circuit to see what frequencies other than ω_0 can produce zero reactance. By using Eq. 9.27,

$$Z_{\text{in}} = R_{\text{ser}} + j\left(\omega L - \frac{1}{\omega C}\right) + \frac{(\omega M)^2}{R_{\text{ser}} + j\left(\omega L - \frac{1}{\omega C}\right)} \qquad (9.57)$$

Let us first simplify this equation by letting $\omega L - 1/\omega C = X$, the net reactance in either primary or secondary. Also, the mutual impedance $j\omega M$ varies only a small percentage over the passband, so that little inaccuracy will occur if we assume this impedance to be constant at the value $j\omega_0 M$. Making these substitutions and recognizing that $M = kL$,

$$Z_{in} = R_{ser} + jX + \frac{(k\omega_0 L)^2}{R_{ser} + jX} \tag{9.58}$$

Rationalizing the last term,

$$Z_{in} = R_{ser} + jX + \frac{(k\omega_0 L)^2 (R_{ser} - jX)}{R_{ser}^2 + X^2} \tag{9.59}$$

Separating the real and j-parts of the last term,

$$Z_{in} = R_{ser} + jX + \frac{R_{ser}(k\omega_0 L)^2}{R_{ser}^2 + X^2} - \frac{jX(k\omega_0 L)^2}{R_{ser}^2 + X^2} \tag{9.60}$$

Notice that the reactance coupled into the primary (the last term in Eq. 9.60) tends to cancel the net primary reactance. Resonance occurs at frequencies that cause the total reactance to be zero. Then, from Eq. 9.60,

$$X - \frac{X(k\omega_0 L)^2}{R_{ser}^2 + X^2} = 0 \tag{9.61}$$

Dividing through by X,

$$1 - \frac{(k\omega_0 L)^2}{R_{ser}^2 + X^2} = 0 \tag{9.62}$$

and

$$(k\omega_0 L)^2 = R_{ser}^2 + X^2 \tag{9.63}$$

$$X^2 = (k\omega_0 L)^2 - R_{ser}^2 \tag{9.64}$$

Let $k = bk_c$, so that b is the ratio of actual coupling k to the critical coupling. Then

$$X^2 = b^2 (k_c \omega_0 L)^2 - R_{ser}^2 \tag{9.65}$$

But $k_c = 1/Q$, so that $k_c \omega_0 L = \omega_0 L/Q = R_{ser}$. Therefore,

$$X^2 = R_{ser}^2 (b^2 - 1) \tag{9.66}$$

and

$$X = \pm R_{ser}(b^2 - 1)^{1/2} \tag{9.67}$$

Note that resonance occurs at $X = 0$ when $b = 1$ or $k = k_c$, but when b is greater than one, resonance will occur at other frequencies where X is not zero. Let us find these frequencies.

$$X = \omega L - \frac{1}{\omega C} = \pm R_{\text{ser}}(b^2 - 1)^{1/2} \tag{9.68}$$

Multiplying through by ωC,

$$\omega^2 LC - 1 = \pm R_{\text{ser}} \omega C (b^2 - 1)^{1/2} \tag{9.69}$$

Since $LC = 1/\omega_0^2$ and from $Q = 1/\omega_0 C R_{\text{ser}}$, $R_{\text{ser}} C = 1/\omega_0 Q$,

$$\left(\frac{\omega}{\omega_0}\right)^2 \pm \frac{\omega}{\omega_0} \frac{(b^2 - 1)^{1/2}}{Q} - 1 = 0 \tag{9.70}$$

Using the quadratic equation to solve for ω/ω_0,

$$\frac{\omega}{\omega_0} = \pm \frac{(b^2 - 1)^{1/2}}{2Q} \pm \left(\frac{b^2 - 1}{4Q^2} + 1\right)^{1/2} \tag{9.71}$$

Since $(b^2 - 1)/4Q^2$ is very small compared to 1, for normal values of b,

$$\frac{\omega}{\omega_0} \simeq \pm \frac{(b^2 - 1)^{1/2}}{2Q} \pm 1 \tag{9.72}$$

The positive sign preceding the 1 is the only sign that gives positive frequencies; therefore, the additional resonant frequencies occur at

$$\omega \simeq \omega_0 \pm \frac{(b^2 - 1)^{1/2}}{2Q} \omega_0 \tag{9.73}$$

When b is greater than 1, maximum power transfer occurs at the frequencies given by Eq. 9.73 but not at ω_0, as shown by the frequency response curves of Fig. 9.17. Notice that the passband splits into two passbands when values of $b > 2.5$ are used. Also observe that values up to about $b = 1.5$ can be used to increase the bandwidth without serious deterioration of the response at ω_0. Values of $b < 1$ result in decreased bandwidth and decreased gain.

The bandwidth of the double-tuned amplifier can be found by the use of considerable additional algebra. Only the final result will be given here

$$B(\text{Hertz}) = (b^2 + 2b - 1)^{1/2} \frac{f_0}{Q} \tag{9.74}$$

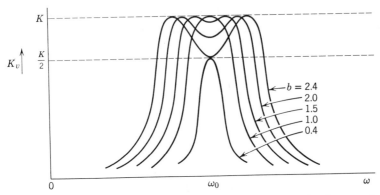

Fig. 9.17.   Frequency response curves of a double-tuned amplifier for several values of b.

This value of bandwidth is valid only for values of b between 1 and 2.5 because maximum power transfer at some frequency is assumed and the band splits into two bands when $b > 2.5$, as was previously shown.

Comparison of Eqs. 9.21 and 9.74 shows that the bandwidth of a double-tuned circuit is increased over a single-tuned circuit by the factor $(b^2 + 2b - 1)^{1/2}$ for the same circuit Q. Thus, since the voltage gain is porportional to the circuit Q, the gain-bandwidth product is also increased by this factor.

We shall now consider the design of a double-tuned amplifier. The design might proceed as follows:

1. Choose the value of b in accordance with the desired frequency response characteristics.
2. Calculate the value of Q using Eq. 9.74.
3. Calculate L using the two relationships.

$$R_o' = \frac{Q_0 \omega_0 L R_o}{Q_0 \omega_0 L + R_o} \text{ and } R_o' = Q \omega_0 L$$

Then

$$Q \omega_0 L = \frac{Q_0 \omega_0 L R_o}{Q_0 \omega_0 L + R_o} \tag{9.75}$$

$$Q = \frac{Q_0 R_o}{Q_0 \omega_0 L + R_o} \tag{9.76}$$

$$Q Q_0 \omega_0 L + Q R_o = Q_0 R_o$$

$$L = \frac{R_o}{\omega_0} \frac{(Q_0 - Q)}{Q Q_0} = \frac{R_o}{\omega_0}\left(\frac{1}{Q} - \frac{1}{Q_0}\right) \tag{9.77}$$

Notice that $1/Q$ appears in this equation in place of $1/2Q$ in the corresponding equation for the single-tuned circuit with matched impedance. This change occurred because the Q of each tuned circuit was defined by considering only the loading on one side of the coupling transformer while the single-tuned circuit included loading from both sides of the coupling circuit.

4. The coefficient of coupling is

$$k = bk_c = \frac{b}{Q} \tag{9.78}$$

5. The tuning capacitors can be determined from the relationship

$$C = 1/\omega_0^2 L$$

6. The tap point on the secondary is $n_2/n_1 = (R_i/R_o)^{1/2}$.

Note that the loading caused by h_{ie} on the secondary-tuned circuit must be the same as the loading of R_o on the primary circuit if the Q of both circuits is to be the same. Thus R_o instead of R_o' appears in the expression for n_2/n.

The design procedure for a double-tuned amplifier will now be illustrated by an example.

EXAMPLE 9.8   Assume that we need to design a 455-kHz amplifier stage with a 10-kHz bandwidth using transistors that have $h_{fe} = 100$, $R_i = 1\ \text{k}\Omega$, and $R_o = 20\ \text{k}\Omega$. This is a typical IF amplifier in a standard broadcast radio receiver.

1. We shall choose $b = 1.2$ from observation of Fig. 9.17.
2. Then $Q = \sqrt{2.84}(45.5) = 78$ from Eq. 9.71.
3. Assuming $Q_0 = 100$, $L = 2 \times 10^4(0.0128 - 0.01)/6.28 \times 4.55 \times 10^5 = 19.6\ \mu\text{H}$.
4. The coefficient of coupling $k = 1.2/78 = 0.0154$.
5. The tuning capacitance $C = 1/(6.28 \times 4.55 \times 10^5)^2(1.96 \times 10^{-5}) = 6.4 \times 10^{-9}\ \text{F}$.
6. The tap point for the 1 kΩ secondary load is $n_2 = n/\sqrt{20} = 0.224n$. The voltage gain to the primary of the coupling transformer can be calculated from the relationship $K_v = \beta R_o'/2h_{ie}$ (Eq. 9.39). But $R_o' = Q\omega_0 L = 78 \times 6.28 \times 4.55 \times 1.96 = 4.25\ \text{k}\Omega$, so $K_v = (4.25 \times 10^5)/(2 \times 10^3) = 212$. The voltage gain to the secondary, which is the stage gain, is $212 \times 0.224 = 47$.

You may have observed from Example 9.8 that L and, hence, the voltage (or power) gain would be considerably higher if the coil Q_0 were higher. The following problem will illustrate this point.

PROBLEM 9.8   Calculate L, C, the tap point n_2/n, and the voltage gain of the amplifier in the example above if the coil Q_0 is 150 instead of 100. *Answer:* $L = 42.7 \times 10^{-6}$ H, $C = 2.94 \times 10^{-9}$ F, $n_2/n = 0.224$, $K_v = 107$.

Double-tuned amplifiers are sometimes coupled by a capacitance, as shown in Fig. 9.18. In this amplifier, the coupling capacitance C_m provides the mutual admittance that couples the two circuits. This coupling circuit is the *dual* of the inductively coupled circuit of Fig. 9.15c. Therefore, the relationship between C_m and C in Fig. 9.18 is the same as the relationship between M and L in Fig. 9.15c. Thus, since $M = k_c L = L/Q$ for critical coupling, as was previously shown, $C_m = C/Q$ for critical coupling in the circuit of Fig. 9.18. Then, in general,

$$C_m = \frac{bC}{Q} \tag{9.79}$$

for the capacitively coupled amplifier. Otherwise, the design proceeds in the same manner as for the inductively coupled amplifier. However, the coupling is more easily adjusted, or changed, in the capacitively coupled amplifier.

PROBLEM 9.9   Determine the value of coupling capacitance required for the amplifier of Prob. 9.8 if capacitive, rather than inductive, coupling is used.   *Answer:* $C_m = 45$ pF.

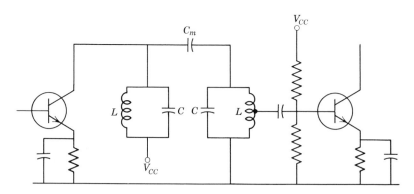

Fig. 9.18.   A capacitively coupled, double-tuned amplifier.

9.5   TUNED-AMPLIFIER STABILITY

Amplifier stability is a term used to indicate the freedom from oscillation in an amplifier. A tuned amplifier is particularly susceptible to oscillation because of its normally high gain and its LC-coupling circuits. These coupling circuits can cause a relative phase shift of 180° at frequencies below resonance and thus cause regenerative (or in-phase) feedback. You will recall that the common-emitter type amplifier causes a voltage polarity reversal as the signal goes through the amplifier. Therefore, if a feedback path that provides another polarity reversal exists between the output and the input, the feedback signal causes regeneration and perhaps oscillation in the amplifier. The regeneration narrows the bandwidth of the amplifier even though the feedback may not be sufficient to cause oscillation.

The undesired feedback in a tuned amplifier may come from three causes. The first is improper circuit layout or inadequate shielding, the responsibility of the technician who constructs the circuit. Unwanted coupling can exist between coupling coils that are widely separated on a chassis unless the coils are enclosed in metal cans that confine the coils' magnetic and electric fields. The magnetic fields are confined at high frequencies because of eddy currents induced in the shield can by the magnetic fields from the coil. These eddy currents cause opposing magnetic fields that reduce the external magnetic fields to almost zero. Capacitive and inductive coupling can also exist between the signal carrying output leads and input leads of an amplifier unless these leads are very short and well separated. Therefore, careful chassis, or circuitboard, layout is essential for a stable amplifier.

The second cause of instability is coupling between amplifier stages through the power supply leads. This coupling results from either long-power supply leads or high-power supply impedance. Decoupling filters between stages are used to reduce this type of coupling. Decoupling will be discussed in Chapter 11.

The third cause of instability is the capacitive coupling due to the collector junction capacitance in the transistor. This capacitance is shown as C_{jc} in the hybrid-π circuit of Fig. 9.19. This coupling, or feedback current, causes regeneration and instability only when both the load and source impedances are inductive as shown. In a tuned amplifier *these impedances are inductive* at frequencies below resonance.

The instability or oscillation occurs because the output voltage V_o causes a feedback current to flow through C_{jc}. This current causes a voltage drop V_i across the source impedance Z_s. The voltage V_i can add to, or substitute for, the source voltage if the source and load impedances are

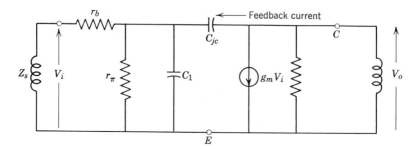

Fig. 9.19. A hybrid-π equivalent circuit showing feedback through the collector-junction capacitance.

inductive, as was previously mentioned, and cause regeneration and instability. The effects of this feedback current can be cancelled or neutralized by feeding a current of equal magnitude but opposite polarity into the base circuit, as shown in the simplified equivalent circuit of Fig. 9.20a. The circuit including C_n and R_n that provides the opposing current is known as the *neutralizing circuit*. The neutralizing current must be obtained from a voltage source of opposite polarity to the amplifier output voltage V_2. This voltage may be obtained either from the secondary of a coupling transformer as shown in Fig. 9.20a or from a tapped-tuned circuit, which has a tap point at V_{CC} (or ground) signal potential as shown in Fig. 9.20b.

Complete neutralization can be obtained if $I_n = I_f$ (Fig. 9.20a) in both magnitude and phase; then the current through Z_s is zero and $V_i = 0$. When $V_i = 0$, the feedback current I_f is

$$I_f = \frac{V_2}{R + 1/j\omega C_{cj}} \tag{9.80}$$

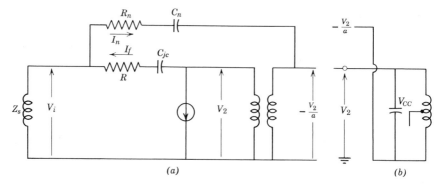

(a)   (b)

Fig. 9.20. A simplified equivalent circuit showing the principle of neutralization.

also

$$I_n = \frac{-V_2/a}{R_n + 1/j\omega C_n} \tag{9.81}$$

Now, if I_n is equal to I_f and opposite in direction, $I_f + I_n = 0$, and

$$\frac{V_2}{R + 1/j\omega C_{cj}} = \frac{V_2/a}{R_n + 1/j\omega C_n} \tag{9.82}$$

Then

$$R_n = \frac{R}{a} \tag{9.83}$$

and

$$C_n = aC_{jc} \tag{9.84}$$

Notice that a is the ratio of primary voltage to secondary voltage and is usually greater than 1 in transistor circuits where stepdown transformer ratios are used.

The resistance R in the simplified circuit of Fig. 9.20a is the equivalent resistance in the feedback path of the more detailed circuit of Fig. 9.19. Some algebraic manipulation of the circuit equations will show the relationship between R and the circuit parameters of Fig. 9.19.[1]

$$R = \frac{r_b(C_{cj} + C_1)}{C_{jc}} \tag{9.85}$$

Therefore, in terms of r_b,

$$R_n = \frac{ar_b(C_{jc} + C_1)}{C_{jc}} \tag{9.86}$$

EXAMPLE 9.9 Let us consider an example of a tuned amplifier using a transistor with $r_b = 100\ \Omega$, $\beta = 100$, $C_{jc} = 5\ \text{pF}$, $C_1 = 100\ \text{pF}$, and $g_m = 0.1\ \text{mho}$ at the chosen q-point. We assume that the voltage stepdown ratio of the output-coupling transformer is 4 and the neutralizing voltage is obtained from the secondary. Then neutralization will be obtained when $C_n = 4 \times 5 = 20\ \text{pF}$ and $R_n = 100(105)/20 = 525\ \Omega$.

Many transistors that are designed especially for high-frequency amplification have small values of r_b and C_1 and can be approximately neutralized with only C_n in the neutralizing circuit. Let us assume that the transistor

---

[1]C. L. Alley and K. W. Atwood, *Electronic Engineering*, Second Edition, John Wiley and Sons, Inc., pp. 321–322.

of the preceding example is to be used at $f_0 = 1$ MHz. Then $X_{cn} = 8000\ \Omega$ and the ratio $X_{cn}/R_n = 8000/525 = 15$. Therefore, if R_n were omitted from the neutralizing circuit at this frequency, the amplitude of feedback current would be essentially unchanged and the phase would be changed by only about $4°$. However, the ratio of X_{cn} to R_n should be checked for a given transistor at the desired frequency before R_n is omitted. If the ratio of X_{cn} to R_n is 5 or higher, R_n can usually be neglected. Note that the voltage stepdown ratio a does not enter into the ratio X_{cn}/R_n.

A field-effect transistor (including MOSFETS) may be neutralized in the same manner as a bipolar transistor except that R_n is always zero. A circuit diagram for a neutralized FET amplifier is drawn in Fig. 9.21. For example, if C_{gd} is 6 pF and the V_{DD} tap point is one-third of the total turns from the bottom of the coil, $C_n = 2C_{gd} = 12$ pF.

Neutralization is frequently called *unilateralization* because the transfer of energy from the output circuit back into the input circuit is reduced to zero. Complete unilateralization is difficult to obtain, however, because of the following reasons:

1. The transistor parameters are functions of the collector (or drain) voltage and thus unilateralization occurs only at the specified q-point.

2. Parameters vary among transistors of a given type. Therefore, the parameters of each transistor must be obtained by measurement at the specified q-point.

3. The neutralizing voltage must be precisely $180°$ out of phase with the transistor output voltage.

Of course, partial neutralization may stabilize an amplifier.

The need for neutralization may be avoided if the voltage gain of the amplifier stage is limited to a reasonably small value by either resistive

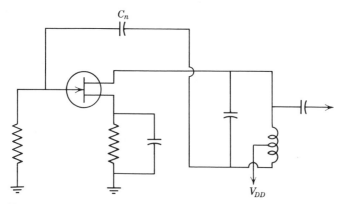

Fig. 9.21.   A neutralized FET amplifier.

loading or mismatching impedances. A relationship between the load conductance G'_L and the source conductance G'_s at resonance (including bias resistance R_b), the junction capacitance C_{jc} (or C_{gd}) and the transistor transconductance g_m required for stable operation can be derived[2] and is given here.

$$g_m < \frac{2G'_i G'_o}{\omega_0 C_{jc}} \tag{9.87}$$

An equivalent circuit is given in Fig. 9.22 to assist in the visualization of the parameters included in Eq. 9.87. We shall use an example to illustrate the use of Eq. 9.87 in stabilizing a tuned amplifier without resorting to neutralization.

EXAMPLE 9.10   Let us assume that a transistor, with $C_{jc} = 5$ pF, $h_{ie} = 1$ kΩ, and $h_{oe} = 50\ \mu$mhos at the q-point $I_C = 5$ mA and $v_{CE} = 10$ V, is to be used as a tuned amplifier at $f_0 = 10$ MHz. Next assume the coil Q_0 is twice the circuit Q so that $R'_o = R_0/2 = 1/2h_{oe} = 10$ kΩ. Let us first check to see if the amplifier will be stable if impedance matching is used. In this case $G'_o = 2/R'_o$ since $G_L = R'_o$, and $G'_i = 2/h_{ie}$ since $G_s = 1/h_{ie}$ (matching occurs in both directions). Testing for stability, g_m must be less than $(8 \times 10^{-7})/(6.28 \times 10^7 \times 5 \times 10^{-12}) = 2.5 \times 10^{-3}$ mho. But the value of g_m at $I_C = 5$ mA is 0.2 mho, which is 80 times too high for stability. We now have the choice of either drastically reducing the q-point collector current or increasing the product $G'_o G'_i$ by a factor greater than 80. A drastic reduction of collector current will cause a major increase of h_{ie} and R_o, which will not only cause a serious mismatch but will also seriously reduce the beta-cutoff frequency

[2]C. L. Alley and K. W. Atwood, *op. cit.*, pp. 332–335.

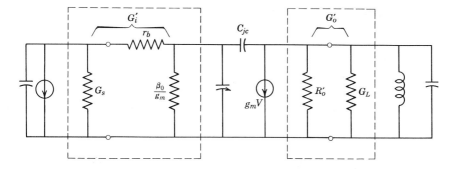

Fig. 9.22.   High-frequency equivalent circuit used in determining stability.

of the transistor. Therefore, stability can be obtained with less sacrifice of power gain if the product $G_o'G_i'$ is increased. Let us increase this product by a factor of 100. This can be done by placing a 560-Ω resistor in parallel with the primary of the coupling transformer so that G_o' will be increased from 2×10^{-4} mho to approximately 2×10^{-3} mho. This resistor will require a lower value of L to provide the same bandwidth, but if the ratio L_2/L_1, or n_2/n is maintained at the initial value so that the coupling circuit impedance ratio does not change, the source resistance R_s of the following transistor is reduced by the same ratio as the load resistance. Then G_i' is increased by approximately the same ratio as G_o', if all the coupling circuits in the amplifier are similarly loaded, and the product $G_o'G_i'$ is increased by a factor of approximately 100 as desired.

Let us gain additional experience by considering one more example.

EXAMPLE 9.11   We shall use the transistor of the example above at the same q-point for a 3.0-MHz amplifier consisting of several identical stages of amplification. The bandwidth of each stage is to be 120 kHz. Single-tuned inductive coupling is chosen and the Q_0 of each tuned circuit is 100. We shall stabilize the amplifier by mismatching and allow a maximum value of transconductance equal to twice the value we expect to use. This safety factor will help to maintain the desired bandwidth. The design procedure might be as follows.

1. Choose the impedance ratio of the coupling circuit to be R_o'/R_i where R_o' includes the tuned circuit loss, as was previously discussed. R_o' can be determined assuming the tuned circuit will not be loaded by the secondary. Then Eq. 9.74 will apply as in the double-tuned case. Since $Q = f_0/B = 25$ and $Q_0 = 100$, $L_1 = 31.8 \, \mu\text{H}$, and $R_o' = 15 \, \text{k}\Omega$. Thus the impedance transformation ratio will be 15.

2. We now need to express G_i' in terms of G_o'. But since the impedance looking into the transformer secondary will be small compared to R_i, G_i' is approximately equal to $G_s = 15G_o'$. Then, by using Eq. 9.87 and assuming $g_m = 0.4$ (including the safety factor), $G_o' = 1.1 \times 10^{-3}$.

3. The loading resistance can now be determined. This resistance in parallel with R_o provides a new value of R_o which, when combined with the coil loss gives a new value of $R_o' = 1/G_o'$. But since the ratio of Q_0/Q must remain the same as before

$$\frac{R_o}{R_o'}(\text{new}) = \frac{R_o}{R_i'}(\text{old}) \text{ and } R_o(\text{new}) = R_o(\text{old})\frac{R_o'(\text{new})}{R_o'(\text{old})} \qquad (9.88)$$

Then, for this example, R_o (new) $= (20 \text{ k}\Omega)(900/15 \text{ k}\Omega) = 1.2 \text{ k}\Omega$ and the loading resistor $R = [(20 \times 1.2)/(20 - 1.2)] \text{k}\Omega = 1.5 \text{ k}\Omega$.

4. The primary inductance can now be determined from the relationship $R_o' = Q\omega_0 L_1$. Thus (new) $L_1 = R_o'/Q\omega_0 = 2 \times 10^{-6} \text{ H}$. The tuning capacitance and secondary inductance can be determined for a given coefficient of coupling, as was discussed in Section 9.2.

The voltage gain of the mismatched amplifier stage can be determined in the usual way. The current gain from base to collector is approximately β since R_o' is small in comparison with R_o. Notice that the value of β appropriate for the frequency f_o must be used. Also the coupled resistance is large compared to R_o', so that $R_L' \simeq R_o'$. For the transistor of this example $\beta = 100$ at $f = 3$ MHz. Then the voltage gain to the primary is approximately, $\beta R_o'/R_{\text{in}} = 90$ and the voltage gain of the stage (to the following base) is $K_v = 90/\sqrt{15} = 23$.

PROBLEM 9.10 A given transistor has $\beta_0 = 120$, $f_T = 300$ MHz, and $C_{ob} = 2.0$ pF at $I_C = 2.0$ mA, $V_{CE} = 10$ V. Using this q-point, this transistor is used in a tuned amplifier with $f_0 = 10$ MHz. $G_o' = 2 \times 10^{-4}$ mho and $G_i' = 2 \times 10^{-3}$ mho at $f = 10$ MHz.

(a) Determine a value of loading resistance that will provide amplifier stability with a safety factor of 5, if only the collector circuit is loaded.
Answer: $R \simeq 1 \text{ k}\Omega$.

(b) The output of the amplifier is inductively coupled to a following amplifier with a coupling coil that has $n_2 = 0.3n$ and $k = 0.83$. Draw the diagram of a neutralizing circuit, using the secondary winding, and determine the value of C_n. Assume $R_n \simeq 0$.   *Answer:* $C_n = 8$ pF.

(c) Calculate the approximate voltage gain for the neutralized matched amplifier (base-to-base).   *Answer:* $K_v = 37$.

PROBLEM 9.11 The intermediate-frequency amplifier of a given television receiver has $f_0 = 40$ MHz. Since several stages of amplification are used, the transformed input resistance of one amplifier stage is the load resistance for the preceding stage. You are assigned the task of designing a transistor amplifier for this purpose using a 2N5208 transistor. The transistor characteristics at $I_C = 2.0$ mA and $V_{CE} = 10$ V (approximately optimum q-point) are given by the manufacturer at $f = 40$ MHz as $g_{ie} = 1$ mmho, $g_{oe} = 0.1$ mmho, C_{cb} (use as C_{jc}) $= 0.5$ pF, $f_T = 625$ MHz, and $\beta_0 = 80$ (low frequency). Use an inductively coupled, single-tuned circuit as a coupling device.

(a) Will the amplifier be stable (assuming proper construction) if impedance matching is used for maximum power transfer?   *Answer:* No.

(b) Assume that your decision for stabilizing the amplifier is to use the
same turns ratio n/n_2, which would provide maximum power gain, but
then load the collector-tuned circuit with a resistance that will provide a
stability safety factor of approximately 5. Note that this technique will
increase both G_0' and G_i' by the same factor. What value of loading resistance is needed assuming the coil Q_0 is very high compared to the circuit Q?
Answer: 710 Ω.

(c) What will be the voltage gain of the stage if the magnitude of h_{ie},
or $|h_{ie}|$ is 250 Ω at 40 MHz?   *Answer:* 12.

(d) What voltage gain can be obtained if the amplifier is neutralized?
Answer: 99.

PROBLEM 9.12   The cascode-connected FET (Section 7.8) was found to
have negligible coupling between output and input, and therefore should
be unconditionally stable if properly constructed. Design a double-tuned
coupling circuit for a cascode amplifier using 2N4416 FET's with $g_m =$
5 mmho, $r_i = 10^4 \Omega$, and $r_o = 10^4 \Omega$ at $f = 40$ MHz, $V_{DS} = 10$ V, $I_D = 3.0$ mA,
and $V_{GS} = -0.2$ V. Assume coil $Q_0 = 100$ and design the amplifier for
maximum power gain. Determine all the circuit components for either
inductive or capacitive coupling and calculate the voltage gain of the stage.
Answer: $K_v = 23$.

CHAPTER 10

DIRECT-COUPLED
AMPLIFIERS

The need frequently arises for an amplifier that will faithfully reproduce very slowly varying signals. The very low-cutoff frequency required for such an amplifier may eliminate capacitive or transformer coupling from practical consideration and leave only direct coupling as a feasible solution. The main disadvantage of direct coupling is that thermal currents generated in the amplifier are amplified along with the signal currents. Thus, thermal stability problems are increased and thermal currents may mask signal currents. Therefore, particular attention must be paid to thermal stability in a dc amplifier. Three amplifier types will be studied in this chapter. They are the Darlington Connection, *npn–pnp* arrangements, and differential amplifiers.

10.1 THE DARLINGTON CONNECTION

One method of direct coupling, known as the Darlington Connection, is shown in Fig. 10.1. In this arrangement the emitter current of transistor T_1 is the base current of transistor T_2. If R_L is small, then $i_{E1} = (\beta_1 + 1)i_{B1}$ and $i_{C2} = \beta_2 i_{B2}$; the ratio $i_{C2}/i_{B1} = (\beta_1 + 1)\beta_2$. The current i_{C1} adds to i_{C2} in the load resistor, but if β_2 is large i_{C1} is negligible, and the total amplification factor is approximately the product $\beta_1\beta_2$. Three transistors are sometimes used in the Darlington Connection to produce a current gain approximately equal to $\beta_1\beta_2\beta_3$.

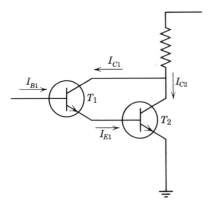

Fig. 10.1.   The Darlington connection.

The thermal currents are also amplified in the Darlington Connection. The discussion in Chapter 5 showed that the thermal current amplification is equal to the current stability factor in any given stage. Resistors may be used in the Darlington circuit to reduce the thermal currents, as shown in Fig. 10.2. The thermal current $S_{I1}I_{CO1}$ from transistor T_1 is divided between resistor R_{B2} and the input of transistor T_2. Also, part of the thermal current I_{CO2} of transistor T_2 may flow through R_{B2} to reduce the stability factor S_{I2}. Notice that R_E is not by-passed because the expected signal frequencies are too low for effective bypassing. Therefore, for good voltage gain, R_E must be small in comparison with R_L. The signal gain is also reduced because part of the signal current is shunted through the stabilizing resistors. In fact, a change in thermal current is indistinguishable from

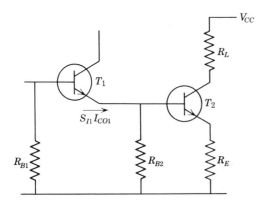

Fig. 10.2.   A Darlington-connected amplifier with linear thermal stabilization.

a signal current. Therefore, with this linear type stabilization, high gain with adequate thermal stability can be achieved only by the use of silicon transistors, which have very small values of I_{CO}.

We shall now consider the design of a Darlington amplifier.

EXAMPLE 10.1   A circuit is connected as shown in Fig. 10.2. The load resistor R_L is determined by the amplifier application. Assume that $R_L = 100\,\Omega$ and $V_{CC} = 20\,\text{V}$. We shall next choose $R_E = 10\,\Omega$. The transistor selected for T_2 has $h_{FE} = 100$ and $r_b = 12\,\Omega$. Since the maximum value of $i_C = 20\,\text{V}/(R_L + R_E) \simeq 180\,\text{mA}$, the average value of collector current should be about 90 mA. Then $(h_{fe} + 1)r_e = \beta_0/g_m = 100/3.6 = 28\,\Omega$ and $h_{ie} = 50\,\Omega$ at this average collector current. Thus, the input resistance of transistor T_2 is $h_{ie} + (h_{fe} + 1)R_E = 50 + (101)10 = 1060\,\Omega$. A sensible choice of value for R_{B2} might be about the same as the input resistance of T_2. Then half of the signal current will be shunted through R_{B2}. We shall choose $R_{B2} = 1\,\text{k}\Omega$. Then the total resistance in the emitter circuit of T_1 (Fig. 10.2) is approximately $1.06\,\text{k}\Omega \times 1\,\text{k}\Omega/2.06\,\text{k}\Omega = 515\,\Omega$. We shall now select transistor T_1 with $h_{FE} = 120$ and $r_b = 100\,\Omega$. The average base current of transistor T_2 is $i_{C2}/\beta_2 = 90\,\text{mA}/100 = 0.9\,\text{mA}$. Assuming v_{BE} of transistor T_2 to be 0.6 V and recognizing that the voltage drop across R_E at the average value of emitter current is $0.09\,\text{A}\,(10\,\Omega) = 0.9\,\text{V}$, the voltage across R_{B2} is 1.5 V and the current through R_{B2} is $1.5\,\text{V}/1\,\text{k}\Omega = 1.5\,\text{mA}$ at this average value of voltage. Then the average emitter current of transistor T_1 is $1.5 + 0.9 = 2.4\,\text{mA}$ and the average base current of T_1 is $i_{E1}/h_{fe1} = 2.4/120 = 20\,\mu\text{A}$. Also, $h_{ie1} = 100 + 120/(0.04 \times 2.4) = 1.35\,\text{k}\Omega$.

The average input impedance of transistor T_1 is about $h_{fe1}/R_{eq} + h_{ie1} = 120(515) + 1350 = 63\,\text{k}\Omega$. Note that this high input impedance is due to the impedance in the emitter circuit R_{eq}. Let us again sacrifice about one-half of the signal current and select $R_{B1} = 68\,\text{k}\Omega$. Then the total current gain for the Darlington amplifier is $K_i \simeq \beta_1\beta_2/4 = 3000$ and the voltage gain $K_v = v_o/v_1 = K_i R_L/R_i = 3000 \times 100/33\,\text{k}\Omega = 9$. Note that transistor T_1 acts as an emitter follower and provides current gain but not voltage gain.

Diodes can be used to stabilize the Darlington amplifier, as shown in Fig. 10.3. If the reverse saturation current I_{S2} of diode D_2 is equal to the thermal current I_{CO2} of transistor T_2, the current stability factor of transistor T_2 is unity. Similarly, if the saturation current I_{S1} is equal to the thermal current I_{CO1} the current stability factor of transistor T_1 is one. The only problem arises in finding diodes that match the transistors in thermal currents. Since the resistance of a reverse-biased diode is very high, the approximate current gain of the amplifier of Fig. 10.3, in terms of the transistor betas, is $\beta_1\beta_2$.

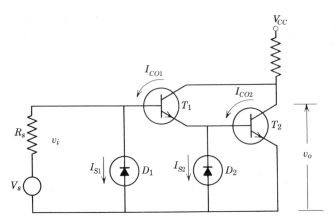

Fig. 10.3.   A diode-stabilized Darlington amplifier.

EXAMPLE 10.2   Let us assume that the Darlington amplifier of the preceding example has the stabilizing resistors replaced by diodes. Then at the average currents previously determined the total current gain is approximately $K_i = h_{fe2}h_{fe1} = 120 \times 100 = 12{,}000$, the approximate input resistance of transistor T_1 is $\beta_1 h_{ie2} + h_{ie1} \simeq 6\,\text{k}\Omega + 3.4\,\text{k}\Omega = 9.4\,\text{k}\Omega$ and the approximate voltage gain $K_v = v_o/v_i = i_o R_o/i_i R_i = K_i(R_o)/R_i = 1.2 \times 10^6/9.4\,\text{k}\Omega = 126$.

We have assumed that the signal source provided forward bias for the Darlington amplifier. If the signal source does not have a dc component to provide this bias, a resistor must be connected between the base of transistor T_1 and V_{CC} to provide the required bias. Some modern devices contain a Darlington configuration in a single container.

PROBLEM 10.1   A given germanium power transistor, which has $h_{ie} = 10\,\Omega$, $\beta = 120$, and $I_{CO} = -80\,\mu\text{A}$ at the desired q-point, is driven by a Darlington-connected germanium transistor with $h_{ie} = 1500\,\Omega$, $\beta = 150$, and $I_{CO} = -5\,\mu\text{A}$ at its desired q-point. Assume h_{re} and h_{oe} are negligible. The amplifier is diode stabilized. If the load resistance in the collector circuit of the power transistor is $30\,\Omega$ and $V_{CC} = -20\,\text{V}$, determine the low-frequency input resistance and the voltage gain of the amplifier.

Answer: $R_i = 2700\,\Omega$, $K_v \simeq 200$.

PROBLEM 10.2   If the Darlington-connected amplifier of Prob. 10.1 is capacitively coupled to its driving source, draw a circuit diagram of the amplifier and determine the value of bias resistance if $I_{C2} = 300\,\text{mA}$. What should be the values of I_S for the diodes? Assume all values are given at $T = 25°C$.   *Answer:* $R \simeq 1.2 \, M\Omega$, $I_{S1} = 5 \, \mu A$, $I_{S2} = 80 \, \mu A$.

10.2   *npn–pnp* COMBINATIONS

A dc amplifier can be constructed by alternating *n-p-n* and *p-n-p* types, as shown in Fig. 10.4. This amplifier is diode stabilized and the input voltage is assumed to provide forward bias for the transistors. Observe that the collector current of transistor T_1 is the base current of transistor T_2. Therefore, the input impedance of transistor T_1 is much lower and the voltage gain much higher for this amplifier as compared with the Darlington amplifier.

EXAMPLE 10.3   Let us use the same transistor T_1 as in the Darlington amplifiers and a *p-n-p* transistor with similar characteristics to T_2 and compare the current gain, input resistance, and voltage gain to the Darlington amplifier values at the same average current values. Recalling that $h_{ie2} = 50 \, \Omega$, $\beta_2 = 100$, $R_L = 100 \, \Omega$, $h_{ie1} = 3.4 \, k\Omega$, and $\beta_1 = 120$, $K_i \simeq \beta_1\beta_2 = 120 \times 100 = 1.2 \times 10^4$, $R_{in} \simeq h_{ie1} = 3.4 \, k\Omega$, and $K_v = K_i R_o/R_i = 1.2 \times 10^6/3.4 \, k\Omega = 353$. Note that the voltage gain is increased by the same ratio as the reduction of input resistance. Observe also that the dc potential in the output is zero when the dc potential of the input is zero. Sometimes this is a distinct advantage.

The alternating *npn–pnp* arrangement may be extended to include any desired number of transistors. For example, Fig. 10.5 uses three transistors. Observe that the zero signal output and input potential can be the same only if an even number of transistor stages is used in the amplifier.

As was previously mentioned, the biggest problem in building a diode-

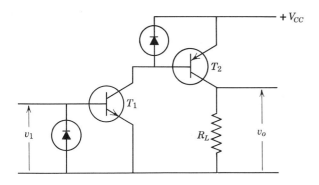

Fig. 10.4.   An *npn-pnp* dc amplifier.

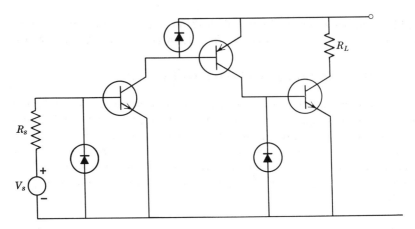

Fig. 10.5.   A three-stage dc amplifier.

stabilized amplifier is finding diodes with the required reverse saturation currents. Sometimes a transistor is used in place of a diode as shown in Fig. 10.6. Since the thermal current in the collector circuit of a transistor is $S_I I_{CO}$, and S_I is approximately equal to R_B/R_E as discussed in Chapter 5, the thermal current in transistor T_3 can be controlled by adjusting R_B. In fact, if I_{CO} in transistor T_3 is large enough, this transistor can compensate for the thermal currents in both T_1 and T_2 and no diodes are required in the two-stage amplifier.

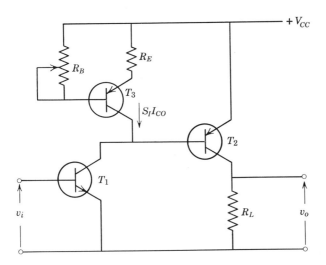

Fig. 10.6.   Thermal stabilization by the use of a transistor.

PROBLEM 10.3 Assume that the amplifier of Fig. 10.6 has $R_L = 30 \, \Omega$ and $V_{CC} = 20 \, \text{V}$. The characteristics of T_2 are $h_{ie} = 10 \, \Omega$, $\beta = 120$, and h_{re} and h_{oe} are negligible. The characteristics of T_1 are $h_{ie} = 1500 \, \Omega$, $\beta = 150$, and h_{re} and h_{oe} are negligible at the respective q-points. What is the voltage gain and what is the input resistance of the amplifier?

Answer: $K_v = 360$, $R_i = 1500 \, \Omega$.

PROBLEM 10.4 Referring to the amplifier of Prob. 10.3, transistor T_2 has $I_{CO} = 80 \, \mu\text{A}$; transistor T_1 has $I_{CO} = 1 \, \mu\text{A}$ and $S_I = 50$; transistor T_3 has $I_{CO} = 10 \, \mu\text{A}$. What must be the approximate ratio R_B/R_E to thermally stabilize the amplifier.

Answer: 13.

10.3 DIFFERENTIAL AMPLIFIERS

The final type of dc amplifier to be considered is the differential amplifier. There are two basic types, balanced and unbalanced. The circuit diagram of a balanced differential amplifier is given in Fig. 10.7. In this amplifier the input signal v_i is balanced with respect to ground. With this type of signal the forward bias of transistor T_1 is increased while the forward bias of transistor T_2 is decreased. If the transistors are matched and linear, the emitter current of one transistor increases by the same amount as the emitter current of the other transistor decreases, and the current through the common-emitter resistor R_E remains constant. Therefore, the voltage across R_E remains constant, and no degeneration is

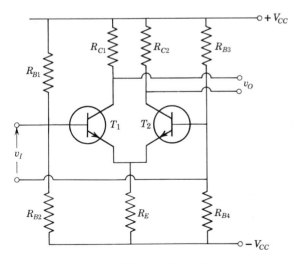

Fig. 10.7.   A balanced differential amplifier.

caused by this resistor. If the transistors are not well matched, a small voltage will appear across R_E; this voltage will tend to degenerate the higher gain transistor and regenerate the lower one and thus improve the balance. Since the impedance between the base and the emitter of each transistor is h_{ie}, the input impedance from base to base is $2h_{ie}$.

The collector potential of one transistor (Fig. 10.7) increases, while the collector potential of the other transistor decreases. The output voltage, which is the difference between the collector potentials, may be fed to a balanced load or to another balanced amplifier. The balanced amplifier that follows may conveniently use the opposite type transistors (p-n-p to follow n-p-n) as shown in Fig. 10.8. However, the transistors may all be of the same type. Note that the two transistors in a balanced amplifier appear to be in series when viewed from the output terminals as they do when viewed from the input terminals. Therefore, the output resistance of the balanced amplifier is twice that of a single-transistor amplifier at the same q-point. However, the current gain and voltage gain of the balanced amplifier are the same as a single transistor similarly biased and with the emitter resistor perfectly bypassed.

The main advantage of the balanced amplifier is that in-phase input signals, which are applied to the two bases, do not produce an output signal, which is the difference between the two collector potentials. These in-phase signals are called *common-mode* signals and include thermally generated currents, changes in v_{BE} due to temperature changes, and extraneous signals such as hum and noise induced in the input leads.

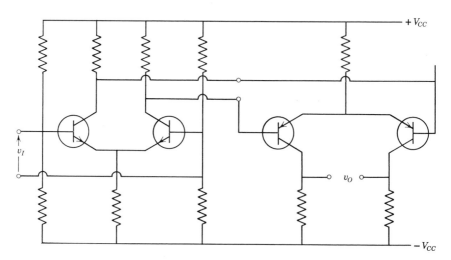

Fig. 10.8.   A two-stage balanced amplifier.

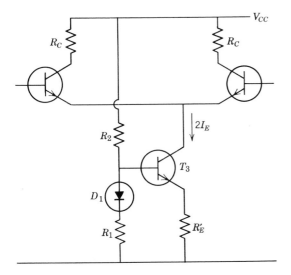

Fig. 10.9.   A differential amplifier with improved q-point stabilization.

These signals are not transferred from one stage to the next providing the two transistors are well matched and maintained at the same temperature. Also, the individual transistors may have excellent q-point stability because of the high permissible value of emitter circuit resistance R_E.

The effective value of R_E can be increased greatly and yet allow the desired value of emitter current to flow if a transistor is used to replace the resistor R_E (Fig. 10.9). Since the collector current of the emitter-circuit transistor T_3 is determined almost entirely by the stabilized bias circuit d_1, R_1, R_2, and R_E', the sum of the emitter currents $2I_E$ of the differential amplifier is held constant; therefore, the q-point collector currents and voltages of the differential amplifier are held constant. The diode d_1 in the stabilized bias circuit compensates for the temperature variation of v_{BE} in transistor T_3.

When an input signal is applied to a differential amplifier each collector voltage varies with respect to V_{CC} or ground, as was previously noted. Therefore, an output signal that is referenced to V_{CC} or ground can be obtained from the differential amplifier to drive a single-ended amplifier or load. The amplifier then has a balanced input and an unbalanced output, and the voltage gain is decreased by a factor of two.

The differential amplifier can be used with both the input and the output unbalanced, as shown in Fig. 10.10. In this circuit the base of one transistor is used as the ground reference. At first glance, it may appear that the unbypassed emitter resistor R_E will cause serious degeneration and low

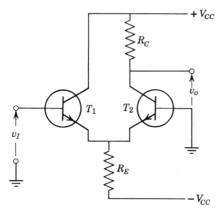

Fig. 10.10.   A differential amplifier with unbalanced input and output.

gain in the amplifier. However, the common-base input impedance h_{ib} of transistor T_2 is in parallel with R_E. Since h_{ib} is normally very small in comparison with R_E, the input impedance of transistor T_1 is

$$R_{\text{in}} \simeq h_{ie1} + (h_{fe1} + 1)h_{ib2} \tag{10.1}$$

But if the two transistors are alike, $(h_{fe1} + 1)h_{ib2} = h_{ie1}$. Therefore,

$$R_{\text{in}} \simeq 2h_{ie1} \tag{10.2}$$

When the output of a differential amplifier is unbalanced the common-mode signals are not cancelled in the output because the output voltage is referenced to a fixed point, V_{CC}, or ground. However, the amplifier gain for common-mode signals may be very small because h_{ib2} is not in parallel with R_E when in-phase signals are applied to both bases. These in-phase signals could be extraneous signals in a balanced input or thermally induced signals ΔI_{CO} and ΔV_{BE} in either a balanced or unbalanced input. Thus, an index of the goodness of a differential amplifier is the *common-mode rejection ratio* (CMRR), defined as follows.

$$\text{CMRR} = \frac{\text{Voltage gain for difference signals}}{\text{Voltage gain for common mode signals}} \tag{10.3}$$

Since the voltage gain for difference signals is

$$K_v = \frac{K_i R_L}{R_{\text{in}}} \simeq \frac{\beta R_L}{2h_{ie}} \tag{10.4}$$

And the voltage gain for common-mode signals is approximately $R_L/2R_E$, since both emitter currents flow through R_E (which doubles the effectiveness of R_E),

$$\text{CMRR} \simeq \frac{(\beta R_L/2h_{ie})}{(R_L/2R_E)} = \frac{\beta R_E}{h_{ie}} \tag{10.5}$$

Therefore, a high common-mode rejection ratio is obtained when R_E is very large compared to $h_{ie}/\beta \simeq h_{ib}$. For example, if $R_E = 2.0 \text{ k}\Omega$, $h_{ie} = 1 \text{ k}\Omega$, and $\beta = 100$, CMRR $= 200$. Note that the unbalanced difference amplifier of Fig. 10.10 is actually a common-collector amplifier directly coupled to a common-base amplifier. It can also be viewed as a cascode configuration. Therefore, its high-frequency response is better than either the balanced configuration or a single common-emitter amplifier.

Transistor manufacturers provide two transistors made on the same chip. These transistors are closely matched and maintain essentially the same temperature. Therefore, they provide very good thermal characteristics in differential amplifier applications. Balanced amplifiers are very commonly used in integrated circuits because of their universal application and excellent thermal characteristics. Integrated circuits are discussed in Chapter 16. An example will be used to illustrate one use of a balanced differential amplifier.

EXAMPLE 10.4 Let us consider the design of an electronic thermometer to replace the glass fever thermometer. One approach is to use a thermistor bridge circuit and a differential amplifier, as shown in Fig. 10.11. The temperature indicator is a basic 0-1 mA meter, which has a temperature scale ranging from 90° F to 110° F. This meter has a very small resistance in comparison with the collector load resistors R_C. The potentiometers P are used to calibrate the thermometer and compensate for differences in transistor characteristics. The transistors are assumed to be matched.

We must obtain a Thevenin's equivalent circuit for the bridge before we can design the amplifier. This can be accomplished by obtaining an expression for the potential difference between points A and B at the output of the bridge and calculating the resistance between these two points. First, the potential at points A and B will be obtained with respect to the negative side of V_{CC}, which will be used as a reference. Using the voltage-divider technique, the open-circuit (base disconnected) voltage at point A is

$$V_A = \frac{R_{th}}{R_{th} + R} V_{CC} \tag{10.6}$$

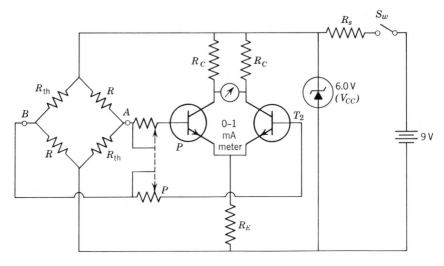

Fig. 10.11.   The circuit diagram of a proposed electronic fever thermometer.

where R_{th} is the resistance of the thermistor and R is a fixed resistance. Similarly,

$$V_B = \frac{R}{R_{th} + R} V_{CC} \tag{10.7}$$

The voltage applied to the differential amplifier is the difference between V_A and V_B. Then

$$V_{AB} = V_A - V_B = \frac{R_{th} - R}{R_{th} + R} V_{CC} \tag{10.8}$$

But V_{AB} must be zero at the lowest temperature (90°F) so that the meter in the collector circuit will read zero. Then $R_{th} = R$ at 90°F. We can thus let $R_{th} = R + \Delta R$ where ΔR is the change in resistance of the thermistor, which results from the temperature departure from 90°F. Making this substitution in Eq. 10.8,

$$V_{AB} = \frac{\Delta R}{2R + \Delta R} V_{CC} \tag{10.9}$$

If a thermistor with a modest temperature coefficient is chosen, ΔR is small compared to $2R$ over the 20°F temperature range. Then

$$V_{AB} \simeq \frac{\Delta R}{2R} V_{CC} \tag{10.10}$$

This expression is the open-circuit voltage for the Thevenin's equivalent circuit.

The resistance of the bridge between points A and B can be obtained by observing that there are two parallel paths between these points and each path has resistance $R + R_{th}$. Then, since $R_{th} \simeq R$, the internal resistance of the bridge is approximately R.

The thermometer circuit of Fig. 10.11 exhibits some practical features such as a 9.0-V power source, which is the value used in almost all miniature portable radios and is, therefore, inexpensive and readily available. The reference diode is used to stabilize V_{CC} so that the calibration will not change appreciably as the battery ages. Since the switch is a push-button type, the circuit will draw current only while the reading is being taken. The design of the amplifier circuit is left as an exercise for the student.

PROBLEM 10.5   Complete the design of the electronic thermometer of Example 10.4. The current through the bridge elements should be large in comparison with the base bias currents, but the power dissipation in the thermistors should be 0.1 W, or less, so that the thermistor temperature will remain essentially equal to the ambient temperature. Assume that the thermistor has a temperature coefficient of 1 percent per °F over the desired temperature range and can be obtained with any desired resistance value at 90°F. Select the transistor type for your circuit and determine the potentiometer resistance that will permit a variation in h_{fe} from h_{fe}min to $2h_{fe}$min.

PROBLEM 10.6   Draw the circuit diagram of a two-stage differential amplifier that has unbalanced input, unbalanced output and uses two pair of *n-p-n* transistors.

PROBLEM 10.7   A given transistor amplifier has a current stabilizing transistor in the emitter circuit, as shown in Fig. 10.9. What is the CMMR if $h_{ie} = 1$ kΩ and $\beta = 100$ for transistors T_1 and T_2, and the output resistance of the stabilizing transistor T_3 is 200 kΩ?   *Answer:* 2×10^4.

PROBLEM 10.8   What is the approximate voltage gain of the amplifier of Prob. 10.7 if the output is unbalanced and the load resistance is 2 kΩ?
                    Answer: 100.

CHAPTER 11

MULTISTAGE AMPLIFIERS

Most amplifier applications, such as radio tuners, phonograph amplifiers, and microphone amplifiers, require more gain than a single stage can provide. Consequently, it is common practice to feed the output of one amplifier stage into the input of the next stage, as has been suggested in the preceding work. When amplifiers are connected in this fashion they are called *cascaded amplifiers*. The thoughtful circuit designer will raise some questions concerning the multistage amplifier such as:

1. How does the gain and bandwidth of the amplifier relate to the gain and bandwidth of a stage?

2. What instability problems might arise?

3. When does amplifier noise become an important factor and how can noise be minimized?

11.1 GAIN AND BANDWIDTH

Two stages of a multistage RC-coupled transistor amplifier are presented by the equivalent circuit of Fig. 11.1. This midfrequency h-parameter circuit will be used to determine the reference (midfrequency) gains of the amplifier. The calculation of voltage gain is quite complicated because of the interaction between stages unless h_{re} is negligibly small. Fortunately, as was previously mentioned, adequate accuracy can usually

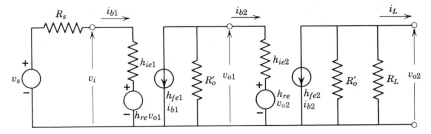

Fig. 11.1.   A midfrequency equivalent circuit for two stages of a multistage amplifier.

be obtained in RC-coupled circuits if h_{re} is assumed to be zero. We shall make this assumption.

Notice in Fig. 11.1 that the current gain of the second stage is $G_{i2} = i_L/i_{b2}$ and the current gain of the first stage is $G_{i1} = i_{b2}/i_{b1}$. Therefore the total current gain of the two-stage amplifier is

$$G_{ia} = \frac{i_L}{i_{b1}} = \frac{i_L}{i_{b2}}\frac{i_{b2}}{i_{b1}} = G_{i2}G_{i1} \tag{11.1}$$

Similarly, the voltage gain of the second stage is $v_{o2}/v_{o1} = G_{v2}$, and the voltage gain of the first stage is v_{o1}/v_i; so the voltage gain of the two-stage amplifier is

$$G_{va} = v_{o2}/v_i = G_{v2}G_{v1} \tag{11.2}$$

Therefore, we need to find the current gain and the voltage gain for each stage, as we have previously done, and use the product of the stage gains to find the amplifier gains. To illustrate these ideas, let us consider the current and voltage gains of a two-stage amplifier.

EXAMPLE 11.1   A two-stage amplifier is connected as shown in Fig. 11.1. The circuit parameters are $R_s = 2$ kΩ, $h_{ie1} = h_{ie2} = 2$ kΩ, $h_{fe1} = h_{fe2} = 100$, $R_o' = 5$ kΩ, and $R_L = 2$ kΩ. (Let us assume $h_{re} = 0$ for this example.) Determine the current gain and voltage gain for each stage and for the total amplifier.

The midfrequency current gain of the second stage is $K_{i2} = i_L/i_{b2} = h_{fe}R_o'/(R_o' + R_L) = 100$ (5 kΩ)/7 k$\Omega = 70.7$. Since the first and second stages are symmetrical, $K_{i1} = 70.7$ also. Then, from Eq. 11.1, the total current gain is $K_{ia} = K_{i1}K_{i2} = 70.7 \times 70.7 = 5000$. (The letter K is used here to denote the midfrequency or reference gain. The symbol G is used as a symbol for gain that is a function of frequency.)

The reference voltage gain for each stage of the amplifier is $K_{v2} = V_{o2}/V_{o1} = K_{i2}R_L/R_i = 70.7 \times 2$ kΩ/2 kΩ = 70.7. However, the voltage gain from the source v_s to the output of the first amplifier v_{o1} is usually preferred over the voltage gain of the first stage only, since the open circuit voltage v_s of devices such as microphones is frequently given by the manufacturer. For the amplifier of Fig. 11.1, $K_{v1} = v_{o1}/v_s = 70.7$ (2 kΩ)/4 kΩ = 35.35 and the amplifier voltage gain $K_{va} = v_{o2}/v_s = 70.7 \times 35.35 = 2500$.

You have previously learned that the bandwidth of an amplifier is the frequency difference between half-power points or frequencies at which the voltage and current gains have dropped to 0.707 times the reference or midfrequency gains. Therefore, we would expect that the bandwidth of two identical amplifiers in cascade would be less than the bandwidth of each individual amplifier because the half-power frequencies of each stage would be the fourth-power frequencies of the amplifier ($0.5 \times 0.5 = 0.25$).

Figure 11.2 shows a hybrid-π equivalent circuit for the amplifier of Fig. 11.1. This circuit is more useful than the h-parameter circuit in determining the upper-cutoff frequency of an amplifier, as was previously discussed. The upper-cutoff frequency of each stage will first be determined; then a relationship will be developed between the upper-cutoff frequency of the amplifier and the cutoff frequencies of the individual stages.

The task of finding the upper-cutoff frequency of the first stage of the amplifier of Fig. 11.2 appears foreboding at first glance because the load on the generator $g_m v_1$ includes the entire amplifier that follows, and this load impedance is complex and difficult to determine. However, it can be shown[1] that an approximately correct value of upper-cutoff frequency can be determined by assuming that the load on the generator $g_m v_1$ is only the resistance to the left of points C-D. In other words, we may assume that the

[1]SEEC Vol. 5, John Wiley and Sons, Inc., New York.

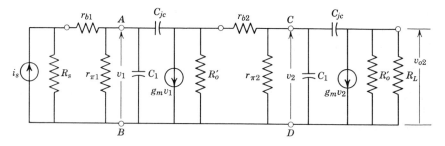

Fig. 11.2.   A hybrid-π equivalent circuit for the amplifier of Fig. 11.1.

amplifier is severed at the points C-D and the part of the circuit to the right of C-D can be neglected when determining the upper-cutoff frequency of the first stage. This problem is then identical to the single-stage problems that we have previously solved. For the single stage, you may recall, the upper-cutoff frequency in radians per second is the reciprocal of the time constant as seen at the points A-B. In Chapter 7 this frequency was shown to be

$$\omega_h = \frac{1}{[r_\pi \| (r_b + R_s)][C_1 + (1 + g_m R'_L)C_{jc}]} \qquad (7.50)$$

where the parallel bars $\|$ mean *in parallel with*, and R'_L is $R'_o \| (r_{b2} + r_{\pi2})$.

The upper-cutoff frequency of the second stage can be determined by considering the time constant between the points C-D when the influence of the first stage to the left of R'_o is neglected. We shall now consider a numerical example to help clarify this concept.

EXAMPLE 11.2   Determine the upper-cutoff frequency for the first stage in Fig. 11.2. The values of the parameters are $R_s = 2$ kΩ, $r_{b1} = r_{b2} = 200$ Ω, $r_{\pi1} = r_{\pi2} = 1800$ Ω, $C_1 = 200$ pF, $C_{jc} = 10$ pF, $R'_o = 5$ kΩ, $g_m = 0.055$, and $R_L = 2$ kΩ.

The value of R'_L in Eq. 7.50 is $R'_L = 5$ k$\Omega \times (200 + 1800)/(5$ k$\Omega + 200 + 1800) = 1.43$ kΩ. When these values are substituted back into Eq. 7.50 the upper-cutoff frequency for the first stage is $\omega_h = 1/[1800 \times 2200/(1800 + 2200)][2 \times 10^{-10} + (1 + 0.055 \times 1.43 \times 10^3)10^{-11}] = 10^6$ rad/s or 1.6×10^5 Hz.

PROBLEM 11.1   Determine the upper-cutoff frequency for the second stage in Fig. 11.2 if the parameters have the values given in Example 11.2.
Answer: $f_h = 1.2 \times 10^5$ Hz.

To develop a relationship for the gain and bandwidth of a multistage amplifier, let us assume that the amplifier has n stages, each with an upper-cutoff frequency ω_{h1}. An expression for the gain of a single-stage amplifier at middle and upper frequencies was derived in Chapter 7 and is repeated below.

$$G = -\frac{K}{1 + j\omega/\omega_{h1}} \qquad (11.3)$$

where G and K may represent either voltage gain or current gain. Since we are interested only in the gain magnitude at the moment, and not the phase, we shall replace the complex denominator of Eq. 11.3 with its magnitude and ignore the negative sign, which indicates a polarity reversal. Then Eq. 11.3 becomes

$$G = \frac{K}{\sqrt{1 + (\omega/\omega_{h1})^2}} \qquad (11.4)$$

Since the gain of the amplifier is the product of the stage gains, the voltage or current gain of the amplifier with n identical stages is

$$G_a = \left[\frac{K}{\sqrt{1 + (\omega/\omega_{h1})^2}}\right]^n = \frac{K^n}{[1 + (\omega/\omega_{h1})^2]^{n/2}} \qquad (11.5)$$

The upper-cutoff frequency of the amplifier ω_{ha} occurs when the gain (voltage or current) of the amplifier decreases to $K^n/\sqrt{2}$. Then, at frequency ω_{ha}, the denominator of Eq. 11.5 must be equal to $\sqrt{2}$. Thus

$$(2)^{1/2} = \left[1 + \left(\frac{\omega_{ha}}{\omega_{h1}}\right)^2\right]^{n/2} \qquad (11.6)$$

and

$$2^{1/n} = 1 + \left(\frac{\omega_{ha}}{\omega_{h1}}\right)^2 \qquad (11.7)$$

Thus

$$\left(\frac{\omega_{ha}}{\omega_{h1}}\right)^2 = 2^{1/n} - 1 \qquad (11.8)$$

and

$$\omega_{ha} = \omega_{h1}\sqrt{2^{1/n} - 1} \qquad (11.9)$$

Equation 11.9 can be used to either determine the upper-cutoff frequency of an amplifier if the upper-cutoff frequency of one stage is known or to determine the upper-cutoff frequency of a single stage if the required upper-cutoff frequency of the amplifier is known. Let us consider an example.

EXAMPLE 11.3   Two identical stages are cascaded to form an amplifier. The midfrequency current gain of each stage is 70.7 and the upper-cutoff frequency of each stage is 10^6 rad/s. Determine the midfrequency current gain of the amplifier and the upper-cutoff frequency of the amplifier.

The number of stages is $n = 2$ so the midfrequency current gain is $K_a = K^n = 70.7^2 = 5000$. The upper-cutoff frequency of the amplifier is

$\omega_{ha} = \omega_{h1}(2^{1/n} - 1)^{1/2} = 10^6(2^{1/2} - 1)^{1/2} = 10^6(1.414 - 1)^{1/2} = 10^6(0.414)^{1/2}$
$= 6.4 \times 10^5 \, \text{rad/sec or } 1.02 \times 10^5 \, \text{Hz}.$

PROBLEM 11.2 If a four-stage amplifier is required to have an upper-cutoff frequency of 1 MHz, determine the upper-cutoff frequency of each identical stage. *Answer:* $f_{h1} = 2.3$ MHz.

The individual cutoff frequencies can be averaged when these frequencies are nearly equal but not identical. However, when the upper-cutoff frequencies differ by a factor of three or more, the cutoff frequency of the amplifier is nearly equal to the lowest of the stage-cutoff frequencies. For example, if an amplifier has three stages with cutoff frequencies of 10^5 Hz, 5×10^5 Hz and 2×10^6 Hz, the upper-cutoff frequency of the amplifier is approximately 10^5 Hz.

The low and midfrequency gain (current or voltage) for a single RC- or transformer-coupled stage can be expressed as

$$G = -\frac{K}{1 - j(\omega_{l1}/\omega)} \tag{11.10}$$

This is identical to the high-frequency gain expression except that the ratio ω_{l1}/ω appears in Eq. 11.10 in place of ω/ω_{h1} in Eq. 11.3. Therefore the relationship between the amplifier low-frequency cutoff ω_{la} and the stage low-frequency cutoff ω_{l1} can be written by comparison with Eq. 11.9 as

$$\omega_{la} = \frac{\omega_{l1}}{\sqrt{2^{1/n} - 1}} \tag{11.11}$$

PROBLEM 11.3 If each stage of a two-stage RC-coupled amplifier has a low-frequency cutoff $f_{l1} = 30$ Hz, determine the low-frequency cutoff of the amplifier. *Answer:* $f_{la} = 47$ Hz.

Table 11.1 gives the relationships between the lower- and upper-cutoff frequencies of the stage and the lower- and upper-cutoff frequencies of the amplifier for n identical stages.

Table 11.1

n	1	2	3	4	5	6	7	8
f_{l1}/f_{la}	1	0.644	0.510	0.435	0.387	0.348	0.333	0.30
f_{h1}/f_{ha}	1	1.55	1.96	2.3	2.58	2.88	3.00	3.33

Since the stage bandwidth must increase as n increases, for a fixed amplifier bandwidth, the stage reference gain will decrease as n increases because increased bandwidth is always obtained at the price of reduced reference gain, as discussed in Chapter 7. Therefore, there is a limit to the amount of gain that can be obtained by the cascading technique for a fixed bandwidth.

The bandwidth of a tuned amplifier that uses identical single-tuned stages has the same relationship between amplifier bandwidth B_a and stage bandwidth B_s as expressed for the RC-coupled amplifier. Thus

$$B_a = B_s \sqrt{2^{1/n} - 1} \qquad (11.12)$$

Therefore, if a two-stage IF amplifier is required to have a 10-kHz bandwidth, the bandwidth of each of the identical stages is $B_s = 15$ kHz.

It has been proven[2] that the relationship between amplifier bandwidth and stage bandwidth for a critically coupled double-tuned amplifier is expressed as follows:

$$B_a = B_s (2^{1/n} - 1)^{1/4} \qquad (11.13)$$

Thus, if a 10-kHz bandwidth, two-stage amplifier uses critically coupled double-tuned amplifiers, the bandwidth of each stage should be 12.5 kHz. There is even less difference between the stage and amplifier bandwidths in a double-tuned amplifier with coupling greater than critical.

11.2   GAIN PHASE OR BODE PLOTS

The gain expressions of Eq. 11.3 and Eq. 11.10 will be used to plot the magnitude of gain and the phase shift of an amplifier as functions of frequency. These plots will not be made by a tedious point-by-point technique but will be approximated by a quick and easy method developed by *Bode* of the Bell Telephone Labs. These plots will be useful in solving frequency response and feedback stability problems. Equation 11.10 is repeated here for convenience.

$$G = -\frac{K}{1 - j(\omega_{l1}/\omega)} = -\frac{K}{1 - j(f_{l1}/f)} \qquad (11.10)$$

This is the gain equation for low and midfrequencies that was developed and discussed in Chapter 7. A plot of gain and phase was also given. How-

[2]Thomas Martin, *Electronic Circuits*, Prentice Hall, Inc.

ever, we shall now make a straight-line approximation of the stage gain in the low and midfrequency region. This plot will be made on log paper, as shown in Fig. 11.3. Notice from Eq. 11.10 that the gain (either current or voltage) is proportional to frequency when the frequency is below about $0.1 f_l$. Therefore, the slope of the curve is 1 below $0.1 f_l$. Similarly, for frequencies above $10 f_l$ where the j-part of the denominator is negligible, the gain is essentially constant and the slope of the curve is zero. If these two straight-line portions are continued until they intersect, the inter-section occurs at f_l. This straight-line approximation is the dashed line of

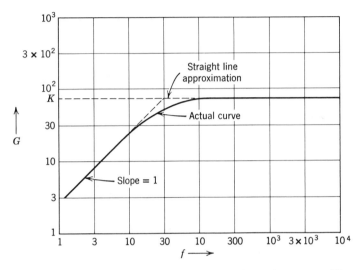

Fig. 11.3.   A Bode plot of low-frequency gain of a single-stage amplifier with $K = 70$ and $f_l = 30$ Hz.

Fig. 11.3. The actual gain curve is the same as the straight-line approxima-tion except for the frequency range between $f = 0.1 f_l$ and $f = 10 f_l$. The maximum departure of the actual curve from the straight-line approxima-tion occurs at $f = f_l$, where the actual gain is $0.707K$ instead of K, which the straight line approximation indicates. The actual gain curve can thus be sketched as a smooth curve that blends into the straight-line approxima-tion at $0.1 f_l$ and $10 f_l$ but passes through the point $0.707 K$ at f_l. The gain could have been expressed in decibels (dB), in which case the departure of the actual curve at $f = f_l$ is $20 \log 0.707 = 3$ dB. Also, the slope of the straight-line approximation, expressed in decibels per decade (increase of frequency by a factor of 10) is $20 \log 10 = 20$ dB/decade.

An amplifier with two identical stages has a gain magnitude that increases proportional to the square of the frequency for frequencies below 0.1 f_l, since the gain amplitudes are multiplied. Therefore, since log scales are used for the plot, the slope of the straight-line approximation below f_l has a slope of 2 or 40 dB/decade (Fig. 11.4), and the departure of the actual curve from the straight-line approximation at f_l is $(0.707K)^2 = 0.5K^2$ or 6 dB. Carrying this idea further, the straight-line gain approximation for an amplifier with n identical stages has a slope of n or $20n$ dB/decade below f_{l1} and the actual gain curve is $3n$ dB below the straight-line approximation at the break point f_{l1}.

A gain magnitude plot can be readily drawn for an amplifier that has

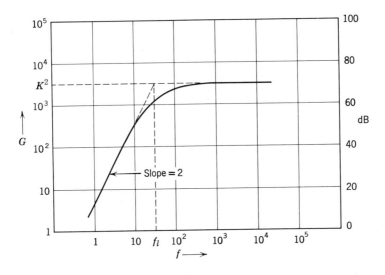

Fig. 11.4.   A gain magnitude plot for two identical stages with $K = 70$ and $f_l = 30$ Hz.

stages with different low-frequency cutoff or break points. To illustrate this idea, let us consider an example.

EXAMPLE 11.4   Consider an amplifier with two stages that have $f_{l1} = 10$ Hz and $f_{l2} = 100$ Hz. Again, the total gain magnitude plot can be obtained by adding the individual stage magnitude plots on log paper. Then the straight-line approximation has a slope of 2 or 40 dB/decade below f_{l1}, a slope of 1 or 20 dB/decade above f_{l1} but below f_{l2}, and zero slope above f_{l2}. Assuming the reference voltage gain of each of these stages is 100, the straight-line approximation of the gain magnitude is sketched in Fig. 11.5. Since the two break frequencies (f_l) are separated by a ratio of 10, the departure of the actual gain curve from the straight-line approximation at each break point is 3 dB. However, the departure of the actual gain curve

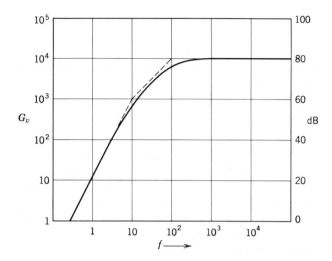

Fig. 11.5. A gain magnitude plot for two stages with $f_{l1} = 10$ Hz and $f_{l2} = 100$ Hz, $K_1 = K_2 = 100$.

at a point midway between f_{l1} and f_{l2} may be difficult to estimate accurately without additional information. The needed information is the calculation of the departure of the actual gain curve from the straight-line approximation at a few strategic frequencies. For example, we already know that the departure is approximately 3 dB at f_l and 0 dB at $0.1\, f_l$ for a single stage. The departure for three additional points each side of f_l is given in Table 11.2.

Table 11.2

f	$0.3\,f_l$	$0.5\,f_l$	$0.7\,f_l$	f_l	$1.4\,f_l$	$2\,f_l$	$3\,f_l$
Ratio	0.96	0.9	0.82	0.7	0.82	0.9	0.96
Departure	0.5 dB	1 dB	2 dB	3 dB	2 dB	1 dB	0.5 dB

The total departure for two stages is the sum of the individual departures in decibels. Therefore, returning to the two-stage amplifier with $f_{l1} = 10$ Hz and $f_{l2} = 100$ Hz, 31.6 Hz (half-way between the 10 Hz and 100 Hz on the log plot) is about $3\,f_{l1}$ and $0.3\,f_{l2}$. Therefore, the departure of the actual gain curve at this point is 0.5 dB + 0.5 dB = 1 dB. Using this information, the actual gain curve is sketched in Fig. 11.5.

PROBLEM 11.4 Consider two stages that have $f_{l1} = 50$ Hz and $f_{l2} = 100$ Hz. Determine the departures of the actual curve at 50 Hz, 70 Hz,

100 Hz, and at 140 Hz. Sketch the straight-line approximation and the actual gain curve.

The gain of a stage at middle and high frequencies can be expressed (from the gain Eq. 7.64 in Chapter 7) as

$$G = -\frac{K}{1 + j(f/f_h)} \tag{11.14}$$

where f_h is the high-frequency cutoff of the amplifier. This equation is identical to the low-frequency equation except for the sign of the operator j and the fact that the variable frequency f is in the numerator of the fraction f/f_h. Therefore, a straight-line approximation of the high-frequency gain magnitude has zero slope to f_h and -1 (downward) slope for frequencies above f_h on a log plot. Thus, all that you have learned concerning slopes and break points and departure of the actual curve from the straight-line approximation can be applied to the high-frequency response as well as the low-frequency response. To illustrate this concept, an example will be given.

EXAMPLE 11.5 A three-stage RC-coupled amplifier has $K_1 = 50$, $K_2 = 100$, $K_3 = 20$, $f_{h1} = 5 \times 10^4$, $f_{h2} = 10^5$, and $f_{h3} = 3 \times 10^5$ Hz. Draw the straight-line approximation and sketch the actual gain curve for the middle- and high-frequency range of this amplifier in Fig. 11.6.

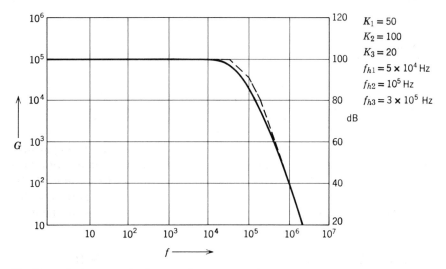

Fig. 11.6.   Gain magnitude plot for a three-stage amplifier.

The midfrequency gain is $K_a = K_1K_2K_3 = 50 \times 100 \times 20 = 100{,}000$. As shown in Fig. 11.6, the straight-line approximation is flat at $K_a = 10^5$ up to $f_{h1} = 5 \times 10^4$ Hz. The slope then becomes -20 dB/decade from 5×10^4 Hz to 10^5 Hz and -40 dB/decade from 10^5 Hz to 3×10^5 Hz. At frequencies above 3×10^5 Hz, the slope decreases at the rate of -60 dB/decade.

The actual gain curve follows the straight-line approximation below $0.1f_{h1}$ or 5×10^3 Hz and above $10f_{h3}$ or 3×10^6 Hz. Between these frequencies, the location of the actual curve can be determined from Table 11.2. Thus, at $f = 10^5$ Hz we have $f/f_{h1} = 10^5/5 \times 10^4 = 2$. The curve is 1 dB below the straight-line approximation due to the break point at f_{h1}. We also have $f/f_{h2} = 10^5/10^5 = 1$. Therefore, the curve is 3 dB below the straight-line approximation due to the break point at f_{h2}. Finally, we have $f/f_{h3} = 10^5/3 \times 10^5 = 0.3$. The actual curve is 0.5 dB below the straight-line approximation due to the break point at f_{h3}. Consequently, the actual curve is 1 dB $+$ 3 dB $+$ 0.5 dB $=$ 4.5 dB below the straight-line approximation.

PROBLEM 11.5   Determine the decibel difference between the straight-line approximation and the actual curve for $f = 5 \times 10^4$ and $f = 3 \times 10^5$ Hz in Example 11.5.

Answer: About 4.25 dB at 5×10^4 Hz and 3.75 dB at 3×10^5 Hz.

Phase plots will now be made using straight-line approximations. We already know from Chapter 7 that a common-emitter (or common-source) stage has a polarity reversal or $180°$ phase shift in the mid-frequency range, a $45°$ phase shift at f_l, and a $225°$ phase shift at f_h. These phase angles were taken with reference to the input signal. Let us use the output voltage of the amplifier as the phase reference of the amplifier in this chapter. The phase will then shift ahead of this reference at low frequencies and behind this reference at high frequencies, as shown in Fig. 11.7 for a single-stage amplifier with $f_l = 10$ Hz and $f_h = 10^5$ Hz. Observe that the actual phase plot is symmetrical about the break frequencies f_l and f_h when the frequency scale is logarithmic and the phase scale is linear. Also notice that a first approximation to the phase plot is the dashed line that shows the entire $90°$ phase change at the break frequencies f_l and f_h. The departure of the actual curve from this first approximation is given in Table 11.3 for a few typical frequencies with reference to f_l or f_h. When a

Table 11.3

Frequency	$0.1f_l$	$0.3f_l$	$0.5f_l$	$0.7f_l$	f_l	$1.4f_l$	$2f_l$	$3.3f_l$	$10f_l$
Departure	$5.7°$	$17.5°$	$26.6°$	$35.3°$	$+45°$	$-35.3°$	$-26.6°$	$-17.5°$	$-5.7°$

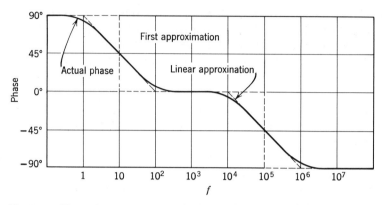

Fig. 11.7.   Phase plot for a single-stage amplifier with $f_l = 100$ Hz and $f_h = 10^5$ Hz.

phase versus frequency curve is plotted through these points (the solid line in Fig. 11.7), the curve is fairly near a straight line with slope -1, which can be drawn through the f_l and $45°$ intercept or the f_h and $-45°$ intercept (the dotted line in Fig. 11.7). Since the phase angles add when complex numbers are multiplied, the total phase of an amplifier can be sketched as a function of frequency by adding the relative phases of the amplifier stages. The phase addition can be performed quite easily by initially drawing the first approximation for the entire amplifier, then sketching the linear approximation for each amplifier, and then adding these linear approximations to obtain the linear approximation for the entire amplifier. Let us consider a two-stage amplifier example.

EXAMPLE 11.6   Draw the straight-line phase approximation for an amplifier with $f_{l1} = f_{l2} = 10$ Hz, $f_{h1} = 10^5$ Hz, and $f_{h2} = 3 \times 10^5$ Hz. The first approximation and straight-line approximations for the amplifier phase are given in Fig. 11.8. The departures from the first approximation, given in Table 11.3, could be used to sketch in the actual phase curve.

Gain and phase plots are usually placed in line vertically so that they can be correlated easily. This arrangement also eliminates some effort in making the phase plot because the break frequencies in the gain plot are obvious and may also be used for the phase plot. This arrangement will be used for the study of feedback amplifier stability in Chapter 13.

PROBLEM 11.6   A two-stage RC-coupled amplifier has $f_{l1} = 10$ Hz, $f_{l2} = 20$ Hz, $f_{h1} = 10^5$ Hz, and $f_{h2} = 5 \times 10^5$ Hz. Draw the straight-line gain and linear-phase approximations; then sketch the gain and phase curves using the departures given in Tables 11.2 and 11.3.

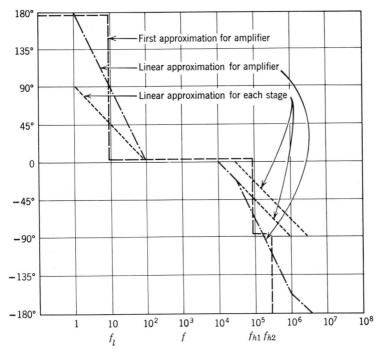

Fig. 11.8. A sketch of relative phase for a two-stage RC-coupled amplifier with $f_{l1} = f_{l2} =$ 100 Hz, $f_{h1} = 10^5$ Hz, and $f_{h2} = 3 \times 10^5$ Hz.

11.3 DECOUPLING FILTERS

You learned in Chapter 9 that instability or oscillation may occur in a tuned amplifier unless signal feedback from the amplifier output to the input is minimized by keeping the signal leads short and well separated. Also, shielding and neutralization may be required to insure stability.

Although instability is not a problem in a single RC-coupled stage because the feedback is usually not in proper phase to reinforce the input signal, instability can readily occur in a multistage amplifier if the opportunity exists for signal feedback from the output of one amplifier to the input of a preceding amplifier. One common feedback path is through a common power supply lead (V_{CC}), as shown in Fig. 11.9. In this figure, two stages are shown in block diagram form. Each stage has a current gain of 100 so that the total current gain $K_i = i_{c2}/i_i = 10,000$. The power supply always has some internal impedance, which is shown as R_p; so the output current i_{c2} causes a signal voltage v_p to appear on the power supply lead. This signal component $v_p = i_{c2}R_p$ causes a feedback current to flow

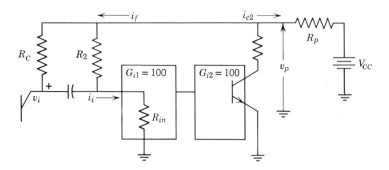

Fig. 11.9. A semiblock diagram showing how feedback can occur through a power supply lead.

through R_C and R_2 into the input of the first amplifier. Since the resistance of R_C is usually much lower than the resistance of R_2, let us assume that the parallel combination of R_C and R_2 is essentially equal to the resistance R_C. Then the feedback current is

$$i_f \simeq \frac{v_p}{R_C + R_{\text{in}}} \tag{11.15}$$

Since $v_p = i_{c2} R_p$, this feedback current can be expressed in terms of the output current i_{c2}.

$$i_f = \frac{i_{c2} R_p}{R_C + R_{\text{in}}} \tag{11.16}$$

Since each of the two common-emitter amplifiers produces a polarity reversal, the feedback current is in phase with the assumed input current and increases the total current of the amplifier. In fact, if the feedback current is equal to the assumed input current without feedback, the feedback current will provide the required input current and the output current will be maintained with no external signal. This is the condition for oscillation or instability. Therefore, if the ratio of i_f/i_{c2} is equal to or greater than the reciprocal of the current gain $1/K_i$, the amplifier will oscillate. The power supply resistance R_p, which will cause oscillation, thus can be determined from Eq. 11.16 by letting $i_f/i_{c2} = 1/K_i$ and solving for R_p.

$$R_p = \frac{i_f}{i_{c2}} (R_C + R_{\text{in}}) = \frac{R_C + R_{\text{in}}}{K_i} \tag{11.17}$$

Of course, R_p must be less than this value for stable operation. For example, let us assume that $R_C = 10\,\text{k}\Omega$ and $R_{\text{in}} = 2\,\text{k}\Omega$ for the amplifier of Fig. 11.9. The value of R_p that will cause oscillation is $12\,\text{k}\Omega/10^4 = 1.2\,\Omega$. Therefore, R_p should not be larger than about $1\,\Omega$. This very small permissible value of R_p can be realized only by using a well-regulated power supply and keeping the power supply leads short. Regulated power supplies will be discussed in Chapter 15. Although a battery is shown as the power supply in Fig. 11.9, the battery only represents the various types of power supplies that might be used. The rectifier-filter type power supplies frequently used in equipment intended for ac input power usually have a

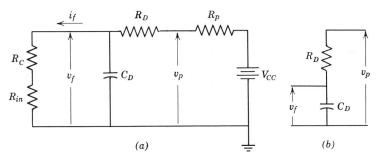

Fig. 11.10. A decoupling filter used to reduce the feedback current through the power supply leads.

capacitor in the output of the power supply. This capacitor has low impedance at the higher signal frequencies. Since its capacitive reactance is inversely proportional to frequency, there will be some low frequency where the feedback is adequate for oscillation unless the low-frequency gain of the RC-coupled amplifier is restricted by the judicious choice of the lower-cutoff frequency of the amplifier.

Instability due to signal feedback through the power supply leads is usually prevented in a multistage amplifier by using decoupling filters. Figure 11.10 shows a decoupling filter consisting of R_D and C_D. The other elements in this figure are extracted from the amplifier circuit of Fig. 11.9. The decoupling filter acts as a voltage divider, as shown in Fig. 11.10b, to reduce the signal component v_p across the power supply to the value v_f. Since v_f is now the signal voltage that forces the feedback current i_f through R_C and R_{in}, the decoupling filter reduces the feedback current by the ratio v_f/v_p. The voltage divider will be effective if X_{CD} is small in comparison with R_D at the low-frequency cutoff f_l of the amplifier. But the dc component of collector current for the lower-level stage (or stages) must pass through R_D, so the value of R_D is determined by the permissible dc voltage drop across R_D. Sometimes R_D is used to reduce V_{CC} by a specified amount.

EXAMPLE 11.7   Let us assume that the desired V_{CC} for the higher-level stage is 25 V and the desired V_{CC} for the lower-level stage is 20 V. Then, if the q-point collector current of the lower-level stage is 1 mA, the desired value of R_D is 5 kΩ. Assume that the amplifier is RC-coupled with $f_l = 16$ Hz and that we wish to reduce the signal feedback by a factor of 0.01. Then $X_{CD} = 0.01R_D$ at 16 Hz and the required value of $C_D = 1/2\pi \times (16)(50) = 200\,\mu\text{F}$.

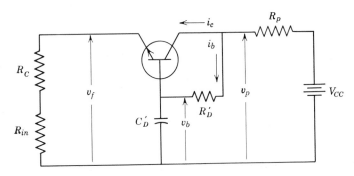

Fig. 11.11.   An *active* decoupling filter.

The impressive reduction of feedback realized in the preceding decoupling filter was obtained at the expense of a 200-μF capacitor. Although this value of capacitor is not prohibitively large or expensive, a saving in cost and space may be realized by using a transistor in the decoupling filter circuit, as shown in Fig. 11.11. This filter is known as an *active* filter because the transistor is known as an *active* element in contrast with a passive R, L, or C element. The transistor amplifier appears to be in the common-collector (or emitter-follower) configuration because R_C and R_{in} are in the emitter circuit. Therefore, the feedback voltage v_f is essentially equal to the signal voltage v_b between the base and ground. But v_b is the voltage across the filter capacitor C_D'. Thus, R_D' and C_D' serve as the decoupling filter in the active circuit. But R_D' may be much larger than the R_D in the passive circuit because i_b instead of i_c flows through R_D'. Therefore, C_D' may be much smaller than C_D.

EXAMPLE 11.8   Consider an active filter as a replacement for the passive filter of the preceding example. Also, assume that the transistor has $\beta_0 = 100$ so $I_B = 10\,\mu\text{A}$ and V_{BE} (dc value) = 0.5 V. Then $R_D' = 4.5\,\text{V}/10\,\mu\text{A}$

$= 4.5 \times 10^5 \, \Omega$ and the 0.01 reduction in feedback current can be obtained at 16 Hz with $C'_D = 1/2\pi(16)(4.5 \times 10^3) = 2.2 \, \mu\text{F}$.

Observe that the required value of filter capacitance for the active filter is reduced by a factor almost equal to β_0. For Example 11.8, the value of C'_D might be increased to $10 \, \mu\text{F}$ with little increase in cost or size to provide improved decoupling. A Darlington connected pair of transistors is often used in high-current applications where I_C may be hundreds of milliamperes or perhaps amperes.

PROBLEM 11.7   A two-stage amplifier is connected as shown in Fig. 11.9. If the current gain of each stage is 150, $R_C = 8 \, \text{k}\Omega$, $R_2 = 200 \, \text{k}\Omega$, and $R_{\text{in}} = 3 \, \text{k}\Omega$, determine the maximum value of power supply resistance R_D for stable operation.   *Answer:* $R_D < 0.47 \, \Omega$.

PROBLEM 11.8   The lower-cutoff frequency of the amplifier of Problem 11.7 is 10 Hz and $R_D = 10 \, \Omega$. Design an active decoupling filter that will reduce the feedback current to 0.2 the value that will cause oscillation. Use a silicon transistor with $\beta_0 = 100$ and allow $V_{CE} = 4.0 \, \text{V}$ for the filter transistor. The current through the filter is 1.0 mA.
              Answer: $R'_D = 350 \, \text{k}\Omega, C'_D = 5 \, \mu\text{F}$.

11.4   AMPLIFIER NOISE

The maximum usable gain of a cascade amplifier is usually determined by the noise generated in the amplifier. High gain is required when the available input signal is very weak. However, if the input signal is not stronger than the noise generated in the first stage of a cascaded amplifier, the noise may make the signal unusable. The signal source always generates some noise and therefore has a finite ratio of signal-to-noise. Ideally, the amplifier would not generate noise, and therefore the signal-to-noise ratio in the output of the amplifier would be the same as the signal-to-noise ratio out of the source. However, the amplifier always adds noise of its own and thus the signal-to-noise ratio in the output of the amplifier is always lower than the signal-to-noise ratio of the source.

Two fundamental sources of noise in a transistor amplifier are:

1. Diode noise, which results from the random injection of charge carriers across the depletion region.

2. Resistor noise, which results from the random motion of electrons in a resistance at temperatures above 0°K.

In both of these noise sources, the average currents are very predictable when a known constant voltage is applied to the device, but small random fluctuations about the average value result from the random motion of electrons or charge carriers. This random motion is due to the kinetic energy of the carriers, which is proportional to the absolute temperature of the material. These random variations are called noise because they are amplified along with the signal and cause background noise radiation from a loudspeaker or headset in an audio system. The noise signals also cause fuzzy oscillograph tracings or "snow" on a TV picture tube. The noise will not be audible or the snow noticeable if the signals are very large in comparison with the noise. Therefore, the *signal-to-noise ratio* determines whether the noise will be disturbing. The tolerance level of the user will depend on the program material, of course. For example, a noisy but intelligible long-distance voice communication may be satisfactory, but a noticeable background noise accompanying a musical concert may be annoying.

The signal-to-noise ratio of the amplifier output depends primarily on the ratio of signal power applied to the amplifier input to the noise power generated in the source and in the first amplifier. The noise that is generated at the input of the first amplifier is amplified by that amplifier and is then large compared to the noise generated in the following amplifier, assuming these amplifiers have similar noise figures.

We shall now investigate the generation of noise in an amplifier and relate this noise to the *noise figure*, *noise temperature*, and signal-to-noise ratio, which are familiar noise criteria used in the industry. First, we shall determine the noise voltage of a resistor. This noise voltage fluctuates randomly and, therefore, does not have discrete frequency components such as one might obtain from the Fourier analysis of a periodic fluctuation. Instead, the noise voltage or current has a continuous noise spectrum over a very broad band of frequencies extending well beyond the upper-frequency limits of transistor amplifiers. Therefore, the noise power in the output of an amplifier is proportional to the bandwidth of the amplifier in Hz. Since Boltzmann's constant k relates temperature to energy, a reasonable relationship between noise power w_n, temperature T, and bandwidth Δf is

$$w_n = kT\Delta f \qquad (11.18)$$

The Δf is used instead of B for bandwidth because the effective noise power bandwidth is different than the half-power bandwidth we have previously defined. However, these bandwidths are nearly the same and we shall consider them to be equal. Nyquist[3] postulated that the noise

---

[3]Nyquist, "Thermal Agitation of Electric Charge in Conductors," *Physical Review*, Vol. 32, p. 110, 1928.

power given by Eq. 11.18 is the maximum power that can be transferred from a noisy (normal) resistor to a noiseless resistor. Since maximum power is transferred when the load resistance is equal to the source resistance, a Thevenin's equivalent circuit can be drawn for the noisy resistor (Fig. 11.12). A simple calculation will show that this equivalent circuit will transfer the maximum power $w_n = kT\Delta f$. Of course, if two normal resistors are connected together, they both generate noise power so that there is no net transfer of energy if they are at the same temperature. The voltage $\bar{v}_n$ is not an rms voltage in the usual sense because it cannot represent a

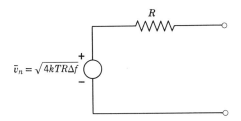

Fig. 11.12.   A Thevenin's equivalent circuit for the noisy resistor.

frequency component. Therefore, the bar is placed over v_n to indicate that it is a fictitious voltage that will yield the proper noise power transfer. Therefore, $\bar{v}_n$ can be treated as an rms voltage.

PROBLEM 11.9   Determine the effective noise voltage of a 100-kΩ resistor at 300°K, as measured with a noiseless voltmeter with a 1-MHz bandpass ($k = 1.38 \times 10^{-23}$ J/°K).   *Answer: 91 μV.*

The Thevenin's equivalent noise source can be transformed to a Norton's or current source, as shown in Fig. 11.13. A simple calculation will show that the short circuit current of the Thevenin's circuit (Fig. 11.12) is

$$\bar{i}_n = \frac{\sqrt{4kTR\Delta f}}{R} = \sqrt{4kTG\Delta f} \tag{11.19}$$

A semiconductor diode can be represented by the Norton's equivalent circuit of Fig. 11.13. If the diode and a resistance at the same temperature are connected together with no external voltage applied, no net noise power will flow between them. Otherwise one of them must get hotter and the other cooler. Therefore, the noise current flow resulting from noise generated in the diode must be equal to the noise current flow resulting from noise generated in the resistor. But the noise in a diode, exclusive of the noise generated in the ohmic resistance of the doped semiconductor, is due to the random variations in current across the junction. But in a diode with no external connection, the saturation current flows in one direction and an injection current equal to the saturation current flows in the opposite direction, as was discussed in Chapter 3. Although the average current is zero, the noise components of the two oppositely directed currents are additive because the random variations depend only on the magnitude of current and not on the direction. Thus the magnitude of *noise-producing* current in the unbiased diode is $2I_s$.

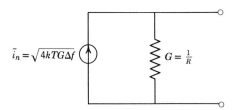

Fig. 11.13.   A Norton's equivalent noise source.

As was discussed in Chapter 3, the dynamic conductance of a diode with no forward bias is

$$G = \frac{q}{kT}I_s \qquad (11.20)$$

If this value of conductance is substituted into Eq. 11.19, the noise component of an unbiased diode becomes

$$\bar{i}_{nd} = \sqrt{4qI_s\Delta f} \qquad (11.21)$$

But the noise-producing current in an unbiased diode is $I = 2I_s$. Therefore, the noise component of diode current can be expressed in terms of the noise-producing diode current by substituting $I_s = I/2$ in Eq. 11.21 to obtain

$$\bar{i}_{nd} = \sqrt{2qI\Delta f} \qquad (11.22)$$

This relationship holds for any value of noise producing diode current I. When reverse bias is applied, $I = I_s$. When forward bias is applied, $I = I_s + I_i$, where I_i is the total injection current. But I_i is normally very large compared with I_s, so I is essentially the q-point or average value of diode current when the diode is forward biased.

Since the noise bandwidth of a diode usually exceeds the useful frequency range of the diode, the noise bandwidth Δf is usually the noise bandwidth of the amplifier or measuring instrument that follows the diode.

PROBLEM 11.10   Determine the short-circuit noise current of a diode at 300°K with 1 mA average current, as measured with a meter with a 1-MHz bandwidth.   *Answer:* $\bar{i}_n = 1.8 \times 10^{-8}$ A.

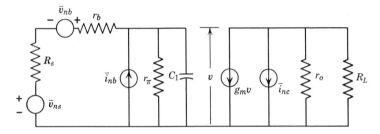

Fig. 11.14.   A noise equivalent circuit for the transistor.

We are now prepared to identify the noise sources in a transistor and to make some noise voltage (or equivalent resistance) calculations. Thermal currents such as I_{CO} have noise components and, therefore, increase the noise of a transistor. Thus, silicon transistors have less noise than germanium transistors if other characteristics are similar. Consequently, we shall assume the transistor that is chosen for low noise is silicon and the thermal currents may be neglected. The main noise sources in a silicon transistor are then the forward-biased emitter junction (diode noise) and the ohmic resistance in the base r_b. A hybrid-π equivalent circuit with these noise sources shown is given in Fig. 11.14. The emitter-junction diode noise has two effects. First it produces a noise voltage across the junction. This voltage is equal to either $\bar{i}_{ne}r_e$ or $\bar{i}_{nb}(h_{fe}+1)r_e$. Since the equivalent resistance $(h_{fe}+1)r_e = r_\pi$ is shown in the equivalent circuit, the equivalent current $\bar{i}_{nb}$ is placed in parallel with it to produce the proper contribution to the noise voltage $\bar{v}$ across the junction. The second effect of the emitter-junction noise current is the direct transmittal of this current, reduced by the ratio α, to the collector circuit. This component, which is not amplified by the transistor, is represented by the noise current source $\bar{i}_{nc}$ in the collector circuit. The resistors r_o and $(h_{fe}+1)r_e$ are noiseless resistors because they are fictitious resistors that account for the transistor characteristics but do not represent ohmic resistance.

The following conclusions can be drawn from the foregoing discussions.

1. The collector circuit–noise current $\bar{i}_{nc}$ can be small if the q-point collector current is small.

2. The noise current $\bar{i}_{bn}$ can be minimized for a given q-point collector current if β is large.

3. The contribution of the base-resistance noise is small if r_b is small in comparison with the driving source resistance R_s.

4. The noise in the output of the amplifier can be minimized by restricting the bandwidth of the amplifier to that required by the signal.

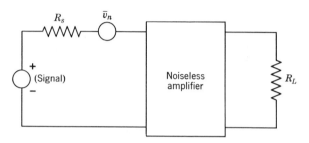

Fig. 11.15.   The representation of amplifier noise by an equivalent noise source.

Thus, a silicon transistor with high β at low values of I_C will have a low noise figure when driven by a source with $R_s \gg r_b$.

A third type of noise is known as $1/f$ noise because its magnitude is approximately inversely proportional to frequency. This type of noise is effective only at the lower frequencies, hundreds of Hertz or less. This type of noise varies quite widely among transistors of the same type and a theoretical basis for its existence has not been firmly established. Recent improvements in transistor surface treatment have greatly reduced the $1/f$ noise in comparison with earlier models. However, in applications which require a very low noise amplifier, the transistor should be hand picked for low $1/f$ noise.

A noise figure of merit for an amplifier is the *spot noise figure, F*. This figure is defined as

$$F = \frac{\text{Noise power delivered by an amplifier to a load}}{\text{Noise power delivered if } R_s \text{ (at 290°K) is the only noise source}}$$

This ratio is expressed for a narrow frequency band at a specific frequency because the amplifier noise is a function of frequency. The noise figure is usually given in decibels at 1 kHz.

The noise produced by both the driving source and the amplifier can be represented by a noise voltage $\bar{v}_n$ at the input of the amplifier, as shown in Fig. 11.15. The equivalent noise voltage $\bar{v}_n$ can be represented as the noise voltage that will be produced by a resistance R_n at 290°K (the standard reference temperature) in the input of the amplifier. Then, the noise figure F is equal to R_n/R_s, as shown by the following relationships. The input noise current to the amplifier is

$$\bar{i}_n = \frac{\bar{v}_n}{R_s + Z_i} = \frac{(4kTR_n\Delta f)^{1/2}}{R_s + Z_i} \tag{11.23}$$

where Z_i is the input impedance of the transistor at the specified frequency. The noise output voltage is then

$$\bar{v}_{no} = \bar{i}_n G_i R_L = \frac{(4kTR_n\Delta f)^{1/2}G_iR_L}{R_s + Z_i} \tag{11.24}$$

Similarly, the noise voltage in the output, which results only from the noise voltage of the source, is

$$\bar{v}_{nos} = \frac{(4kTR_s\Delta f)^{1/2}G_iR_L}{R_s + Z_i} \tag{11.25}$$

The noise power in the load is $\bar{v}_{no}^2/R_L$, whereas if the amplifier were noiseless the noise power in the load would be $\bar{v}_{nos}^2/R_L$. By definition, the spot noise figure is the ratio of these two powers. Then, by using Eqs. 11.24 and 11.25, the spot noise figure is

$$F = \frac{4kTR_n\Delta fG_i^2R_L^2}{4kTR_s\Delta fG_i^2R_L^2} = \frac{R_n}{R_s} \tag{11.26}$$

Observe that the noise figure F is independent of the load resistance. Also, the input impedance Z_i does not appear explicitly in Eq. 11.24. However, the equivalent noise resistance R_n is a function of the transistor input impedance.

We shall now find the equivalent noise resistance R_n in terms of the transistor parameters given in Fig. 11.14. Since R_n represents the total noise resistance including the source resistance, let us define a noise resistance R_a which, at 290°K, will produce noise equivalent to the noise generated by the amplifier. Then $R_n = R_s + R_a$, and in terms of R_a

$$F = \frac{R_s + R_a}{R_s} = 1 + \frac{R_a}{R_s} \tag{11.27}$$

We shall first find the noise voltage $\bar{v}_{na}$ which, if placed in the input of the transistor in series with the driving source, will produce the same output noise current as the transistor. Figure 11.14 shows that there are

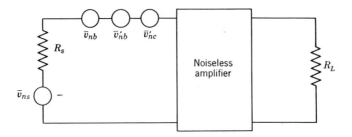

Fig. 11.16.   An equivalent noise circuit for the transistor.

three noise sources in the transistor: one voltage source $\bar{v}_{nb}$ and two current sources $\bar{i}_{nb}$ and $\bar{i}_{nc}$. The voltage $\bar{v}_{nb}$ is already in series with the driving source and we need to convert the two current sources to equivalent voltage sources in series with $\bar{v}_{nb}$ as shown in Fig. 11.16. The total noise produced by these three noise sources will be equal to the noise produced by the fictitious resistance R_a.

The equivalent voltage $\bar{v}'_{nb}$ results from the current source $\bar{i}_{nb}$. This current source can be easily transformed to the equivalent voltage source $\bar{v}'_{nb}$ with the aid of the equivalent circuit given in Fig. 11.17, wherein all noise sources except $\bar{i}_{nb}$ have been turned off. The voltage $\bar{v}'$ in the Thevenin's circuit (Fig. 11.17b) is the open circuit voltage at terminals b'-E of the Norton's circuit (Fig. 11.17a).

The equivalent noise voltage $\bar{v}'_{nc}$ can be found from the equivalent circuit given in Fig. 11.18. Since $\bar{i}_{nc} = g_m \bar{v}$, $\bar{i}_{nc}$ can be expressed in terms of $\bar{v}'_{nc}$ since $\bar{v} = \bar{v}'_{nc} z_\pi / (R_s + r_b + z_\pi)$.

$$\bar{i}_{nc} = \frac{g_m \bar{v}'_{nc} Z_\pi}{R_s + r_b + Z_\pi} \tag{11.28}$$

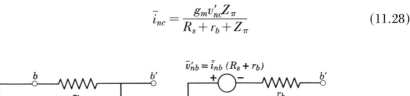

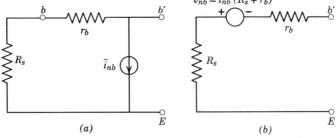

Fig. 11.17.   Equivalent circuits (a) Norton and (b) Thevenin for the base current noise component $\bar{i}_{bn}$.

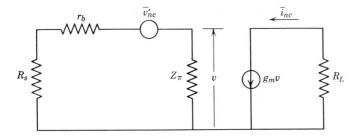

Fig. 11.18.   An equivalent circuit used to determine $\bar{v}'_{nc}$ in terms of $\bar{i}_{nc}$.

Note that Z_π is the impedance of r_π and $j\omega C_1$ in parallel. By using Eq. 11.28 to obtain $\bar{v}'_{nc}$ explicitly,

$$\bar{v}'_{nc} = \bar{i}_{nc} \frac{(R_s + r_b + Z_\pi)}{g_m Z_\pi} \tag{11.29}$$

Now, the effective noise voltage $\bar{v}_{na}$ that we are seeking is the effective sum of the three noise components we have found. But these component noise voltages are uncorrelated; that is, their phase or frequency components are unrelated. However, the noise power of each component is proportional to the square of the component voltage, and these noise powers can be added directly. Therefore

$$\bar{v}_{na}^2 = \bar{v}_{nb}^2 + (\bar{v}'_{nb})^2 + (\bar{v}'_{nc})^2$$

or

$$\bar{v}_{na}^2 = \bar{v}_{nb}^2 + \bar{i}_{bn}^2 (R_s + r_b)^2 + \frac{\bar{i}_{cn}^2 |R_s + r_b + Z_\pi|^2}{g_m^2 |Z_\pi|^2} \tag{11.30}$$

We now need to express these voltage components in terms of the base resistance r_b and the q-point diode current components I_B and I_C that cause them. Letting $\bar{v}_{na}^2 = 4kTR_a\Delta f$ and $\bar{i}_n^2 = 2qI\Delta f$, as was previously suggested,

$$4kTR_a\Delta f = 4kTr_b\Delta f + 2qI_B\Delta f (R_s + r_b)^2 + \frac{2qI_C\Delta f |R_s + r_b + Z_\pi|^2}{g_m^2 |Z_\pi|^2} \tag{11.31}$$

and, dividing through by $4kT\Delta f$,

$$R_a = r_b + \frac{q}{2kT}I_B (R_s + r_b)^2 + \frac{qI_C}{2kT} \frac{|R_s + r_b + Z_\pi|^2}{g_m^2 |Z_\pi|^2} \tag{11.32}$$

This expression for R_a can be simplified if the relationships $I_B = I_C/h_{FE}$ and $qI_C/kT = g_m$ are used. Then

$$R_a = r_b + \frac{g_m(R_s + r_b)^2}{2h_{FE}} + \frac{|R_s + r_b + Z_\pi|^2}{2g_m|Z_\pi|^2} \qquad (11.33)$$

Equation 11.33 shows that R_a, and therefore, the noise figure F, is a function of frequency because Z_π decreases as the frequency increases. We have found that $\beta_0 = g_m r_\pi$, and in general $\beta = g_m Z_\pi$; therefore, Z_π is equal to β/g_m and $h_{ie} = r_b + Z_\pi$, so that Eq. 11.33 can be written

$$R_a = r_b + \frac{g_m(R_s + r_b)^2}{2h_{FE}} + \frac{g_m(R_s + h_{ie})^2}{2\beta^2} \qquad (11.34)$$

This expression points up the importance of small I_C and, hence, small g_m with large β and large h_{FE} (dc beta) for low noise at low frequencies.

PROBLEM 11.11 A transistor with $r_b = 200\ \Omega$, $\beta = 100$, and $h_{FE} = 80$ at $I_C = 0.1$ mA is used in an amplifier with $R_s = 10$ kΩ. Determine the equivalent noise resistance R_a and the noise figure F.

Answer: $R_a = 5250\ \Omega$, $F_a = 1.52$.

Sometimes the amplifier noise is expressed as a *noise temperature T_a* instead of a noise figure. The noise temperature is the temperature which the source resistance R_s must have in order to produce the same noise as the amplifier. Then

$$(4kT_aR_s\Delta f)^{1/2} = (4k(290°)R_a\Delta f)^{1/2} \qquad (11.35)$$

and

$$T_a = 290\frac{R_a}{R_s}\ °K \qquad (11.36)$$

PROBLEM 11.12 Determine the noise temperature of the amplifier in Problem 11.11.   *Answer: $T_a = 151\,°K$.*

The noise figure increases at frequencies above f_β because the signal gain decreases while the transistor noise remains constant. This effect shows up in the last term of Eq. 11.34 because β and h_{ie} are both inversely proportional to frequency for frequencies above f_β. However, as Eq. 11.34 shows, R_a does not increase appreciably with frequency until the magnitude of h_{ie} becomes smaller than R_s.

PROBLEM 11.13 Determine the noise figure of the amplifier in Problems 11.11 and 11.12 at $f = 20 f_\beta$. *Answer: F = 2.13.*

The proper choice of q-point collector current depends on the required frequency of operation. At very small values of I_C, f_β and f_τ increase almost proportionally to collector current. Therefore, a better noise figure may be obtained in a high-frequency amplifier with values of I_C of the order of a few milliamperes, rather than microamperes. The noise temperature of a typical common-emitter amplifier is given as a function of frequency for two different values of I_C in Fig. 11.19. This figure shows that the preferable source resistance and q-point collector current is $R_s = 500\ \Omega$ and $I_C = 1.0\ \text{mA}$ at $f = 10^7$ Hz.

Manufacturers frequently give noise-figure contours similar to the one shown in Fig. 11.20 for their low-noise transistors. These contours greatly simplify the problem of selecting the best value of I_C for a fixed source resistance. For example, if $R_s = 2\ \text{k}\Omega$, a good value of I_C is 50 μA, which will yield a noise figure less than 3 dB at $f = 1$ kHz.

Field-effect transistors have unusually low noise figures because the channel resistance is the main noise source. The thermal saturation current is also a diode noise source that may be very small. Therefore, noise figures smaller than 0.5 dB may be obtained with a FET when the driving source resistance is high.

The signal-to-noise (power) ratio may be determined at both the input and the output of the amplifier providing:

1. The open-circuit signal voltage of the source and the source resistance are known.

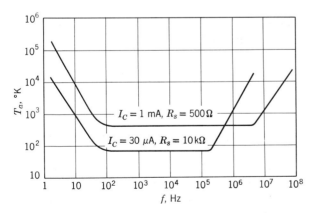

Fig. 11.19. Noise temperature as a function of frequency for a typical common-emitter amplifier at $T = 295°$K.

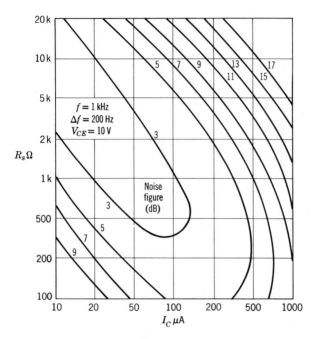

Fig. 11.20.   Constant noise-figure contours for a typical low-noise transistor (2N2443).

2. The noise figure or noise temperature of the amplifier is known.

Since the signal and source noise are in series and, therefore, work into the same impedance, the signal-to-noise ratio of the source is v_s^2/v_{ns}^2.

EXAMPLE 11.9   A microphone is rated as having a -60-dB (1 mV) open-circuit voltage, a 25-kΩ impedance (resistance), and a frequency response from 50 to 15,000 Hz (assumed noise bandwidth). Determine its signal-to-noise ratio (SNR). (Assume $T = 290°$K.) SNR $= 10^{-6}/4 \times 1.38 \times 10^{-23}(290)(2.5 \times 10^4)(1.5 \times 10^4) = 1.65 \times 10^5$. This ratio is usually expressed in dB. Thus SNR $= 10 \log (1.65 \times 10^5) = 52$ dB.

The SNR at the output of the amplifier is reduced by the noise figure F of the amplifier, since the output noise, referred to the input, is represented by $(R_s + R_a)$. Then, if $1/f$ noise is neglected, the signal-to-noise ratio of an amplifier with $F = 2$, or 3 dB, when used with the microphone of Problem 11.13 is 8.3×10^4 or 49 dB. Note that the noise figure in decibels is subtracted from the input SNR in decibels to obtain the output SNR in decibels.

PROBLEM 11.14 Determine the noise figure both as a ratio and in decibels for the transistor of Fig. 11.19 at $I_C = 30$ μA, $R_s = 10$ kΩ, and $f = 10^5$ Hz.   *Answer: $F \simeq 1.28 = 1.1$ dB.*

PROBLEM 11.15 Determine the noise temperature of the 2N2443 amplifier (Fig. 11.20) at $I_C = 50$ μA, $R_s = 500$ Ω, and $f = 1$ kHz.
Answer: $T_a = 290°$K.

PROBLEM 11.16 A given strain gauge has a 10-kΩ internal resistance and produces a 100-μV rms open-circuit signal. This signal is amplified by a bipolar silicon transistor which has $r_b = 200$ Ω, $\beta_0 = 100$, $h_{fe} = 80$, and $f_T = 5$ MHz at $I_C = 100$ μA, $V_{CE} = 10$ V. Determine the theoretical signal-to-noise ratio both into the amplifier and out of the amplifier, neglecting $1/f$ noise, if the noise bandwidth of both the source and the amplifier is 10 kHz. Assume $T = 290°$K.   *Answer: 6250 (38 dB), 4800 (36.8 dB).*

PROBLEM 11.17 A three-stage RF amplifier must have a total voltage gain of 6400 and a 200-kHz bandwidth at a center frequency of 5.0 MHz. If the transistor of Fig. 11.19 is used for each stage and all stages are identical, determine the gain and bandwidth of each stage. Which values of I_C and R_s listed in Fig. 11.19 give the best noise figure, and what is the value of this noise figure?
Answer: $K_v = 40$, $B = 392$ kHz, $F = 2.38$ (3.8 dB) approximately.

PROBLEM 11.18 A given dynamic microphone is rated -57 dB output (open circuit), 20 kΩ resistance, and 40–15 kHz frequency response. Choose a suitable transistor, or transistors, and design a low-noise amplifier that will provide approximately 1 V rms output. Determine the approximate SNR in the output of your amplifier.

PROBLEM 11.19 A given condenser microphone is rated -74 dB and has $C = 10^{-9}$ F as an internal capacitance. This microphone will provide uniform frequency response from 20 to 20,000 Hz if it feeds into an amplifier having adequate input resistance. Design an amplifier that will provide approximately 1 V rms output over the 20–20,000 Hz range when used with this microphone. You choose the transistors. Use a decoupling filter in the power supply lead if appropriate. Determine the approximate SNR in the output of your amplifier.

CHAPTER 12

POWER AMPLIFIERS

A typical amplifier consists of several stages of amplification. Most of these stages are small-signal, low-power devices. For these stages efficiency is usually unimportant, distortion is negligible, and the equivalent circuits accurately predict the amplifier behavior. In contrast, the final stage of an amplifier (and in some cases an additional driver stage) is usually required to furnish appreciable signal power to its load. Typical loads include loudspeakers, antennas, positioning devices, and so on. These are commonly called power amplifiers. Because of this relatively high-power level, the efficiency of the power amplifier is important. Also, distortion becomes a problem because the amplifier parameters vary appreciably over the signal cycle. Therefore, the equivalent circuits are only rough approximations, and graphical methods assume increased importance. Heat dissipation also becomes a problem. This chapter will discuss these problems that are peculiar to power amplifiers, namely, distortion, efficiency, push-pull configurations, and thermal conduction.

12.1 AMPLIFIER DISTORTION

The power amplifier is usually expected to deliver the maximum power for which it is capable to the load. Therefore, it may be driven from approximately cutoff on one half cycle to saturation on the next half cycle. Also, the load line will probably pass near the maximum dissipation curve, as shown in Fig. 12.1. Therefore, h_{fe} and h_{ie} may vary widely over the signal cycle. Fortunately, these two nonlinearities tend to cancel, as shown in

286

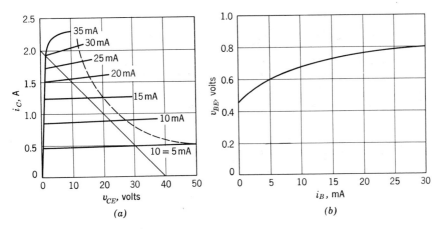

Fig. 12.1.   Characteristics of a typical silicon-power transistor.

Fig. 12.2. The upper part of this figure was constructed by plotting i_C as a function of i_B from values taken along the load line in Fig. 12.1a. The lower curve in Fig. 12.2 is the inverted input characteristic obtained directly from Fig. 12.1b.

If v_{BE} is known, the value of i_B can be determined directly from the lower curve in Fig. 12.2. However, one usually knows the waveform of the driving source voltage v_S. In this case, we note that $v_S = i_B R_s + v_{BE}$, where R_s is the internal resistance of the driving source. As we have already noted, equations of this type represent straight lines on the characteristic curves. In fact, notice that a value v_S may be obtained by drawing a line with slope R_s from a value of v_{BE} to the $i_B = 0$ axis. To help clarify this concept, let us consider an example.

EXAMPLE 12.1   The transistor whose characteristics are given in Fig. 12.2 is connected to a voltage source whose open-circuit output voltage v_S is $v_S = 0.75 + 0.25 \sin \omega t$. This voltage source has an internal impedance equal to 6.7 Ω. Let us determine the waveform of i_C for this input signal.

When $t = 0$, the signal component (the sinusoid) is zero. Thus, $v_S = 0.75$ V, which is the q-point for the transistor in this example. This voltage appears on the curve in Fig. 12.2 at point C_1. Now, v_{BE} will be less than v_S because of the drop across R_s, which is equal to $i_B R_s$. Fortunately, we have already graphically solved problems of this type when we drew load lines on the characteristic curves of diodes or transistors. In this example, the intercept with the v_{BE} axis occurs at the voltage v_S. The slope of the dashed line from point C_1 toward the input characteristic i_B versus v_{BE} curve has

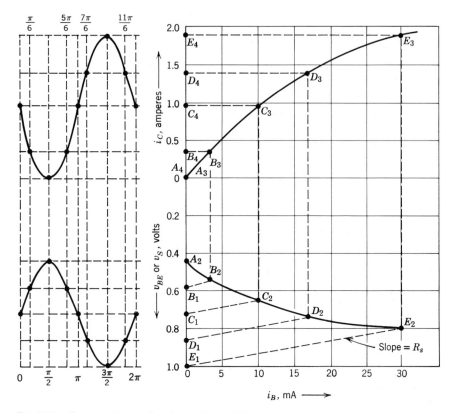

Fig. 12.2.   Construction used to determine the distortion in a transistor amplifier.

a value of $R_s = \Delta v_{BE}/\Delta i_B$, as shown in Fig. 12.2. In this example, $R_s = 6.7\ \Omega$, so that if $\Delta i_B = 15$ mA, the value of $\Delta v_{BE} = 0.015 \times 6.7 = 0.1$ V. A dashed line with this slope is drawn from point C_1 until this line intercepts the input characteristic curve at point C_2. The point C_2 represents the actual voltage v_{BE} when $v_S = 0.75$ V. The q-point value of v_{BE} is $\simeq 0.66$ V.

The value of i_B is found by projecting point C_2 up to the i_B axis. In this case, the q-point value of i_B is 10.5 mA. This same value of i_B can be projected up to the dynamic output curve (point C_3). This point C_3 determines the value of output collector current i_C when projected over to the i_C axis (point C_4). Thus, the quiescent collector current is approximately 0.97 A. Additional values of v_S are projected in this fashion until i_C can be drawn. Since R_s is a constant, all the lines from v_S to the input characteristic curve will be parallel. In this example, we have projected values of v_S for $t = 0$, $t = \pi/6$, $t = 5\pi/6$, $t = \pi$, $t = 7\pi/6$, and so forth, so that we could construct the i_C versus time curve.

PROBLEM 12.1   Repeat Example 12.1 if R_s is changed to 20 Ω.

Note from the construction in Fig. 12.2 that the output signal is not sinusoidal because the positive half cycle of collector current is smaller than the negative half cycle. If the usable portion of the input characteristic has more curvature than the transfer (i_C versus i_B) characteristic, there exists a value of source resistance R_s that will make the positive half cycles of output current equal to the negative half cycles. This equalization and, hence, minimum (or near-minimum) distortion can be obtained because the source resistance dilutes the effect of the non-linearity of the input characteristic.

The value of driving source resistance R_s, which gives minimum distortion, can be determined from the relationships shown in Fig. 12.3.

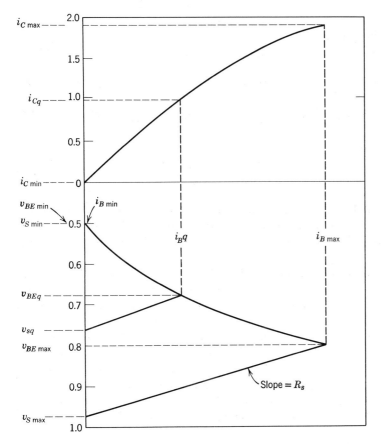

Fig. 12.3.   Arrangement for determining the optimum R_s.

First, the value of i_{Cq} is located midway between i_{Cmax} and I_{Cmin} to give equal successive half cycles of output current. Then, the corresponding values of i_{Bmax}, i_{Bq}, and i_{Bmin} are determined from the transfer curve. Then the corresponding values of v_{BEmax}, v_{BEq}, and v_{BEmin} are obtained from the input characteristics. Also, the corresponding source potentials may be expressed as follows.

$$v_{Smax} = v_{BEmax} + i_{Bmax}R_s \qquad (12.1)$$

$$V_{Sq} = v_{BEq} + i_{Bq}R_s \qquad (12.2)$$

$$v_{Smin} = v_{BEmin} + i_{Bmin}R_s \qquad (12.3)$$

But, v_S is assumed to be sinusoidal or symmetrical. Therefore

$$v_{Smax} - v_{Sq} = v_{Sq} - v_{Smin} \qquad (12.4)$$

Rearranging Eq. 12.4,

$$v_{Smax} + v_{Smin} = 2v_{Sq} \qquad (12.5)$$

Substituting the values of v_S given by Eqs. 12.1, 12.2, and 12.3 into Eq. 12.5 we obtain,

$$v_{BEmax} + v_{BEmin} + (i_{Bmax} + i_{Bmin})R_s = 2(v_{BEq} + i_{Bq}R_s) \qquad (12.6)$$

Solving for R_s

$$R_s = -\frac{v_{BEmax} + v_{BEmin} - 2v_{BEq}}{i_{Bmax} + i_{Bmin} - 2i_{Bq}} \qquad (12.7)$$

PROBLEM 12.2   Determine the value of source resistance that will give minimum (or near minimum) distortion for the amplifier with characteristics given in Figs. 12.1 or 12.2.   *Answer: $R_s = 2.5 \, \Omega$.*

The value of R_s that provides low distortion for the preceding problem is unreasonably small for most sources, such as a driver transistor. Therefore, the distortion will probably not be minimized. Consequently, we need to be able to determine the percent distortion of the amplifier when the source resistance is not optimized in order to make a judgment about the acceptability of the distortion level.

Harmonic (or waveform) distortion is analyzed by assuming the driving source to be sinusoidal and then finding the amplitude coefficients of the first few terms of the Fourier series which represents the distorted output. For example, the output current can be represented by the series

$$i_C = I_0 + I_1 \cos \omega t + I_2 \cos 2\omega t + I_3 \cos 3\omega t \ldots . \qquad (12.8)$$

The series may contain only cosine terms, as given, when the $t = 0$ axis is chosen at a point of even symmetry as shown later. The problem can be simplified further if the output distortion can be identified as being either primarily composed of even harmonics or primarily composed of odd harmonics. For example, if successive half cycles are dissimilar, as shown in Fig. 12.4a, even harmonics must be present and they usually predominate over odd harmonics which may also be present. Note that the total current in Fig. 12.4a very closely resembles the collector current waveform of Fig. 12.2, which resulted from nonoptimum source resistance. However, this waveform can be obtained by adding the dc component I_0, the fundamental component $I_1 \cos \omega t$, and the second harmonic component $I_2 \cos 2\omega t$. Thus, for this special case where successive half cycles are dissimilar, we can assume the distortion to be primarily second harmonic and write the following relationship.

$$i_C = I_0 + I_1 \cos \omega t + I_2 \cos 2\omega t \qquad (12.9)$$

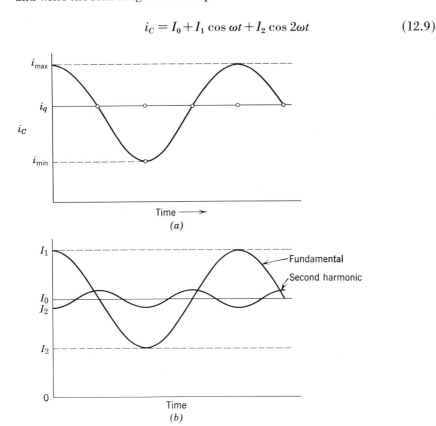

Fig. 12.4.   Second harmonic distortion. (a) Total current and (b) frequency components.

But when $t = 0$, $i_C = i_{Cmax}$. In addition, $\cos \omega t = 1$ and $\cos 2\omega t = 1$. Therefore, by using Eq. 12.9,

$$i_{Cmax} = I_0 + I_1 + I_2 \tag{12.10}$$

Also, when $\omega t = \pi/2$ or $90°$, $i_C = i_{Cq}$, $\cos \omega t = 0$, and $\cos 2\omega t = -1$. Then

$$i_{Cq} = I_0 - I_2 \tag{12.11}$$

And when $\omega t = \pi$ or $180°$, $i_C = i_{Cmin}$, $\cos \omega t = -1$, and $\cos 2\omega t = 1$. Thus

$$i_{Cmin} = I_0 - I_1 + I_2 \tag{12.12}$$

These equations can be solved simultaneously to yield I_0, I_1, and I_2 in terms of i_{Cmax}, i_{Cmin}, and i_{Cq}, which we know. For example, if Eq. 12.12 is subtracted from Eq. 12.10, I_1 is obtained.

$$I_1 = \frac{i_{Cmax} - i_{Cmin}}{2} \tag{12.13}$$

Similarly, if Eq. 12.11 is added to Eq. 12.10, and Eq. 12.13 is substituted for I_1 in the result, I_0 is obtained

$$I_0 = \frac{i_{Cmax} + i_{Cmin} + 2I_{Cq}}{4} \tag{12.14}$$

Then I_2 may be found by substituting Eq. 12.14 into Eq. 12.11,

$$I_2 = \frac{i_{Cmax} + i_{Cmin} - 2i_{Cq}}{4} \tag{12.15}$$

The percentage of second harmonic distortion in the amplifier is

$$\% \text{ second harmonic} = \frac{I_2}{I_1} \times 100 \tag{12.16}$$

PROBLEM 12.3   Determine the peak amplitude of the fundamental, the peak value of the second harmonic, and percentage of second harmonic distortion in the amplifier represented by Fig. 12.2 with the maximum level input signal given.
Answer: $I_1 = 0.97$ A, $I_2 = 0.04$ A, second harmonic $= 2.6\%$.

This value of distortion is not objectionable for many applications but is unacceptable in a high-fidelity amplifier where 1 percent distortion is usually considered to be the upper acceptable limit.

Sometimes successive half cycles of output current or voltage are similar, as they may be when optimum source resistance is used or when balanced push-pull amplifiers, which are discussed in Section 12.3, are employed. However, these half cycles are not usually sinusoidal because the peaks of the output signal tend to be flattened because of the decreased h_{fe} at the extremes of the current excursion. This type of distortion does not contain even harmonics that cause successive half cycles to be dissimilar. Figure 12.5 shows that third harmonics can cause this type of distortion. Usually the third harmonic is much larger than the higher order odd harmonics so that the collector current can be expressed in terms of the dc component, the fundamental, and the third harmonic as follows:

$$i_C = I_0 + I_1 \sin \omega t + I_3 \sin 3\omega t \qquad (12.17)$$

But when $\omega t = 0$, Fig. 12.5 and Eq. 12.17 show that

$$i_q = I_0 \qquad (12.18)$$

Also, when $\omega t = \pi/6$ the driving source voltage is one-half the maximum value, as shown in Fig. 12.2, and with $\sin \pi/6 = 0.5$ and $\sin 3\pi/6 = 1$ substituted into Eq. 12.17, this equation becomes

$$i_{\pi/6} = I_0 + 0.5I_1 + I_3 \qquad (12.19)$$

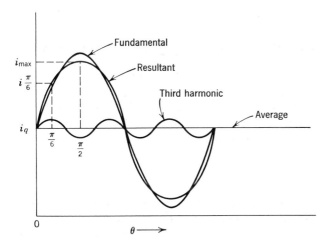

Fig. 12.5.   Typical third-harmonic distortion.

When $\omega t = \pi/2$, Fig. 12.5 and Eq. 12.17 show that

$$i_{max} = I_0 + I_1 - I_3 \qquad (12.20)$$

These equations can be solved simultaneously to obtain I_1 and I_3 in terms of i_q, $i_{\pi/6}$, and i_{max}. The results are

$$I_1 = \frac{2(i_{\pi/6} + i_{max} - 2i_q)}{3} \qquad (12.21)$$

$$I_3 = \frac{2i_{\pi/6} - i_{max} - i_q}{3} \qquad (12.22)$$

EXAMPLE 12.2 We shall determine the value of fundamental component and third harmonic component for a current that has the waveform shown in Fig. 12.3 with $i_q = 0.97$ A, $i_{max} = 1.94$ A, and $i_{\pi/6} = 1.5$ A.

The fundamental component is found from Eq. 12.21. $I_1 = 2(1.5 + 1.94 - 2 \times 0.97)/3 = 3/3 = 1$ A. Similarly, from Eq. 12.22, we find $I_3 = (2 \times 1.5 - 1.94 - 0.97)/3 = 0.03$ A. The percent third harmonic distortion is $I_3/I_1 \times 100 = (0.03/1) \times 100 = 3$ percent.

PROBLEM 12.4 Determine the value of the fundamental component, the third harmonic component, the dc component, and the percent third harmonic distortion for a current with $i_q = 500$ mA, $i_{max} = 975$ mA, and $i_{\pi/6} = 775$ mA.
Answer: $I_0 = 500$ mA, $I_1 = 500$ mA, $I_3 = 25$ mA, percent third harmonic = 5 percent.

12.2   POWER OUTPUT AND EFFICIENCY

We shall now discuss a method of determining the power output and the efficiency of an amplifier. The voltage drop and power loss in emitter circuit stabilizing resistors will be neglected. These losses may be included by increasing V_{CC} to compensate for the IR drop across R_E and then using this higher V_{CC} to calculate the power delivered by the power supply.

Figure 12.6 shows how the power output and efficiency of an amplifier can be determined from the collector characteristics. The maximum power available is of major interest and that power is obtained when the transistor is biased halfway between saturation and collector-current cutoff and is driven from saturation to cutoff. The maximum power output for a sinusoidal driving signal is then $V_c I_c$ where V_c and I_c are the rms values of

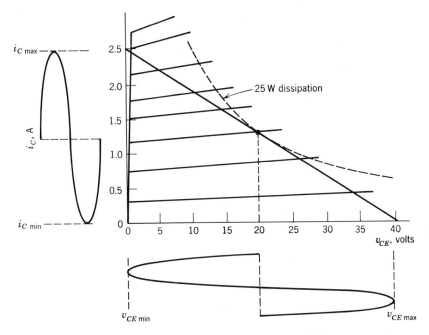

Fig. 12.6.   Current, voltage, and power relationships in an amplifier.

collector voltage and current. But these rms values can be obtained by dividing the peak-to-peak values by two, to obtain peak values, and then dividing by $\sqrt{2}$ to obtain rms values. Therefore,

$$P_o = \frac{(v_{CE\max} - v_{CE\min})(i_{C\max} - i_{C\min})}{8} \tag{12.23}$$

The amplifier of Fig. 12.6 could deliver the same signal power to the load with approximately half as much input power from the power supply if the load is transformer coupled, as shown in Fig. 12.7. The improved efficiency occurs because the dc power loss in the load is eliminated. If the ohmic resistance of the transformer primary is negligible, the dc load line is nearly vertical. The ac signal will swing as high above V_{CC} as it swings below V_{CC}. Thus, V_{CC} is now the average value of v_{CE}. An example will be used to illustrate the contrast between the amplifier with R_L in the collector circuit and a transformer-coupled amplifier.

EXAMPLE 12.3   The collector characteristics of an amplifier with a 16-Ω load is shown in Fig. 12.6. We shall determine the collector circuit

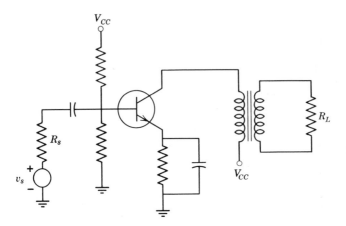

Fig. 12.7.   A transformer-coupled power amplifier.

efficiency if R_L is connected between the collector and V_{CC} and then compare this efficiency to that of a transformer-coupled amplifier (Fig. 12.7), which operates along the same load line.

From Eq. 12.23, the ac signal power is $P_o = (40 - 1)(2.45 - 0)/8 = 11.9$ W. The collector power input for the amplifier with the resistance in the collector circuit is $P_i = V_{CC}I_{C\,ave} = 40 \times 1.225 = 49$ W. The collector circuit efficiency is $\eta_c = (P_o/P_i)\,100 = (11.9/49)\,100 = 24$ percent.

When the transformer-coupled amplifier (Fig. 12.7) is considered, V_{CE} will swing as high above V_{CC} as it swings below V_{CC}. Then, $V_{CC} = (V_{CEmax} + V_{CEmin})/2 = (40 + 1)/2 = 20.5$ V. The ac power output is the same as above. However, the power input is now $P_i = V_{CC}I_{C\,ave} = 20.5 \times 1.225 = 25.1$ W. The collector circuit efficiency is now $\eta_c = (11.9/25.1) = 47.4$ percent, which is essentially twice as great.

The power output would be increased and, hence, the efficiency increased if the collector voltage v_{CE} could be driven to zero on one half cycle and the collector current i_C could be driven to zero on the next half cycle. Then, using Fig. 12.7 and Eq. 12.23,

$$P_{o\max} = \frac{2V_{CC}2I_{Bq}}{8} = \frac{V_{CC}I_{Cq}}{2} \tag{12.24}$$

The maximum efficiency is then

$$\text{max eff} = \frac{V_{CC}I_{Cq}/2}{V_{CC}I_{Cq}} \times 100 = 50 \text{ percent} \tag{12.25}$$

This is the maximum theoretical efficiency of a Class A amplifier with sinusoidal input. The efficiency can approach 100 percent if the input signal is a square wave. A Class A amplifier is defined as an amplifier in which collector current flows during the entire cycle. We have considered only Class A amplifiers to this point in our work.

The Class A amplifier is inefficient because of the relatively large q-point current that flows through the amplifier all the time. This q-point current must be equal to the peak signal current. Otherwise, the collector current is cut off during part of the signal cycle and serious distortion results. Notice that the collector power input to a Class A amplifier is essentially constant. However, the signal output power increases with the signal level. The collector dissipation is the collector power input minus the signal power output. Consequently, the collector power dissipation decreases and the collector efficiency increases as the signal level increases.

PROBLEM 12.5   The transistor whose characteristics are given in Fig. 12.6 is to be operated with $V_{CC} = 40$ V and R_L (connected from collector to V_{CC}) = 20 Ω. Determine the maximum value of ac power output, collector power input, and collector efficiency.

Answer: $P_o \simeq 9.7$ W, $P_i = 39.2$ W, $\eta_c = 24.7$ percent.

PROBLEM 12.6   A circuit is connected as shown in Fig. 12.7. The characteristics of the transistor are given in Fig. 12.6. The maximum collector voltage is 40 V and the ac impedance of the transformer primary is 20 Ω. Determine the maximum value of ac power output, collector power input, and collector efficiency.

Answer: $P_o \simeq 9.7$ W, $P_i \simeq 20$ W, $\eta_c \simeq 48.5$ percent.

12.3   PUSH-PULL AMPLIFIERS

A high efficiency amplifier can be built by using two transistors in an arrangement known as push-pull (Fig. 12.8). This amplifier is similar to the balanced amplifier discussed in Chapter 10 except transformer coupling is used to provide balanced signals of opposite polarity to the two bases. Transformer coupling is also used to add the balanced outputs of the two transistors in a single load R_L. The push-pull amplifier can be operated Class A and will provide about twice the power output obtainable from a single transistor under similar circumstances. Of course, since the total collector input power also doubles (two stages), the collector efficiency remains the same. The distortion of the push-pull connection is lower, however, because the even harmonic distortion is cancelled as a result of

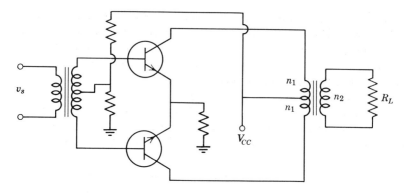

Fig. 12.8. A transformer-coupled push-pull amplifier.

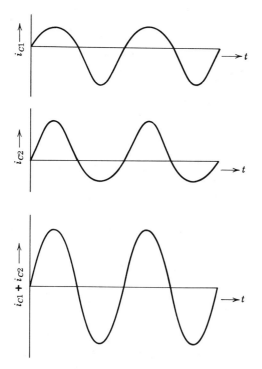

Fig. 12.9.   Illustration of the cancellation of even harmonics in the output of a balanced push-pull amplifier.

the balanced arrangement. The positive half cycle of one transistor adds to the negative half cycle of the other transistor to provide an output with similar successive half cycles (Fig. 12.9).

The main advantage of a push-pull amplifier results from the fact that the q-point currents of the individual transistors can be drastically reduced and the efficiency, therefore, markedly increased. In fact, each transistor can be biased at very nearly cutoff, so that each transistor delivers power to the load only during one-half of the signal cycle. Collector current flows approximately half the time in each transistor. This type of operation is known as *Class B*.

When each transistor is biased precisely at cutoff, distortion occurs in the output because h_{fe} decreases rapidly with collector current at very small values of collector current. This type of distortion is known as *crossover distortion* and is illustrated in Fig. 12.10a. In this figure, the transfer characteristic of one transistor is inverted and reversed as compared to the other transistor so the sum of the two curves is the total transfer curve of the push-pull amplifier. Most of the nonlinearity of the total transfer curve occurs at small values of i_C, as shown.

The crossover distortion can be eliminated by providing a small forward-bias base current, as shown in Fig. 12.10b. The minimum value of bias is determined by extending the relatively straight portion of the transfer

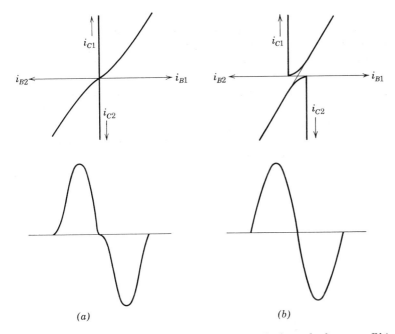

(a)   (b)

Fig. 12.10.   Illustration of the crossover distortion which results from cutoff bias and the reduction of distortion by the use of projected-cutoff bias. (a) Crossover distortion and (b) projected-cutoff bias.

curve of one transistor until the extension crosses the $i_C = 0$ axis. The value of i_B found at this intersection is known as *projected-cutoff bias*. If the transfer curve for the other transistor is inverted and reversed, and the projected transfer curves are made to coincide at the projected cutoff bias value of i_B, the total transfer curve is essentially straight and little distortion occurs. Since the magnetic flux in the output transformer that results from the dc components of collector currents tend to cancel, the net flux, which is effective in producing voltage in the secondary, is proportional to the difference between the individual collector currents. Therefore, the total or *composite* transfer characteristic is the difference between the individual transfer characteristics at any point.

PROBLEM 12.7   The transfer characteristic for a 2N2147 power transistor is given in Fig. 12.11. Determine the projected-cutoff bias for this transistor.   *Answer:* $I_B \simeq 2$ mA.

A set of composite (or total) collector characteristics can be drawn by inverting one set and reversing it as compared with the other and vertically aligning the individual q-points. However, the bottom half of such a composite set contains the same information as the top half, so only the top half, as shown in Fig. 12.12, is needed. The composite (total) load line passes through the composite q-point, which is at $i_C = 0$ and $v_{CE} = V_{CC}$. The operation of the individual transistor follows the composite load line except during the time when both transistors are conducting. Then the total collector current is the difference between the individual collector currents. Since each transistor works into one-half of the output transformer primary, the composite load line represents the impedance of only one-half of the primary, which is one-fourth of the total primary impedance.

The maximum permissible collector current and the maximum power output for the Class B push-pull amplifier can be determined as functions

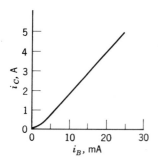

Fig. 12.11.   The transfer characteristic of a typical power transistor.

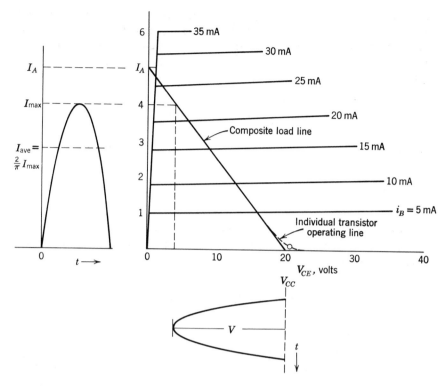

Fig. 12.12.   Basic relationships in a Class B push-pull amplifier.

of the permissible transistor dissipation with the aid of Fig. 12.12. The signal is assumed to be sinusoidal. The total transistor power dissipation is

$$P_d = P_i - P_o \qquad (12.26)$$

where P_i is the collector power input from the power supply and P_o is the signal power output, as was previously discussed. But, assuming cutoff bias,

$$P_i = I_{ave}V_{CC} = \frac{2}{\pi}I_{max}V_{CC} \qquad (12.27)$$

and

$$P_o = \frac{I_{max}^2}{2}R_L \qquad (12.28)$$

Then, substituting Eq. 12.28 and Eq. 12.27 into Eq. 12.26

$$P_d = \frac{2}{\pi}I_{max}V_{CC} - \frac{I_{max}^2}{2}R_L \qquad (12.29)$$

We need to find the value of I_{max} that will give maximum power dissipation $P_{d\,max}$. This can be done by differentiating P_d with respect to I_{max} and equating this derivative to zero. If you do not understand the process of differentiation (differential calculus), the result given in Eq. 12.31 will still serve your needs.

$$\frac{dP_d}{dI_{max}} = \frac{2}{\pi}V_{CC} - I_{max}R_L = 0 \qquad (12.30)$$

and, solving for the I_{max} that gives maximum power dissipation

$$I_{max} = \frac{2}{\pi}\frac{V_{CC}}{R_L} = \frac{2}{\pi}I_A \qquad (12.31)$$

Where I_A is the current axis intercept (V_{CC}/R_L) of the load line as shown in Fig. 12.12. Substituting this value of I_{max} (Eq. 12.31) into Eq. 12.29, the maximum power dissipation can be found in terms of I_A and V_{CC}.

$$P_{dmax} = \frac{4}{\pi^2}I_A V_{CC} - \frac{2}{\pi^2}I_A{}^2 R_L \qquad (12.32)$$

But $I_A R_L = V_{CC}$, so

$$P_{dmax} = \frac{4}{\pi^2}I_A V_{CC} - \frac{2}{\pi^2}I_A V_{CC} = \frac{2}{\pi^2}I_A V_{CC} \qquad (12.33)$$

However, the power dissipation capabilities of the two transistors P_d is usually known or obtainable and the maximum value of I_A can be found from Eq. 12.33.

$$I_A max = \frac{\pi^2}{2}\frac{P_d}{V_{CC}} \simeq \frac{5P_d}{V_{CC}} \qquad (12.34)$$

A method of determining P_d will be discussed in Section 12.5. The intercept current I_A must be reduced as the q-point collector current I_{Cq} is increased above cutoff. An empirical alteration of Eq. 12.34 takes the q-point current of the individual transistor into account

$$I_A max \simeq \frac{5P_d}{V_{CC}} - 3I_{Cq} \qquad (12.35)$$

The minimum load resistance, maximum power output, and maximum efficiency can now be determined if V_{CC} and P_d are known. We shall illustrate this idea by an example.

EXAMPLE 12.4 Consider an amplifier with characteristics given in Fig. 12.12 and $V_{CC} = 20$ V. Assume that $P_d = 20$ W total for the two transistors and $I_{Cq} = 100$ mA. Then, using Eq. 12.35, $I_A = 5 - 0.3 = 4.7$ A. The minimum value of load resistance is $R_L = V_{CC}/I_A = 20/4.7 = 4.25 \ \Omega$. The maximum power output, assuming the saturation voltage is negligible, is $P_o = V_{CC}I_A/2 = 47$ W. At this power output, the collector power input is $P_i = I_{ave}V_{CC} = [I_A(2/\pi) + I_{Cq}]V_{CC} = (4.7 \times 0.636 + 0.1)20 = 62$ W. The maximum collector circuit efficiency is $(47/63)100 = 75$ percent.

Figure 12.13a shows the power output, power dissipation and percent efficiency of a Class B amplifier as a function of peak output signal amplitude I_{max}. The Class B amplifier is assumed to be biased at cutoff. The power output, power dissipation, and efficiency of a Class A amplifier, either push-pull or single ended, is given as a function of I_{max} in Fig. 12.13b. Both amplifiers have the same power dissipation capabilities P_{dmax}. Note the increased power output capability of the Class B amplifier as compared to the Class A. For example, if the power dissipation capability is 10 W in either amplifier, the maximum power output from the Class A amplifier is 5 W and the maximum power output from the Class B amplifier is 25 W, neglecting the saturation voltage.

The balanced source voltage can be obtained from a phase *inverter* or *splitter* instead of a transformer with some saving in cost and weight. A typical phase inverter circuit is shown in Fig. 12.14. Since the currents through resistors R_{L1} and R_{L2} are essentially equal, the voltages applied to the push-pull bases will be essentially equal but of opposite polarity, providing $R_{L1} = R_{L2}$. Since R_{L1} is in the emitter circuit of the phase

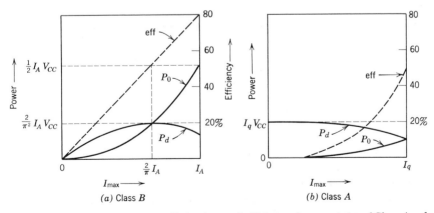

Fig. 12.13. Power output, power dissipation, and efficiency characteristics of Class A and Class B amplifiers.

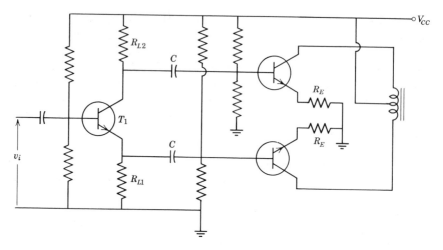

Fig. 12.14. A typical phase-inverter circuit.

inverter, the voltage gain from the input v_i to either base is approximately one. Distortion will occur in the phase inverter if the resistance of R_{L1} and R_{L2} is not small in comparison with the input resistance of the push-pull transistors, because this input resistance varies over the signal cycle and tends to distort the current waveform in the phase inverter. The unbypassed emitter resistor R_E in the push-pull amplifier tends to maintain a high and fairly constant input impedance to this amplifier.

PROBLEM 12.8 Two transistors with a total power dissipation capability of 15 W are used in a Class B push-pull amplifier with $I_{Cq} = 50$ mA and $V_{CC} = 25$ V. Determine the minimum permissible value of ac load resistance and the maximum power output, assuming sinusoidal signals, for this amplifier. Neglect the transistor saturation voltage.

Answer: $R_L = 8.4 \ \Omega, w = 35.6$ W.

12.4 COMPLEMENTARY-SYMMETRY AMPLIFIERS

Both the input and output transformers can be eliminated in a push-pull amplifier type known as a complementary-symmetry amplifier. This amplifier uses one n-p-n and one p-n-p transistor with similar characteristics, as shown in Fig. 12.15. The bias resistors R_1, R_2, R_3, and R_4 provide the desired forward bias, usually projected cutoff bias, for the transistors. When the input signal v_i goes through a positive half cycle the emitter junction of the n-p-n transistor T_1 is forward biased and the

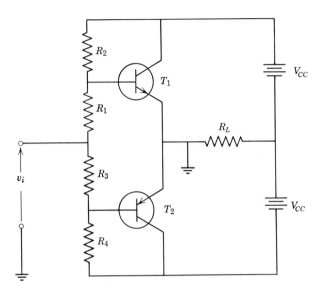

Fig. 12.15.   A complementary symmetry push-pull amplifier.

emitter junction of the *p-n-p* transistor is reverse biased, causing the collector current of T_1 to pass through the load resistor R_L from left to right. During the next half cycle when the input voltage is negative, the *p-n-p* transistor is forward biased and the *n-p-n* transistor is cut off so that the collector current of T_2 flows through the load resistor R_L from right to left. When the signal current is zero, the two transistor currents are nearly equal, so essentially no direct current flows through the load resistor. Notice that this amplifier, like any push-pull amplifier can be operated Class *A* or at any *q*-point between Class *A* and Class *B*. Operation at *q*-points between Class *A* and Class *B* is known as Class *AB*. However, high power amplifiers are usually operated at projected cutoff Class *B* because of the high efficiency of this mode.

The circuit of Fig. 12.15 has a serious limitation. That is, the power supply that is represented by two batteries does not have a ground point. Therefore, the output signal voltage appears between the power supply and ground, and this power supply would not be usable for other amplifiers in the system. The shunt capacitance between the power supply and ground or chassis would also limit the high-frequency response of the amplifier.

The emitter follower version of the complimentary-symmetry amplifier permits the grounding of the power supply as shown in Fig. 12.16. However, as compared with the common-emitter configuration, the input impedance

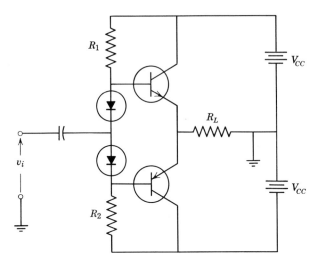

Fig. 12.16. A common-collector configuration of the complementary symmetry amplifier.

is high and the required input voltage v_i is large. The forward bias for the transistors in Fig. 12.16 is provided by the two forward biased diodes. The low dynamic resistance of these diodes and their negative temperature coefficients provide improved thermal stability of the transistors as compared with the circuit of Fig. 12.15. A small resistance can be connected in series with the diodes to provide precise adjustment of the q-point.

The driver transistor for the complementary-symmetry amplifier can be directly coupled (Fig. 12.17). Let us consider the design of a typical amplifier of this type to improve our understanding of its operation.

EXAMPLE 12.5   We assume that V_{CC} (each supply in Fig. 12.17) $= 15$ V and $R_L = 100 \ \Omega$. Then the current axis intercept I_A for either T_1 or T_2 is $I_A = 15$ V/100 $\Omega = 150$ mA. From Eq. 12.34, using projected cutoff bias, the power dissipation requirement for the two transistors T_1 and T_2 is $I_A V_{CC}/5 = 450$ mW. The silicon complementary pair 2N3703 and 2N3705 have adequate current capability, and power dissipation and voltage ratings for this application and will, therefore, be used. These transistors have $\beta_0 \simeq 100$ over the collector current range 10 mA to 150 mA. Projected cutoff bias is approximately $I_C = 1$ mA and $I_B = 15$ μA. Therefore, the forward biasing diodes, with perhaps some additional resistance, should be chosen so that these q-point values are obtained. We should expect the diodes to be low-current silicon diodes to match the characteristics of the emitter junctions of the transistors.

When the input signal v_i is positive, the collector of the driver T_3 goes

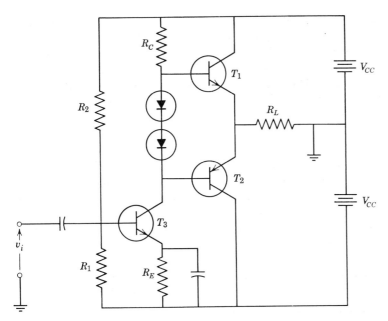

Fig. 12.17.   A complementary symmetry amplifier with a direct-coupled driver.

negative, so that the transistor T_1 is cut off and the emitter end of the load resistor has approximately the same potential as the collector of the driver. When the driver transistor T_3 is driven to saturation, the potential of the emitter end of the load resistor R_L differs from the potential of the negative end of the lower V_{CC} by the drop across R_E, plus the saturation voltage of transistor T_1, plus the voltage V_{BE} of transistor T_2. Therefore, the voltage $I_E R_E$ of the driver transistor should be fairly small; we shall use 1 V.

When the input signal goes negative, the collector of the driver transistor T_3 goes positive, and transistor T_1 conducts and transistor T_2 is cut off. Then the emitter end of the load resistor approaches the potential of the positive end of the upper V_{CC}. However, both the collector current of T_3 and the base current of T_1 must flow through the collector load resistor R_C. Therefore, at the peak negative input voltage v_i the transistor T_3 is cut off and the maximum-signal base current of transistor T_1 must flow through R_C. Thus, the difference between the maximum-positive load potential and the positive power supply potential is the voltage $i_{Bmax} R_C$ plus the voltage V_{BE} of transistor T_1. So the voltage $i_{Bmax} R_C$ should not exceed a volt or so. We shall allow 1.5 V. Then with i_C (of T_1) max $= 150$ mA and $\beta_0 = 100$, $i_{Bmax} = 1.5$ mA and $R_C = 1.5$ V/1.5 mA $= 1$ kΩ.

We can now design the driver stage with $R_C = 1$ kΩ. At quiescent conditions, the voltage drop across the driver load R_C is approximately $V_{CC} = 15$ V. Therefore, the q-point collector current is 15 mA and the value of R_E required for 1 V drop is $1/0.015 = 67\,\Omega$. The driver operates as a Class A amplifier with $V_{CC} = 30$ V and maximum dissipation $P_d = 15$ V (15 mA) $= 225$ mW. The 2N3704 transistor, which is similar to the 2N3705 except $\beta_0 \simeq 200$, has adequate voltage and dissipation ratings for this application. The bias resistors R_1 and R_2 can be determined by the techniques discussed in Chapter 5. We assume that a current stability factor of 25 is adequate for this silicon transistor. Then $R_b = 25 \times 67 = 1.7\,\text{k}\Omega$, $V_B = I_B R_b + V_{BE} - I_E R_E = 1.63V$, $R_1 = (30/28.37)1.7\,\text{k}\Omega = 1.8\,\text{k}\Omega$ and $R_2 = (30/1.63)1.7\,\text{k}\Omega = 32\,\text{k}\Omega$. The coupling capacitor C and the emitter bypass capacitor C_E can be determined when the driving source resistance is known. The bypass capacitor C_E is frequently omitted to provide higher input resistance and better linearity (less distortion) at the expense of lower power gain.

You may have observed that the power dissipation rating of the Class A driver must be about the same as either of the Class B pair. Therefore, a

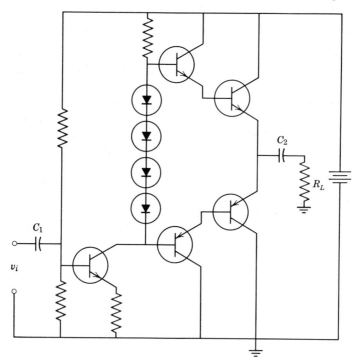

Fig. 12.18.   A high-power complementary symmetry amplifier using Darlington intermediate amplifiers.

high-power amplifier will require a high-power driver unless a Class B intermediate amplifier is used as shown in Fig. 12.18. This figure also shows that a single power supply can be used if the load resistor R_L is capacitively coupled to the amplifier. The amplifier design can proceed in the same manner as was previously discussed, except the effective β of the Darlington pair is treated as $\beta_1\beta_2$. The capacitor C_2 is calculated in the usual manner. Since the output resistance of the (emitter-follower) amplifier is very low, the coupling capacitance can be determined from the values of f_l and R_L. For example, if $R_L = 10\ \Omega$ and $f_l = 32$ Hz, determined by C, $C = 1/\omega_l R_L = 500\ \mu\text{F}$.

The availability of complementary-symmetry pairs of transistors is limited, especially in the high-power types. This problem is resolved by using the *quasi-complementary symmetry* circuit of Fig. 12.19. Observe that this amplifier has two transistors of the same type in the output. However, the lower transistor is driven by the collector current of the driver amplifier instead of its emitter current. In other words, the lower amplifier consists of a *pnp–npn* combination instead of a Darlington connection. The polarity reversal in this arrangement permits the transistor T_2 to be an *n-p-n* instead of a *p-n-p* type. At first glance one might suspect that the input resistance of the driver T_4 is much lower than the

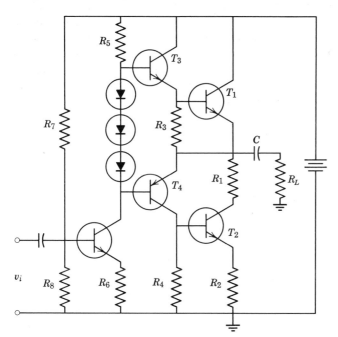

Fig. 12.19. A quasi-complementary symmetry amplifier.

input resistance of T_3. This is not true because the output voltage $\Delta i_{C2} R_L$ is between the emitter of T_4 and ground, and the base-to-ground signal voltage v_{B4} of transistor T_4 is essentially equal to the output voltage. Therefore,

$$R_{\text{in}} = \frac{\Delta v_{B4}}{\Delta i_{B4}} \simeq \frac{\Delta i_{C2} R_L}{i_{B4}} = \beta_2 \beta_4 R_L \qquad (12.36)$$

Thus, the input resistance of transistor T_4, with $R_L = 10\ \Omega$, $\beta_2 = 100$, and $\beta_4 = 200$ is $R_{\text{in}} = 200{,}000\ \Omega$. The input resistance of transistor T_3 is also approximately $200{,}000\ \Omega$.

The resistors R_1, R_2, R_3, and R_4 (Fig. 12.19) are needed to provide adequate thermal stability when germanium transistors are used. The resistors R_1 and R_2 should be small in comparison with R_L to avoid serious loss of output power. For example, if $R_L = 8\ \Omega$, R_1 and R_2 should not exceed about $1\ \Omega$. The resistors R_3 and R_4 should be of the same order of magnitude as the input resistance of the power transistors T_1 and T_2, as was discussed in the Darlington amplifier section of Chapter 10. Thus, these resistors should be about $100\ \Omega$ in a high-power amplifier. Note that the input resistance of T_2, as seen from the viewpoint of R_4, is $h_{ie2} + (\beta_2 + 1)R_2$. Although h_{ie2} varies widely over the signal cycle, it is usually small compared to $(\beta + 1)R_2$. The same relationship holds for the input resistance of transistor T_1, as seen from the viewpoint of R_3. Notice that the output resistances of transistors T_3 and T_4 are in parallel with R_3 and R_4, respectively. The diodes between the bases of T_3 and T_4 cause this output resistance to be quite low, as long as emitter current flows in transistors T_3 and T_4. Thus, the current stability factors of transistors T_1 and T_2 are much lower than the ratio R_3/R_1 or R_4/R_2 when current flows in transistors T_3 and T_4. The stabilizing resistor may be omitted when silicon transistors are used. Also, diodes, instead of resistors, may be used for thermal stabilization.

The complementary-symmetry amplifier can be arranged in a common-emitter configuration by using two pnp–npn combinations, as shown in Fig. 12.20. The capacitor C_E maintains the emitters of transistor T_3 and T_4 at ac ground potential and the resistors R_3 and R_4 compensate for differences between transistors T_3 and T_4, and transistors T_1 and T_2. Either R_3 or R_4 can be adjusted until the V_{CE} of transistor T_1 is equal to the V_{CE} of transistor T_2. Lack of this balance in the output transistors decreases the available power output. The available power output from this configuration is somewhat higher than the common-collector configurations previously considered because the peak-to-peak output voltage is equal to the total power supply voltage minus the saturation voltages of transistors T_1 and T_2. These saturation voltages may be only a few tenths of a volt.

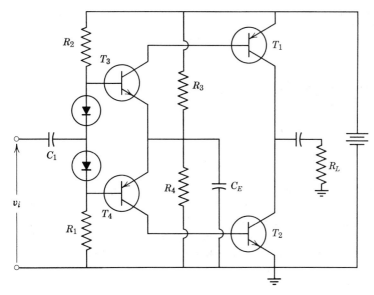

Fig. 12.20.   A common-emitter complementary symmetry amplifier.

The power gain of the common-emitter configuration is also higher and, therefore, the required voltage and power gain can be obtained with fewer transistors than the common-collector configurations. The current through resistors R_1 and R_2 should be several times the maximum base current of transistors T_3 and T_4. Projected cutoff bias is desirable.

PROBLEM 12.9   If the β_0 of all transistors in Fig. 12.20 are approximately 100, $R_L = 8\,\Omega$, and the power supply voltage = 40 V, determine R_1 and R_2 values so that the current through them is 3 times i_{b3} max; determine, also, R_3 and R_4 so that their current is $0.3i_{e3}$ max.

Answer: $R_1 = R_2 = 27\,\text{k}\Omega, R_3 = R_4 = 2.7\,\text{k}\Omega.$

12.5   THERMAL CONDUCTION AND THERMAL RUNAWAY

As was noted in Chapter 5, small-signal transistors rely on air convection and connecting lead conduction to remove heat from the transistor to the surrounding atmosphere. Therefore, the heat dissipation rating is based on the ambient temperature, usually 25°C, and this rating must be reduced as the ambient temperature is increased above the reference temperature. You may recall that the reciprocal of the derating factor is known as thermal resistance.

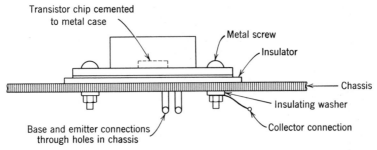

Fig. 12.21.   Heat sink arrangement for a typical power transistor.

Power transistors (above about 1 W) are usually fastened securely to a large metal plate or chassis that serves as a heat sink. The collector of the power transistor is usually connected electrically and mechanically to the transistor case. Maximum thermal conductivity is obtained when the transistor case is electrically and mechanically fastened to the heat sink. However, an insulator is usually required to electrically isolate the collector from the heat sink (Fig. 12.21).

Every transistor has a maximum permissible junction temperature $T_{j\mathrm{max}}$, which ranges from about 85°C to 110°C for germanium and about 130°C to 175°C for silicon. Thermal resistance θ_T is defined as the ratio of temperature rise in degrees centigrade (or Kelvin) to the power conducted in watts. Therefore, the junction temperature can be related to the power being dissipated P_d, the thermal resistance θ_T, and the ambient temperature T_a by the following equation, previously given in Chapter 5.

$$T_j = \theta_T P_d + T_a \tag{12.37}$$

The thermal resistance θ_T consists of three parts:

1. The thermal resistance between the collector junction and the transistor case θ_{jc}.
2. The thermal resistance between the transistor case and the heat sink θ_{cs}.
3. The thermal resistance between the heat sink and the ambient surroundings θ_{sa}.

Heat conduction is comparable to electrical conduction. Thus, the equivalent circuit of Fig. 12.22 represents the *thermal relationships* of the transistor and its heat sink. The capacitors shown represent the thermal capacitance of the three parts of the circuit. The thermal capacitance C_j of the collector junction is very small because the mass of the transistor chip and hence its heat capacity is very small. Therefore, the

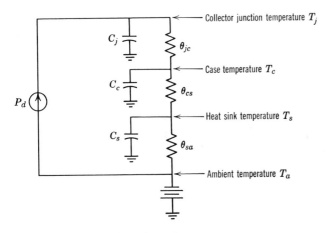

Fig. 12.22.   An equivalent thermal circuit.

thermal time constant $\theta_{jc}C_j$ is of the order of milliseconds. The significance of this time constant is that the load line can pass through the area above the maximum dissipation curve *providing the excessive dissipation does not continue for more than a few milliseconds.* This situation may occur in a Class B or AB amplifier. The thermal capacitance C_c of the case is much greater than that of the junction. Therefore, a transistor amplifier that is designed to operate with a heat sink can operate at least several seconds without the heat sink before the transistor is damaged. Similarly, the thermal capacity of the heat sink is usually much greater than that of the transistor case, so that the amplifier may operate for several minutes with an inadequate heat sink.

The thermal resistance θ_{jc} from the collector junction to the case is usually given by the manufacturer for each type of power transistor. Also thermal resistance θ_{cs} data are usually available for the various transistor mounting systems.[1] The thermal resistance θ_{sa} of a $\frac{1}{8}$-in. thick sheet of bright aluminum is given as a function of area (both sides) in Fig. 12.23. When the heat sink is horizontal, the thermal resistance increases by about 10 percent because of the reduced convection. On the other hand, black painting or amodizing of the heat sink will lower its thermal resistance because of increased heat radiation.

We can now determine the dissipation capability of a power transistor with a specific heat sink system by applying "Ohm's law" for thermal circuits wherein power dissipation P_d is comparable to current, temperature difference is analogous to potential difference, or voltage, and thermal

[1]See *Motorola Power Transistor Handbook*, first edition, Motorola Semiconductor Products Division, Phoenix, Arizona, p. 23.

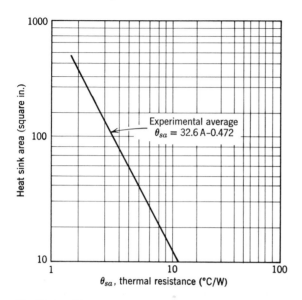

Fig. 12.23. Heat sink area versus thermal resistance. Courtesy Motorola Semiconductor Products Division. Units mounted in the center of square sheets of ⅛-inch thick bright aluminum. Heat sinks were held vertically in still air. (Heat sink area is twice the area of one side.)

resistance compares with electrical resistance, of course. An example will be used to illustrate this procedure.

EXAMPLE 12.6   Let us consider a transistor with $\theta_{jc} = 1°C/W$ mounted on a ⅛-in. aluminum 10-in. by 10-in. sheet. A mica washer coated with silicone grease is used to insulate the transistor from the chassis. The thermal resistance of this washer is 0.5°C/W. The thermal resistance of the aluminum heat sink is $\theta_{sa} = 2.5°C/W$, as seen in Fig. 12.23, if mounted vertically and 2.75°C/W if mounted horizontally. The total thermal resistance from junction to ambient is $\theta_T = 1.0 + 0.5 + 2.5 = 4°C/W$ with vertical heat sink. If the maximum junction temperature of the transistor is 175°C and the ambient temperature is 25°C, the maximum transistor dissipation is $P_{d\,max} = (175 - 25)/4 = 37.5$ W. However, if the maximum expected ambient temperature is 75°C, the maximum safe transistor dissipation is $P_{d\,max} = (175 - 75)/4 = 25$ W. The maximum power output from this transistor in a Class A amplifier with $T_a = 75°C$ is $P_o = 25/2 = 12.5$ W. However, if two of these transistors are mounted on the same heat sink with the same type washers, the total thermal resistance is $\theta_T = (1.5/2) + 2.5 = 3.25°C/W$. The maximum dissipation of both transistors at $T_a = 75°C$ is $P_{d\,max} = 100/3.25 \simeq 30$ W. The maximum sinusoidal power output from these two transistors when operated Class B at 75°C ambient temperature is $P_o = 2.5 \times 30 = 75$ W, as can be noted in Fig. 12.13a.

A power transistor is often given a power dissipation rating with the case temperature held at 25°C. For example, the transistor in the example above can dissipate $(175 - 25)/1 = 150$ W at 25°C case temperature. This case temperature can be maintained by cooling the case with circulating water. Normally, this rating is intended to indicate only the thermal resistance between junction and case. For example, if a transistor is rated as having $T_{j\,\text{max}} = 100$°C and $P_{d\,\text{max}} = 150$ W at 25°C case temperature, we may calculate that $\theta_{jc} = (100 - 25)/150 = 0.5$°C/W.

The transistor is not completely protected when the heat sink is adequate for the required dissipation; the transistor can be destroyed by *thermal runaway*. This thermal runaway results from the increasing collector current caused by the increasing I_{CO}. You may recall from Chapter 5 that $\Delta I_C = S_I \Delta I_{CO}$, where S_I is the current stability factor. We were not concerned with thermal runaway in the RC-coupled amplifier because the dc load line did not cross the maximum dissipation curve so the transistor could not be destroyed by heat. In fact, the power dissipation decreases with increasing collector current when the q-point is to the left of the load line center. However, in a transformer-coupled or complementary-symmetry amplifier the dc load line is almost vertical, as shown in Fig. 12.24.

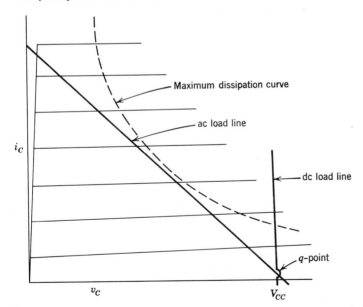

Fig. 12.24. Illustration of the susceptibility of a transformer-coupled amplifier or complementary symmetry amplifier to thermal runaway.

Therefore, the dc collector voltage V_{CE} is almost independent of the collector current at the value V_{CC} and the power dissipation is essentially proportional to the average collector current.

Thermal runaway occurs when the rate of increase of collector junction temperature exceeds the ability of the heat sink to remove the heat. The portion of heat dissipation that results from I_{CO} is

$$P_d = S_I I_{CO} V_{CC} \qquad (12.38)$$

Then the rate of increase of this power dissipation with time is (if you do not understand differential calculus, you can skip the following section to Eq. 12.43 and accept that equation on faith).

$$\frac{dP_d}{dt} = S_I V_{CC} \frac{dI_{CO}}{dt} \qquad (12.39)$$

The rate of heat conduction away from the collector junction to the ambient surroundings is the reciprocal of the thermal resistance, or $1/\theta_T$. Thus, in order to prevent a continual buildup of heat, the following inequality must hold.

$$\frac{1}{\theta_T} > S_I V_{CC} \frac{dI_{CO}}{dt} \qquad (12.40)$$

If I_{CO} doubles for each 10°C temperature increment, the I_{CO} increases by about 7 percent per °C. Then

$$\frac{dI_{CO}}{dt} = 0.07 I_{CO} \qquad (12.41)$$

and

$$\frac{1}{\theta_T} > S_I V_{CC}(0.07 I_{CO}) \qquad (12.42)$$

Note that the current stability factor S_I must be controlled to prevent thermal runaway. By using Eq. 12.42 to obtain S_I explicitly,

$$S_I < \frac{14.3}{\theta_T V_{CC} I_{CO}} \qquad (12.43)$$

The thermal current I_{CO} must be determined at the maximum expected junction temperature.

EXAMPLE 12.7   Let us consider a Class B amplifier with $V_{CC} = 16$ V, $R_L = 8 \, \Omega$, and $I_{co} = 0.1$ mA at 25°C and doubles for each 10°C increase. Then $I_A = 16/8 = 2.0$ A and the maximum dissipation occurs when $I_{c\,max} = 0.636 I_A = 1.27$ A and $I_{ave} = 0.636(1.27) = 0.81$ A. Therefore, the maximum power dissipation is $V_{CC} I_{ave} = 16 \times 0.81 = 13$ W. We assume that θ_T is 4.0°C/W, as in the preceding example, and the maximum ambient temperature is 40°C. Then, from Eq. 12.35, $T_{j\,max} = 13 \times 4 + 40 = 92$°C and I_{co} at this temperature is

$$I_{co} = 0.1 \text{ mA}(2)^{92-25/10} = 0.1 \text{ mA}(2)^{6.7} \simeq 10 \text{ mA} \qquad (12.44)$$

Thermal stability factor must be less than $S_I = 14.3/(4 \times 16 \times 0.01) = 22.3$ to insure that thermal runaway will not occur.

PROBLEM 12.10   If the supply voltage V_{CC} is raised to 20 V and the thermal resistance θ_T is reduced to 3°C/W, all other conditions remaining the same as given in Example 12.7, determine the maximum value of the stability factor S_I to insure thermal stability.   *Answer: $S_I < 12.5$.*

PROBLEM 12.11   Design a transistor power amplifier using an emitter follower type complementary-symmetry arrangement that will provide full power output to an 8-Ω speaker. The driving source is a radio tuner with a 1.0-V rms output voltage, open circuit, and a 1.0-kΩ internal resistance R_g. The total power supply voltage is 40 V. Use transistors of your choice, including those with characteristics given in Appendix 1. Draw a circuit diagram. Calculate the maximum sinusoidal power output, the maximum permissible stability factor using the heat sink of your choice, and determine the voltage gain of your amplifier. You should use a gain, or volume, control something like the one shown in Fig. 12.25 to control the output power of the amplifier.   *Answer: $P_o \simeq 20$ W, $K_v = v_o/v_s = 14$.*

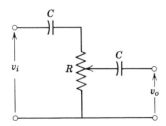

Fig. 12.25.   A typical gain or volume control circuit.

PROBLEM 12.12   Design the amplifier of Prob. 12.11 using a quasi-complementary symmetry arrangement for the power amplifier instead of the complementary-symmetry configuration.

PROBLEM 12.13   Use the common-emitter complementary-symmetry arrangement for the output stage instead of the emitter-follower arrangement in the design of the amplifier of Prob. 12.11.

PROBLEM 12.14   Assuming that a power amplifier with voltage gain of 15 and input resistance of 1 MΩ or higher is needed, replace the input transistor in the amplifier of Fig. 12.19 with a FET and determine the g_m required to provide the required voltage gain if $R_5 = 4.7$ kΩ and the other transistors are as given in the example associated with Fig. 12.19. The power supply voltage is 30 V. Determine the required value of I_D.

Answer: $g_m = 3.3$ mmho, $I_D \simeq 3$ mA.

CHAPTER 13

NEGATIVE FEEDBACK

The performance of an amplifier can be altered by the use of feedback; that is, by adding part or all of the output signal to the input signal. If there are an even number of polarity reversals (or no polarity reversals) between the input and the output of the amplifier, the feedback is said to be positive. This type of feedback is used in oscillator circuits. On the other hand, if there is an odd number of polarity reversals in the amplifier so that the feedback signal tends to cancel the input signal in the midfrequency range, the feedback is said to be negative. This negative feedback, which is the subject of this chapter, can reduce distortion, increase the bandwidth, change the output impedance and the input impedance, and stabilize the gain of an amplifier. All these improvements are gained at the expense of reduced midfrequency gain.

13.1 THE EFFECT OF FEEDBACK ON GAIN, DISTORTION, AND BANDWIDTH

The block diagram of an amplifier with negative feedback is given in Fig. 13.1. The feedback voltage v_f has a polarity opposite to the input voltage v_i, so that the signal v_a that actually drives the amplifier is the difference between v_i and v_f, or

$$v_a = v_i - v_f \qquad (13.1)$$

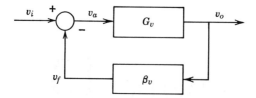

Fig. 13.1.   The block diagram of an amplifier with negative feedback.

The feedback factor $\beta_v{}^1$ is the ratio of the feedback voltage v_f to the output voltage v_o. This ratio is usually obtained by a resistive voltage divider, as shown later. Thus

$$v_f = \beta_v v_o \tag{13.2}$$

We shall let G_v be the voltage gain or amplification of the amplifier without feedback. Thus

$$v_o = G_v v_a \tag{13.3}$$

The voltage gain G_{vf} of the amplifier with feedback is

$$G_{vf} = \frac{v_o}{v_i} \tag{13.4}$$

But, by using Eq. 13.1, $v_i = v_a + v_f$ and substituting the value of v_o from Eq. 13.3,

$$G_{vf} = \frac{G_v v_a}{v_a + v_f} \tag{13.5}$$

Substituting $\beta_v v_o$ for v_f (Eq. 13.2) and then dividing both the numerator and denominator of Eq. 13.5 by v_a, we have

$$G_{vf} = \frac{G_v v_a}{v_a + \beta v_o} = \frac{G_v}{1 + \beta(v_o/v_a)} \tag{13.6}$$

Finally, substituting G_v for v_o/v_a in Eq. 13.6,

$$G_{vf} = \frac{G_v}{1 + G_v \beta_v} \tag{13.7}$$

---

[1]In this chapter β will be used as the feedback factor, not as the transistor current amplification factor, which will be designated as h_{fe}. This β is commonly used in the literature to represent the feedback factor, or ratio.

Equation 13.7 could be written in terms of current gains G_{if}, G_i, and a current feedback ratio β_i instead of their voltage counterparts. Making these substitutions into Eq. 13.7,

$$G_{if} = \frac{G_i}{1 + G_i\beta_i} \tag{13.8}$$

In a given amplifier, the feedback reduces both the current gain and the voltage gain by the same factor, since the relationship $G_v = G_i Z_L/Z_i$ holds for any amplifier, with or without feedback. Therefore, $1 + G_v\beta_v = 1 + G_i\beta_i$ and $G_v\beta_v = G_i\beta_i$. Thus the term $G\beta$ can mean either $G_v\beta_v$ or $G_i\beta_i$. This reduction in gain is the disadvantage of negative feedback, and additional gain must be provided in the amplifier to compensate for the feedback. However, several advantages accrue that make negative feedback very attractive.

One desirable characteristic of negative feedback is improved gain stability. You may observe from Eq. 13.7 or Eq. 13.8 that the product $G\beta$ (meaning either $G_v\beta_v$ or $G_i\beta_i$) may be large in comparison with one. Under this condition

$$G_f = \frac{G}{1 + G\beta} \simeq \frac{G}{G\beta} = \frac{1}{\beta} \tag{13.9}$$

Thus, the gain becomes almost independent of the amplifier characteristics and depends primarily on the resistance ratio of a voltage divider.

Another desirable characteristic of negative feedback is the reduction of harmonic or nonlinear distortion. The reason for this reduction may be seen from Fig. 13.2, where the amplifier is assumed to distort the sinusoidal input voltage by flattening the peaks. The feedback voltage v_f has the same waveform as the output voltage. Therefore, the flattened peaks of the feedback voltage subtract less from the input voltage, thus accentuating the peaks of the amplifier input voltage v_a and predistorting v_a in a manner that will partially compensate for the flattening caused by the amplifier.

The amount of distortion reduction caused by negative feedback can be determined with the aid of Fig. 13.3. The amplifier with gain G has distortion D_o appearing in its output without feedback. After feedback is applied, distortion D_f appears in the output. But distortion D_f is first reduced by the factor β and applied to the input of the amplifier, which in turn amplifies βD_f by the gain G and adds it, in opposite polarity, to the original distortion D_o to obtain the total distortion D_f. Thus,

$$D_f = D_o - G\beta D_f \tag{13.10}$$

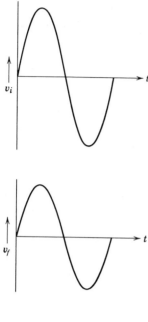

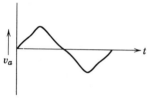

Fig. 13.2. A sketch showing how the feedback voltage v_f predistorts the amplifier input voltage v_a to partially compensate for the amplifier distortion.

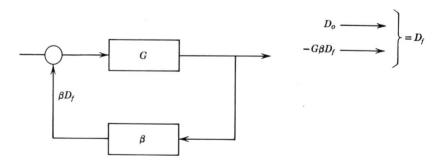

Fig. 13.3. A block diagram illustrating the reduction of distortion in the output of an amplifier.

Solving explicitly for D_f in terms of D_o,

$$D_f = \frac{D_o}{1 + G\beta} \tag{13.11}$$

Notice that the distortion is reduced by the same factor as the gain. Frequently, the feedback factor β is chosen to reduce the distortion a given amount.

The bandwidth of an amplifier is also increased by the use of negative feedback. We shall use the expression of voltage gain for an RC-coupled stage to show this. You may recall that the expression for gain (voltage or current) for the low and middle frequency range is

$$G = \frac{K}{1 - jf_l/f} \tag{13.12}$$

where K is the reference, or midfrequency, gain and f_l is the low-frequency half-power point. Using this expression for G in the feedback formula, Eq. 13.7

$$G_f = \frac{K/(1 - jf_l/f)}{1 + K\beta/(1 - jf_l/f)} = \frac{K}{1 - j(f_l/f) + K\beta} = \frac{K}{1 + K\beta - jf_l/f} \tag{13.13}$$

Then, dividing both numerator and denominator of Eq. 13.13 by $1 + K\beta$,

$$G_f = \frac{K/(1 + K\beta)}{1 - jf_l/(1 + K\beta)f} \tag{13.14}$$

You may observe from Eq. 13.14 that the low-frequency cutoff for the amplifier with feedback is $f_l' = f_l/(1 + K\beta)$.

Similarly, we can determine the effect of feedback on the upper-cutoff frequency of an RC-coupled amplifier, since we know the gain expression for middle and high frequencies is

$$G = \frac{K}{1 + jf/f_h} \tag{13.15}$$

Using this value of G in the feedback Eq. 13.7, G_f is

$$G_f = \frac{K/(1 + jf/f_h)}{1 + K\beta/(1 + jf/f_h)} \tag{13.16}$$

Eq. 13.16 can be rearranged in the same manner as Eq. 13.13 to give

$$G_f = \frac{K/(1+K\beta)}{1+jf/(1+K\beta)f_h} \tag{13.17}$$

Observe that the upper half-power frequency of the amplifier with feedback is $f_h' = (1+K\beta)f_h$. Sometimes negative feedback is used for the specific purpose of increasing the bandwidth of an amplifier. However, when several stages are included in the feedback loop, or transformer coupling is used, the improvement in bandwidth is not so simply related to the feedback factor as in the preceding example.

An example of a feedback amplifier may help clarify some of the foregoing concepts.

EXAMPLE 13.1   An amplifier that is assumed to consist of a single RC-coupled stage is connected as shown in Fig. 13.1. The characteristics of the amplifier without feedback are voltage gain $K_v = 1000$, distortion $= 6\%$, $f_l = 20$ Hz, and $f_h = 200$ kHz. Determine the characteristics of the amplifier with negative feedback when the voltage feedback ratio $\beta_v = 0.01$.

The value of $1+K_v\beta_v$ or $(1+K\beta)$ appears in each equation, so let us evaluate this term $1+K\beta = 1 + 1000 \times 0.01 = 1 + 10 = 11$. From Eq. 13.7, $K_{vf} = K_v/(1+K\beta) = 1000/11 = 90.9$ (if we use the approximation given by Eq. 13.9, $K_{vf} \simeq 1/0.01 = 100$). The distortion with feedback (Eq. 13.11) is $D_f = D_o/(1+K\beta) = 6\%/11 = 0.54\%$. The low-frequency half-power point with feedback f_l' is $f_l/(1+K\beta) = 20/11 = 1.8$ Hz. The upper half-power frequency is $f_h' = (1+K\beta)f_h = 11 \times 200$ kHz $= 2.2$ MHz with negative feedback.

PROBLEM 13.1   Repeat Example 13.1 if $\beta_v = 0.1$.
Answer: $(1+K\beta) = 101$, $K_{vf} = 9.9$, $D_f = 0.059$, $f_l' = 0.198$ Hz, $f_h' = 20.2$ MHz.

PROBLEM 13.2   Assume the acceptable distortion level in the amplifier of Example 13.1 is 1%. Determine the feedback factor β_v that will reduce the distortion from 6% to 1%. What is the voltage gain of the amplifier with this amount of feedback?   Answer: $\beta_v = 0.005$, $K_{vf} = 167$.

13.2   THE EFFECT OF NEGATIVE FEEDBACK ON INPUT IMPEDANCE AND OUTPUT IMPEDANCE

Negative feedback will either increase or decrease the input impedance of an amplifier depending on whether the feedback signal is

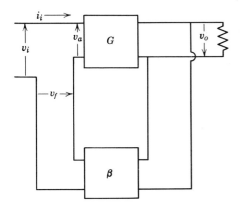

Fig. 13.4. Series input connection for a feedback circuit.

added in series or parallel with the input signal. Let us first consider the series connection shown in Fig. 13.4. The input impedance Z_{if} is the ratio of input voltage v_i to the input current i_i. Thus

$$Z_{if} = \frac{v_i}{i_i} = \frac{v_a + v_f}{i_i} \qquad (13.18)$$

but $v_f = G\beta v_a$, so

$$Z_{if} = \frac{v_a + G\beta v_a}{i_i} = \frac{v_a}{i_i}(1 + G\beta) \qquad (13.19)$$

However, v_a/i_i is the impedance Z_i of the amplifier without feedback. Therefore, in terms of Z_i,

$$Z_{if} = Z_i(1 + G\beta) \qquad (13.20)$$

EXAMPLE 13.2   An amplifier with $Z_i = 1\ \text{k}\Omega$ has a voltage gain $K = 1000$ and $\beta_v = 0.01$. With the feedback connection shown in Fig. 13.4, the input impedance with feedback is $Z_{if} = 1\ \text{k}\Omega(11) = 11\ \text{k}\Omega$.

We shall now consider the parallel arrangement of input and feedback circuits shown in Fig. 13.5. In the parallel circuit the input current i_i is the sum of amplifier input current i_a and the feedback current i_f. Thus, the input admittance with *feedback* is

$$Y_{if} = \frac{i_i}{v_i} = \frac{i_a + i_f}{v_i} \qquad (13.21)$$

But, using the amplifier current gain G_i and the current feedback factor $\beta_i = i_f/i_o$, the feedback current $i_f = i_a G_i \beta_i$. Making this substitution for i_f

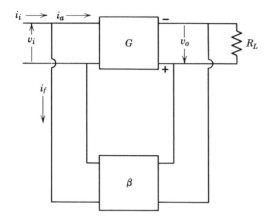

Fig. 13.5.   Parallel input connection for a feedback circuit.

in Eq. 13.21,

$$Y_{if} = \frac{i_a + i_a G_i \beta_i}{v_i} = \frac{i_a}{v_i}(1 + G_i \beta_i) \tag{13.22}$$

Since the input admittance of the amplifier without feedback Y_i is i_a/v_i, Y_{if} can be written in terms of Y_i.

$$Y_{if} = Y_i(1 + G\beta) \tag{13.23}$$

As was mentioned previously, $G_i \beta_i = G_v \beta_v$. We can also show this by recalling that $G_v = G_i R_L/R_i$. Also, $\beta_v = v_f/v_o = i_f R_i/i_o R_L = \beta_i R_i/R_L$. Thus, the product $G_v \beta_v = G_i \beta_i$, providing the input resistance R_i as seen from the feedback network is the same as the input resistance as seen by the driving source. This equality can exist only if the input current is the driving source current and the source resistance R_s is included as part of the input resistance. This subject will be discussed in greater detail and examples will be given in Section 13.3. At the moment let us accept the fact that $G_i \beta_i$ may equal $G_v \beta_v$.

EXAMPLE 13.3   An amplifier has $K_v = 1000$, $\beta_v = 0.01$ and $Z_i = 1000 \ \Omega$ and is connected with parallel feedback (Fig. 13.5). Then, the input admittance without feedback, $Y_i = 1/Z_i = 10^{-3}$ mho and the input admittance with feedback $Y_{if} = 10^{-3}(11) = 1.1 \times 10^{-2}$ mhos. The input impedance with feedback is $Z_{if} = 1/Y_{if} = 90.9 \ \Omega$.

Up to this point, the feedback network has been shown as being connected in parallel with the output terminals. This parallel connection is

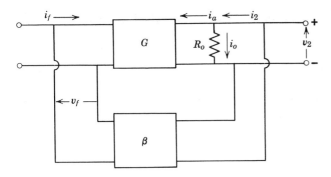

Fig. 13.6.   The parallel output connection, or voltage feedback.

commonly known as *voltage* feedback because the feedback quantity (either current or voltage) is proportional to the output voltage. This voltage feedback is illustrated in Fig. 13.6. The output impedance with feedback is desired and can be determined by turning the input current source off and applying a voltage source v_2 to the output terminals. The current i_2 that flows is the sum of the current i_o through the output resistance, which may include the load resistance, and the current i_a that flows into the amplifier as a result of the feedback. The current flow into the feedback network should be, and usually is, negligible. Then

$$Y_{of} = \frac{i_2}{v_2} = \frac{i_o + i_a}{v_2} \tag{13.24}$$

But with the driving current source off, the feedback current i_f is also the amplifier input current, so $i_a = i_f G_i = i_o \beta_i G_i$. Making this substitution in Eq. 13.24, y_{of} is

$$Y_{of} = \frac{i_o + i_o \beta_i G_i}{v_2} = \frac{i_o}{v_2}(1 + G_i \beta_i) \tag{13.25}$$

But i_o/v_2 is the output admittance Y_o without feedback. Therefore, Y_{of} can be written in terms of Y_o

$$Y_{of} = Y_o(1 + G\beta) \tag{13.26}$$

EXAMPLE 13.4   An amplifier is connected as shown in Fig. 13.6 and has $K_v = 1000$ and $\beta = 0.01$. The output resistance $R_o = 500\ \Omega$ before feedback is applied. Then, with feedback, $Y_{of} = 2 \times 10^{-3}(11) = 2.2 \times 10^{-2}$ mho and $R_{of} = 45\ \Omega$.

You may recall that the loading of the feedback circuit on the output of the amplifier was neglected. If this loading is not negligible it can be included as part of R_o.

You may have observed the regularity with which the factor $(1 + G\beta)$ appears as a modifier in negative feedback circuits. Let us now investigate the final feedback connection—the series arrangement of feedback and load in the amplifier output. This connection, which is known as current feedback because the feedback quantity (either current or voltage) is proportional to the output current, is shown in Fig. 13.7. Again, voltage v_2 is applied to the output terminals and current i_2 flows as a result. But i_2 is the sum of the current i_o through the output resistance and the current i_a that flows into the amplifier as a result of the feedback. Note that the feedback current $\beta_i i_2$ is opposite in direction to the assumed amplifier input current i_i. But with the driving source current turned off, the feedback current *is* the amplifier input current, so $i_i = -\beta_i i_2$. Then

$$i_2 = i_o + i_a = i_o - G_i \beta_i i_2 \qquad (13.27)$$

or

$$i_2(1 + G\beta) = i_o \qquad (13.28)$$

If the voltage drop $i_2 R_f$ across the feedback resistor is small in comparison with v_2, then

$$i_o \simeq \frac{v_2}{R_o} \qquad (13.29)$$

and

$$i_2(1 + G\beta) \simeq \frac{v_2}{R_o} \qquad (13.30)$$

Thus, since the output resistance with feedback R_{of} is v_2/i_2,

$$R_{of} = \frac{v_2}{i_2} = R_o(1 + G\beta) \qquad (13.31)$$

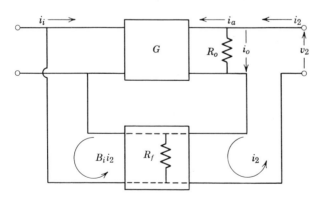

Fig. 13.7.   The series output connection, or current feedback.

EXAMPLE 13.5   The amplifier of the preceding example with $K_v = 1000$, $\beta_v = 0.01$ and $R_o = 500\,\Omega$ has the series output connection, or current feedback. Then, the output resistance with feedback, $R_{of} = 500(11) = 5500\,\Omega$.

PROBLEM 13.3   A given amplifier has $K_v = 10^4$, $R_i = 3\,k\Omega$, and $R_o = 3\,k\Omega$. This amplifier is to be inserted in a transmission line with characteristic impedance $Z_0 = 150\,\Omega$. If feedback is used to make the impedance of the amplifier match the line in both directions, determine the feedback factor, the type connections (series or parallel), and the voltage gain with feedback.   *Answer:* $\beta_v = 1.9 \times 10^{-3}$, $K_{vf} = 500$.

13.3   FEEDBACK CIRCUITS

Many types of feedback circuits can be devised but only a few typical circuits will be discussed here. We shall first consider the circuit of Fig. 13.8, which has the parallel output connection, or voltage feedback, and the series input connection. The feedback is negative because one polarity reversal occurs between the emitter of transistor T_1 where the feedback is applied and the collector of transistor T_2 where the output voltage is obtained. Since $i_f = v_o/(R_f + R_1)$ and $v_f = i_f R_1$, the voltage feedback factor is approximately

$$\beta_v = \frac{v_f}{v_o} = \frac{R_1}{R_f + R_1} \qquad (13.32)$$

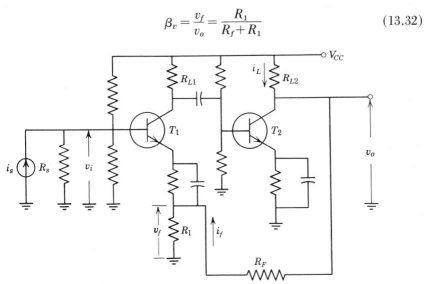

Fig. 13.8.   A feedback circuit with parallel output connection, or voltage feedback, and series input connection.

The impedance looking into the emitter circuit of T_1 is in parallel with R_1 as far as the feedback signal is concerned, and this may cast serious doubts on the accuracy of Eq. 13.32. However, since the feedback voltage v_f is nearly equal to the input voltage v_i when normal amounts of feedback are used, the signal voltage between the emitter and base is small compared to v_f. Therefore, most of the feedback current i_f flows through R_1, and Eq. 13.32 is fairly accurate. The resistor R_1 should be small enough to permit good voltage gain in the first stage and R_f should be large in comparison with the load impedance Z_L so that R_f will not seriously load the output.

EXAMPLE 13.6   Let us use the foregoing assumptions and analyze the circuit shown in Fig. 13.8. Both transistors have $h_{fe} = 150$, $h_{ie} = 2\,\text{k}\Omega$, $h_{re} \simeq 0$, and $h_{oe} \simeq 10^{-5}$ mhos. The value of R_{L2} is $1\,\text{k}\Omega$ and is essentially the total load impedance of transistor T_2. The value of $R_s = 10\,\text{k}\Omega$ and $R_{L1} = 2\,\text{k}\Omega$. The bias circuit resistance $R_b = 12\,\text{k}\Omega$.

We choose $R_1 = 100\,\Omega$ so that the input impedance of transistor T_1 is $Z_{i1} = h_{ie} + (h_{fe} + 1)R_1 = 17\,\text{k}\Omega$. The total shunt resistance at the input of stage 1 is the parallel combination of R_s, R_b, and $R_{sh} = 1/(1/10\,\text{k}\Omega + 1/12\,\text{k}\Omega + 1/17\,\text{k}\Omega) = 4.14\,\text{k}\Omega$. Then, the current gain for stage 1 is $K_{i1} = i_{c1}/i_s = h_{fe}R_{sh}/Z_{i1} = 150 \times 4.14\,\text{k}\Omega/17\,\text{k}\Omega = 36.4$. The input impedance to transistor T_2 is $Z_{i2} = h_{ie2} = 2\,\text{k}\Omega$. The total shunt load on transistor T_1 is the parallel combination of R_{L1}, R_b, and h_{ic2} (the value of $1/h_{oe}$ is much greater than this parallel combination and can safely be ignored). This shunt resistance is about $925\,\Omega$. The current gain for stage 2 is $K_{i2} = i_L/i_{c1} = 150 \times 925/2000 \simeq 69$. Then the total current gain of the amplifier $= K_i = K_{i1}K_{i2} = 36.4 \times 69 = 2510$, and the total voltage gain is $K_v = v_o/v_i = K_iR_{L2}/R_{sh} = 2510 \times 10^3/4.14 \times 10^3 = 606$.

We assume the feedback is to be used to reduce the distortion of this amplifier by a factor of 6. Then the value of $1 + K_v\beta_v = 6$ so that $K_v\beta_v = 5$. Thus, $\beta_v = 5/K_v = 5/606 = 0.00825$. Then, from Eq. 13.32, we have $R_f = (R_1/\beta_v) - R_1 = (100/0.00825) - 100 \simeq 12\,\text{k}\Omega$. Since this value of R_f is more than ten times the load resistance R_{L2}, we assume it is large enough to have negligible loading effect on R_{L2}. The Z_o without feedback is $\simeq R_L$ (since $1/h_{oe} \gg R_L) = 1000\,\Omega$. The output resistance with feedback is $Z_{of} \simeq 1\,\text{k}\Omega$ $(1 + K_v\beta_v) = 1\,\text{k}\Omega/6 = 167\,\Omega$. Also the input impedance to transistor T_1 as $Z_{if} = Z_i(1 + K\beta) = 17\,\text{k}\Omega \times 6 = 102\,\text{k}\Omega$.

PROBLEM 13.4   Repeat Example 13.6 if the h_{fe} of both transistors is 200. Assume we wish to reduce the distortion to 10 percent of the value without feedback.
Answer: $K_{if} \simeq 367$, $K_{vf} \simeq 84$, $\beta_v = 0.0119$, $R_f \simeq 8.3\,\text{k}\Omega$, $Z_{of} \simeq 100\,\Omega$, $Z_{if} = 220\,\text{k}\Omega$.

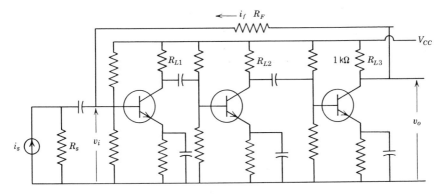

Fig. 13.9.   A feedback circuit with parallel input and output connections.

A typical circuit that uses both parallel input connection and parallel output connection is shown in Fig. 13.9. Observe that an odd number of common-emitter stages must be included in the feedback loop to obtain negative feedback. We shall analyze this circuit in the following example.

EXAMPLE 13.7   A circuit is connected as shown in Fig. 13.9. All three transistors have $h_{fe} = 100$, $h_{ie} = 1\,k\Omega$, $h_{re} \simeq 0$, and $h_{oe} = 10^{-5}$ mhos. The bias circuit resistance of each stage is $R_b = 10\,k\Omega$. The source resistance $R_s = 2\,k\Omega$, $R_{L1} = R_{L2} = 2\,k\Omega$, and $R_{L3} = 1000\,\Omega$. We wish to reduce the distortion of this circuit to one-tenth of its open-circuit value and then determine the characteristics of the circuit with feedback.

First, we must determine the amplifier characteristics with R_f open. We note that the values of R_L for each stage are much less than $1/h_{oe}$, so we can neglect h_{oe}. The input impedance to each transistor is h_{ie}, which is equal to $1\,k\Omega$. The shunt resistance on the input of the first stage is the parallel combination of R_s, R_b, and h_{ie}. Thus $R_{sh} = 1/(1/2\,k\Omega + 1/10\,k\Omega + 1/1\,k\Omega) = 625\,\Omega$. The current gain of the first stage is $K_{i1} = i_{c1}/i_{b1} = h_{fe}R_{sh}/h_{ie} = 100 \times 625/1000 = 62.5$. Notice that if the first transistor is regarded as a current source and R_{L1} is used for R_s, the circuit for the second stage is identical to the circuit for the first stage. Therefore, $K_{i2} = K_{i1} = 62.5$. A similar observation for the circuit driving the third stage indicates $K_{i3} = K_{i1}$ also. Thus, the total open-loop current gain is $K_i = K_{i1} \times K_{i2} \times K_{i3} = 62.5^3 = 2.42 \times 10^5$. Since $R_{L3} \ll 1/h_{oe}$, the output impedance, $Z_o = 1000\,\Omega$. Then, the total open-loop voltage gain K_v is $K_v = K_i Z_o/R_{sh} = 2.42 \times 10^5 \times 10^3/625 = 3.87 \times 10^5$.

Since we desire a reduction of distortion by a factor of 10, the term $(1 + K\beta) = 10$. Then $K\beta = 9$. Now, since the output voltage v_o is very large $(3.87 \times 10^5$ times as large) in comparison with the input voltage v_i, the feedback current i_f is very nearly equal to v_o/R_f. Also, the output signal current $i_o = v_o/R_{L3}$. Therefore, the feedback current factor, or ratio, is

$$\beta_i = \frac{i_f}{i_o} \simeq \frac{v_o/R_f}{v_o/R_L} = \frac{R_L}{R_f} \tag{13.33}$$

Now, $K_i = 2.42 \times 10^5$ and $K_i\beta_i = 9$. Consequently, $\beta_i = 9/2.42 \times 10^5 = 3.71 \times 10^{-5}$. From Eq. 13.33, we note that $R_f = R_L/\beta_i = 10^3/3.71 \times 10^{-5} = 2.69 \times 10^7\ \Omega$. The input impedance of the amplifier, including R_s, can be found from Eq. 13.23 as $Z_{if} = Z_i/(1+K\beta) = Z_i/10 = 625/10 = 62.5\ \Omega$. The output impedance (from Eq. 13.26) is $100\ \Omega$.

Let us now see if $K_i\beta_i$ does equal $K_v\beta_v$ for this circuit. For this amplifier, $\beta_v = Z_i/(R_f+Z_i) = 625/2.69 \times 10^7 = 2.32 \times 10^{-5}$. Then, $K_v\beta_v = 2.32 \times 10^{-5} \times 3.87 \times 10^5 = 9$.

PROBLEM 13.5 Find a value of R_f that will reduce the distortion of the amplifier in Example 13.7 by a factor of 50. Find the value of K_{if}, K_{vf}, Z_{if}, and Z_{of} with this value of R_f connected in the circuit.

Answer: $R_f = 4.95\ \mathrm{M}\Omega$, $K_{if} = 4.84 \times 10^3$, $Z_o = 20\ \Omega$.

Figure 13.10 shows a feedback system with series output connection, or current feedback, and parallel input connection. Since the output current flows through the resistor R_1 in the emitter circuit of the output stage, the voltage across this resistance is proportional to the output current. There-

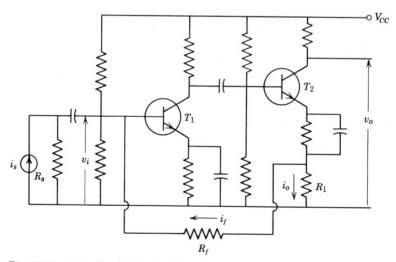

Fig. 13.10. A feedback circuit with series output connection (current feedback) and parallel input connection.

fore, the feedback current is proportional to the output current. If the signal voltage across $R_1(i_oR_1)$ is large in comparison with v_i, the current feedback factor is approximately

$$\beta_i = \frac{i_f}{i_o} = \frac{R_1}{R_f} \tag{13.34}$$

PROBLEM 13.6 Assume that the amplifier of Fig. 13.10 uses the same components as the amplifier of Example 13.6. The unbypassed resistor $R_1 = 100\ \Omega$ has been changed from the input stage to the output stage, however, so that the input impedance as seen by the current source i_s is the parallel combination of R_s, R_b, and h_{ie} or 1.33 kΩ for the values given previously. We assume that the total current gain (i_o/i_s) is the same as before, or 2590. Then if the desired value of $K_i\beta_i = 9$, determine the required value of R_f. Since the output impedance without feedback was assumed to be approximately 1 kΩ, determine the output impedance and the input impedance, as seen by the current driving source I_s, with the feedback specified.   *Answer:* $R_f = 2.88 \times 10^4\ \Omega$, $R_{of} = 10$ kΩ, $R_{if} = 133\ \Omega$.

Negative feedback can be used to provide q-point stability in a dc amplifier, as shown in Fig. 13.11. This amplifier is the quasi-comple-

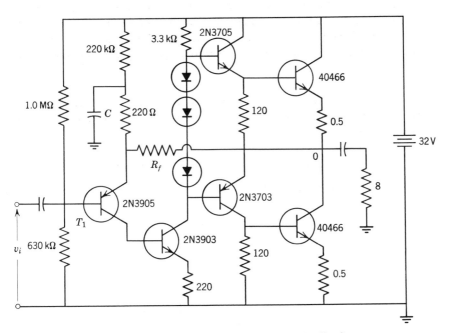

Fig. 13.11.   A quasi-complementary symmetry amplifier with feedback.

mentary-symmetry amplifier given in Fig. 12.19 with an additional input stage T_1 to provide the extra gain needed for adequate feedback. The series input connection and parallel output connection for the feedback network was chosen to provide high input impedance and low output impedance for the amplifier, which is often desirable.

EXAMPLE 13.8   A circuit is connected as shown in Fig. 13.11. The q-point collector current of transistor T_1 is about 72 μA, which is the q-point base current of the following transistor. We shall hold the emitter potential of transistor T_1 at 16 V with respect to ground, with no direct current through the feedback resistor R_f. Then the dc feedback will stabilize the potential at point 0 in the output at approximately 16 V, which will permit maximum output signal amplitude. Therefore, the total emitter-circuit resistance for T_1 needs to be 16 V/72 μA = 220 kΩ. This high resistance is bypassed by capacitor C and 220 Ω additional resistance added in series to provide ac or signal feedback as well as dc feedback. We shall choose the feedback resistor R_f to provide the value of β_v that will give the 15-V peak output signal when a 0.5-volt peak signal is applied at the input. This will require $K_{vf} = 30$. Since the voltage gain K_v without feedback is large in comparison with 30, we can easily approximate $\beta_v = 1/30 = 0.033$ and $R_f \simeq 220 \times 30 = 6.6$ kΩ. The dc feedback factor $\beta_v = 220$ k$\Omega/(220$ k$\Omega +$ 6.6 k$\Omega) \simeq 1$. Thus, the dc voltage at point 0 must be approximately the same as the dc voltage at the emitter of T_1. The voltage gain of the amplifier without feedback is approximately 600 and h_{fe} of T_1 is approximately 70 at $I_C = 72$ μA.

PROBLEM 13.7   Determine the ac input resistance of transistor T_1, Fig. 13.11, using the transistor parameters and the feedback circuit given in Example 13.8.   *Answer:* $R_{if} = 840$ kΩ.

13.4   STABILITY OF FEEDBACK CIRCUITS

The equation for gain with feedback was developed in Section 13.1 and is repeated here for convenience.

$$G_f = \frac{G}{1 + G\beta} \tag{13.35}$$

where G_f is the gain with feedback, or closed-loop gain, G is the gain without feedback, or forward open-loop gain, and β is the feedback factor or ratio. Observe that the denominator of Eq. 13.35 will be equal to zero if the term

$G\beta = -1$. Then the gain G_f will be infinite, or output will occur with no input and the amplifier will oscillate or be unstable. This situation can arise at either a high frequency or a low frequency where the relative phase shift of the amplifier is 180° with respect to the midfrequency phase, or the feedback voltage is actually in phase with the signal voltage. Thus, if the magnitude of total loop gain $G\beta$ is equal to at least *one* at a frequency for which the relative phase shift is 180°, oscillation will occur. Therefore, as was stated previously, the condition for oscillation is

$$G\beta = -1 = 1\underline{|\pm 180} \qquad (13.36)$$

The gain-phase plots, or Bode plots, discussed in Chapter 11, can be used to predict the stability of an amplifier with feedback. In fact, with the aid of a Bode plot, the feedback network can be designed so that the amplifier will not only be stable, but will provide near-optimum frequency and transient response. This design will be the subject of this section.

EXAMPLE 13.9   We now consider the stability of the three-stage RC-coupled amplifier of Fig. 13.12. Both the low and high half-power frequencies are determined for each stage, either by calculation or by measurement, as discussed in Chapters 7 and 11. Let us assume that all three stages have a low half-power frequency at 100 Hz and have three different upper half-power frequencies at 10^5, 3×10^5, and 10^6 Hz. The gain magnitude and phase plots are sketched in Fig. 13.13 by using the asymptote technique discussed in Chapter 11. The midfrequency current gain of the amplifier is assumed to be 10^5 and the total load resistance in the output of the amplifier is assumed to be 1 kΩ. The current gain axis is labeled in decibels as well as numerical gain. The relative phase is plotted on the same frequency scale as the gain magnitude. Observe that the relative phase shift is 180° at two frequencies, one low and the other high. These frequencies are approximately 40 Hz and 6×10^5 Hz. At the lower frequency, the forward current gain G_i of the amplifier is 7×10^3 or 77 dB and at the higher frequency $G_i = 5 \times 10^3$ or 74 dB at 180° relative phase. But the magnitude of the total loop gain $G\beta$ must be less than one at these 180° relative phase frequencies or the amplifier will oscillate when feedback is applied. Therefore, the magnitude of the feedback factor β_i that will cause marginal oscillation, considering each frequency individually, is $\beta_i = 1/(7 \times 10^3) = 1.4 \times 10^{-4}$ or -77 dB at 40 Hz and $\beta_i = 2 \times 10^{-4}$ or -74 dB at 6×10^5 Hz. Note that the smaller amount of feedback is usable at the lower frequency because the half-power frequencies of all the stages are the same. Therefore, if this smaller feedback is used, the amplifier will be stable at the higher frequency, but will marginally oscillate at the lower

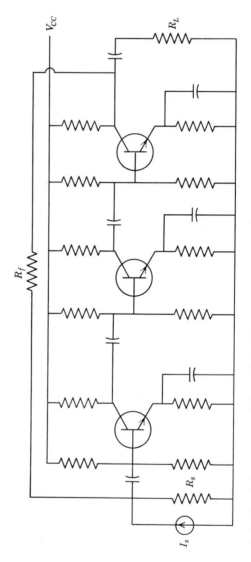

Fig. 13.12.   A three-stage amplifier with feedback.

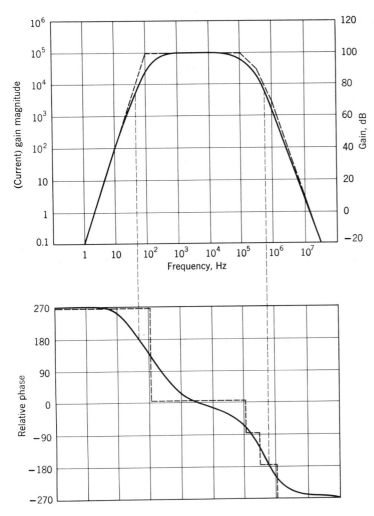

Fig. 13.13.   A Bode plot for the amplifier of Fig. 13.12.

frequency. Of course, since oscillation is intolerable at any frequency, the feedback factor must be reduced not only to provide stability but also to provide satisfactory frequency response and transient response. Theoretical considerations and experiences have shown that the total loop gain magnitude should not exceed about 0.2 or -14 dB at the frequencies that give 180° phase shift. Then, there are no peaks in the frequency response and no more than a 5-percent overshoot in the transient response. This reduction of gain (a factor of 5, or 14 dB) is known as *gain margin*. Smaller values of gain margin may be used if peaks in the response or ringing in the transient response can be tolerated. A sketch of the time and frequency response of a feedback amplifier with a 3-dB gain margin in contrast with the same amplifier with a 14-dB gain margin is given in Fig. 13.14. A step current input is used to excite the transient and only the low-frequency transient is shown. Gain margin values between 6 dB and 14 dB are frequently used.

We shall complete the feedback design for the amplifier of Fig. 13.12 using a 14-dB, or factor of 5, gain margin. The feedback factor required to reduce the loop gain magnitude to 0.2 or -14 dB at 40 Hz is $\beta_i = 0.2/(7 \times 10^3) = 2.8 \times 10^{-5}$ or -77 dB-14 dB $= -91$ dB. The total mid-frequency loop gain with this magnitude of feedback is $K_i\beta_i = 10^5(2.8 \times 10^{-5}) = 2.8$, or 100 dB-91 dB $= 9$ dB. This value of permissible loop gain is disappointingly small because of the modest improvement in the amplifier characteristics. A smaller gain margin would permit somewhat higher loop gain at the expense of peaks in the frequency response or ringing in the transient or time response. A capacitor can be used in the

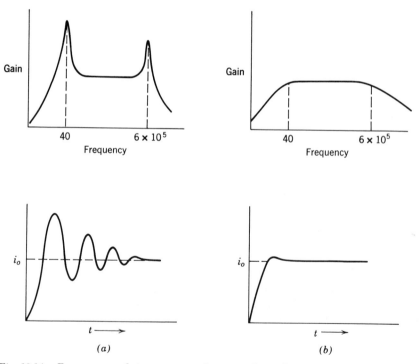

Fig. 13.14.   Frequency and time response for two values of gain margin. (a) 3-dB gain margin and (b) 14-dB gain margin.

feedback circuit to improve the characteristics of this amplifier. This technique is known as phase-lead compensation and will be discussed after we have considered a two-stage amplifier.

EXAMPLE 13.10   The stability of the two-stage RC-coupled amplifier shown in Fig. 13.8 will now be considered. We shall assume that there are two low half-power frequencies at 10 and 100 Hz and two high half-power frequencies at 10^5 and 10^6 Hz. The midfrequency voltage gain was previously calculated to be 833. A Bode plot for this amplifier is given in Fig. 13.15. Observe that the phase plot does not cross the ± 180 values. Therefore, the amplifier will be stable for any finite value of loop gain. Thus, the amplifier is said to be *unconditionally stable*. However, the transient response may be unsatisfactory or severe peaks may occur in the frequency response if the loop gain is not limited. The *gain margin* specification given for the three-stage amplifier is not appropriate because there is no frequency at which the phase is 180°, except zero or infinity,

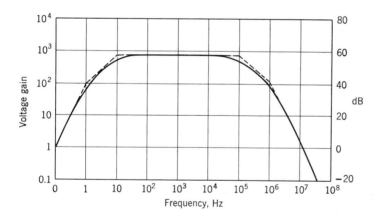

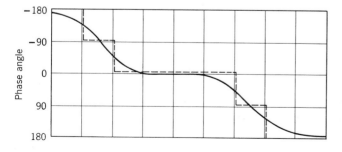

Fig. 13.15.   A Bode plot for the amplifier of Fig. 13.8.

where the gain is zero. Therefore, another margin known as *phase margin* is used to obtain the desired transient characteristics. For example, a 45° phase margin means that the amplitude of the total loop gain is *one* when the phase shift is 45° less than 180°, or ±135°. For the Bode plot of Fig. 13.15, the forward voltage gain of the amplifier is approximately 60 or 36 dB at both the low- and high-frequency 135° phase points. This phase margin compares roughly with the 14-dB gain margin as can be seen in Fig. 13.13. Thus, the value of feedback factor β_v required to reduce the loop gain to *one* at this phase margin is $\beta_v = 1/60 = 0.0167$ or -36 dB. The midfrequency value of loop gain is therefore $K_v\beta_v = 833 \times 0.0167 = 13.9$ or 59 dB $- 36$ dB $= 23$ dB for this 45° phase margin. This is a satisfactory value of loop gain for many applications. Of course, higher loop gain can be realized with a smaller phase margin.

PROBLEM 13.8   The phase margin of the amplifier in Example 13.10 is reduced to 30°. Determine the midfrequency loop gain $K_v\beta_v$.
Answer: 27.8 or 29 dB.

The relatively large amount of feedback that can be applied to the two-stage amplifier in comparison with the three-stage amplifier may lead us to believe that no more than two stages should be included in the feedback loop. However, values of $K\beta$ greater than about 10 are physically unrealizable in a two-stage amplifier. Also, the technique known as phase-lead compensation, mentioned in Example 13.9, can make the phase characteristics of the three-stage amplifier comparable with the uncompensated two-stage amplifier, and thus permit a similar closed-loop gain. You may have observed that the relative phase shift in the amplifier causes the instability and that each term of the form $K/(1 + jf/f_h)$ contributes relative phase shift up to 90°. The two-stage amplifier was unconditionally stable because only two such terms appear at high frequencies because of the shunt capacitance and two appear at low frequencies because of the coupling and bypass capacitance. The phase shift at low frequencies due to the coupling capacitors can be eliminated by using direct coupling. Therefore, the low-frequency phase characteristics of the three-stage amplifier will be similar to the two-stage amplifier if direct coupling is used to eliminate the coupling and emitter bypass capacitors in one stage of the three-stage amplifier. Phase-lead compensation will be used to effectively eliminate one of the high-frequency terms in the amplifier gain expression.
The phase-lead network consists simply of a small capacitance C_f in parallel with the feedback resistor R_f, as shown in Fig. 13.16. Since the impedance (Z_f) of R_f and C_f in parallel is usually very much larger than either R_L or R_i, particularly in a three-stage amplifier, the output current

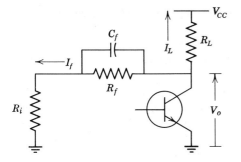

Fig. 13.16.   A phase-compensated feedback network.

is very nearly $I_L = V_o/R_L$ and the feedback current I_f is very nearly V_o/Z_f. Therefore, the current feedback factor is

$$\beta_i' \simeq \frac{I_f}{I_L} = \frac{R_L}{Z_f} \tag{13.37}$$

But

$$Z_f = \frac{R_f(1/j\omega C_f)}{R_f + 1/j\omega C_f} = \frac{R_f}{1 + j\omega R_f C_f} \tag{13.38}$$

Since the time constant $R_f C_f = 1/\omega_f$, where ω_f is the break frequency or crossover frequency of the network $R_f = 1/\omega_f C_f$, then

$$Z_f = \frac{R_f}{1 + j\omega/\omega_f} = \frac{R_f}{1 + jf/f_f} \tag{13.39}$$

and, by using Eq. 13.37,

$$\beta_i' = \frac{R_L(1 + jf/f_f)}{R_f} = \beta_i(1 + jf/f_f) \tag{13.40}$$

where β_i is the midfrequency current feedback factor. We shall now write the complete equation for the high-frequency gain of the three-stage RC-coupled amplifier with phase compensated feedback

$$G_{fi} = \frac{K_1 K_2 K_3/(1 + jf/f_{h1})(1 + jf/f_{h2})(1 + f/f_{h3})}{1 + K_1 K_2 K_3 \beta_i (1 + jf/f_f)/(1 + jf/f_{h1})(1 + jf/f_{h2})(1 + jf/f_{h3})} \tag{13.41}$$

Observe that the term $(1 + jf/f_f)$ in the numerator of the term $G_i \beta_i'$ will cancel one of terms of the form $(1 + jf/f_h)$ in the denominator if f_f is equal to one of the values of f_h. Let us make $f_f = f_{h2}$, then the $G_i\beta_i'$ term is

$$G_{fi}\beta_i' = \frac{K_1K_2K_3\beta_i}{(1+jf/f_{h1})(1+jf/f_{h3})} \tag{13.42}$$

This term has the gain magnitude characteristics of a three-stage amplifier but the relative high-frequency phase characteristics (for the total loop gain) of a two-stage amplifier. However, the forward open-loop gain magnitude can be plotted as though the $(1+jf/f_f)$ term is in the amplifier gain G_i instead of the feedback path β_i'. Then the gain magnitude can also be plotted as though the three-stage amplifier has only two high-frequency break points or poles. This plot will not give the correct high-frequency forward gain but will give the correct feedback design.

EXAMPLE 13.11   We shall now apply phase-lead compensation to the three-stage amplifier with feedback shown in Fig. 13.12, letting $f_f = f_{h2} = 3 \times 10^5$ Hz. Also, direct coupling will be used to couple two of the stages. The modified forward-gain magnitude is sketched in Fig. 13.17. Also, the relative phase is sketched as though the phase compensation is in the amplifier. We shall allow a 45° phase margin, which occurs at approximately $f = 30$ Hz and $f = 10^6$ Hz. Notice that the low-frequency 45° phase margin point has higher forward gain than the comparable high-frequency point. As was previously mentioned, this situation results from the two break points or poles occuring at the same frequency (100 Hz). Therefore, the gain-phase characteristics would be improved and a higher closed-loop gain permitted if one of the coupling circuits had a considerably different value of f_l, for example, 10 Hz. Of course, additional direct coupling could eliminate the pole completely. Therefore, we shall be concerned primarily with the high-frequency gain at the 45° phase margin. This current gain magnitude is 6×10^3 or 77 dB. Therefore, the value of β_i that will give a 45° phase margin is $\beta_i = 1/6 \times 10^3 = 1.67 \times 10^{-4}$ or -77 dB. The mid-frequency closed-loop gain is, therefore, $K_i\beta_i = (10^5)(1.67 \times 10^{-4}) = 16.7$ or 100 dB $- 77$ dB $= 23$ dB, and the amplifier distortion is reduced by the factor $(1+K\beta) = 17.7$. This is a very satisfactory value of loop gain for most applications. The value of R_f (Fig. 13.12) required to give this value of feedback is (with $R_L = 1$ kΩ) $R_f = R_L/\beta_i = 6 \times 10^5$ Ω. The value of C_f required to cancel the pole or break frequency at 3×10^5 Hz is $C_f = 1/R_f\omega_f = 0.88$ pF. This value of capacitance is near the minimum value obtainable from a parts store. The value of R_f could be decreased and the value of C_f increased proportionally if the feedback is taken from a tap in the load circuit (Fig. 13.18). In this circuit R_f would be decreased by a factor of 10 and C_f would be increased by a factor of 10 compared with the values calculated when the feedback is obtained directly across the output.

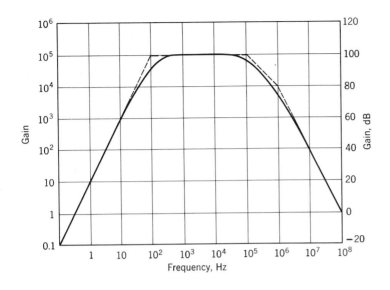

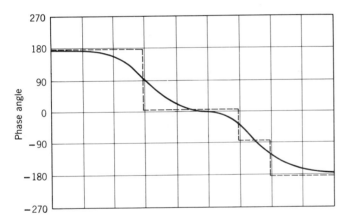

Fig. 13.17.   A Bode plot for the three-stage compensated feedback amplifier.

The value of $K\beta$ determined at a specified phase margin may be larger than the designer desires to use. Smaller values of $K\beta$ can be used, of course, with improved stability (larger gain or phase margins).

PROBLEM 13.9   A given three-stage transistor amplifier has stage upper-cutoff frequencies of 10^5, 5×10^5, and 3×10^6 Hz. The amplifier input resistance $R_i = 10\,\text{k}\Omega$, the load resistance $R_L = 1\,\text{k}\Omega$, and the driving source resistance $R_s = 5\,\text{k}\Omega$. The current gain $I_l/I_s = 10^4$. Direct coupling is used so that stability at low frequencies is not a problem, assuming

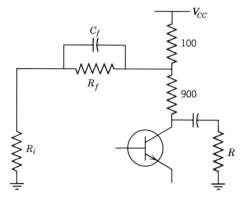

Fig. 13.18. Illustration of a tapped output to reduce R_f and increase C_f.

adequate power supply decoupling is employed. A feedback circuit similar to the one given in Fig. 13.12 is used except that a capacitor C_f is placed in parallel with R_f, and C_f will be chosen so that the zero in the feedback network will cancel the amplifier pole at 5×10^5 Hz. Make a Bode plot similar to Fig. 13.17 and determine the values of R_f, C_f, and $K\beta$ for a 45° phase margin. *Answer:* $R_f = 200$ kΩ, $C_f = 1.6$ pF, $K\beta = 50$.

The gain and phase plot of an amplifier that has been constructed can be readily obtained by the use of a good signal generator and a dual trace oscilloscope; the output and input voltage can be compared as to magnitude and phase. The gain and phase plot for an amplifier design can be easily obtained by using a computer program known as ECAP (Electronic Circuit Analysis Program) if a computer with this program is available.

13.5 LOG GAIN PLOTS

The Bode plots discussed in the preceding section permit the design of stable feedback circuits that provide satisfactory frequency and transient response, but they do not show the frequency characteristics or bandwidth of the amplifier after the feedback has been applied. Therefore, we shall develop in this section a technique that will show the closed-loop, as well as the open-loop, gain characteristics and, as an added bonus, make the feedback design process easier. This proposed technique is just a slight extension and simplification of the Bode plot method previously discussed.

Let us consider the amplifier of Problem 13.9 without the compensating capacitor C_f. The Bode plot for this amplifier is given in Fig. 13.19. Observe that the amplifier gain without feedback is about 2000 at a phase margin of

45°, as shown by the dashed lines. Then the feedback factor that will provide this 45° phase margin is $1/2000 = 5 \times 10^{-4}$, as previously discussed. Also, as previously discussed, the gain of an amplifier *with feedback* is approximately $1/\beta$ as long as $G\beta$ is large compared to 1. Therefore, within the limits of this approximation, the gain of the amplifier *with* feedback is 2000, which is the *same* as the gain without feedback at the frequency that produces $-135°$ relative phase shift. In other words, the straight-line approximation of the closed-loop gain curve (with feedback) intersects the open-loop forward gain curve at the frequency at which $G\beta = 1$. This relationship should be expected if one reviews the rules for making straight-line approximations. At frequencies above the point of intersection $G\beta$ is less than one, so the closed-loop and open-loop gains are equal, within the limits of the straight-line approximations. The *phase margin* is then the relative phase of the amplifier at the frequency of the intersection of the straight-line approximations of the open-loop and closed-loop gains. The upper-cutoff frequency, or bandwidth, of the amplifier with feedback is approximately at the first break frequency in the straight-line approximation of the *closed-loop* gain curve. This break frequency is at the intersection of the closed- and open-loop gain curves, as shown in Fig. 13.19 at $f = 5 \times 10^5$ Hz, when the feedback does not include compensation.

A close relationship exists between the relative phase of the open-loop amplifier and the break frequencies of the straight-line gain plot of the amplifier. For example, the $-135°$ relative phase occurs at approximately the second break frequency of the open-loop amplifier, as shown in Fig. 13.19. Therefore, the two gain curves must intersect at a frequency no greater than the second break frequency if no compensation is used, and the desired phase margin is at least 45°. Thus the phase plot is not required for the feedback design, provided the break frequencies are reasonably well separated.

The effect that compensation in the feedback circuit has on the bandwidth of the closed-loop amplifier can now be shown. Since the gain with feedback $G_f \simeq 1/\beta'$, and $\beta' = \beta(1 + jf/f_f)$ as given in Eq. 13.40, G_f has a break frequency at f_f as shown in Fig. 13.20. The value of f_f was chosen to be $5 = 10^5$ Hz, which is the same as the value used in Problem 13.9. Note that the bandwidth of the amplifier with feedback is the same as f_f because of the break frequency at f_f. The zero in the feedback circuit corrects the phase by 90° approximately, so the intersection of the G and G_f curve can occur at the third break frequency of G instead of the second break frequency. Observe that the difference in the slopes of the two curves at their intersection is the same for both the compensated and uncompensated feedback systems, Figs. 13.19 and 13.20. In other words, the difference in the slopes of G and G_f (straight-line approximations)

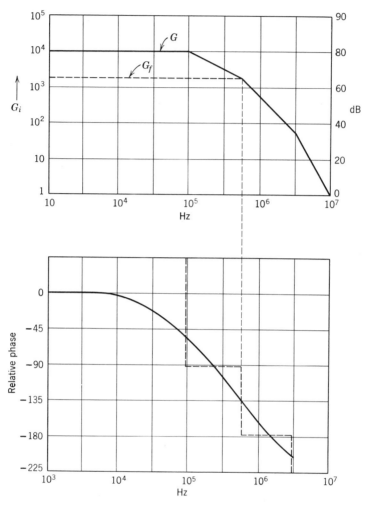

Fig. 13.19.   A Bode plot for the uncompensated amplifier of Problem 13.9.

should not exceed 20 dB per decade, or 6 dB per octave, if the phase margin is to be at least 45°. Note that $K_f = 200$ and β is ten times as high as the uncompensated case with the same phase margin. Also observe that the bandwidth of the feedback amplifier could be increased and the phase margin increased if the frequency f_f were increased to the neighborhood of 10^6 Hz, or perhaps 2×10^6 Hz.

The preceding feedback examples have shown that the stability problems decrease and more feedback can be applied to the amplifier when the half-power frequencies of the amplifier are widely separated, in

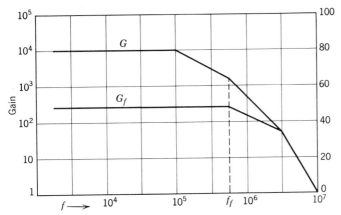

Fig. 13.20.   Straight-line gain approximations for G and G_f for the amplifier of Fig. 13.19 with compensated feedback.

contrast with their occurrence in a narrow frequency range. Therefore, stabilization can be obtained by adding shunt capacitance in one of the stages, so that its upper-cutoff frequency is low compared to the other stages. The benefit of this technique can be seen from the log gain plot of Fig. 13.21. As illustrated, the open-loop gain of the amplifier may decrease greatly before the next higher stage-cutoff frequency is reached. As noted, the requirement for stability is that the closed-loop gain G_f intersect the open-loop gain curve G above the second break frequency when no com-

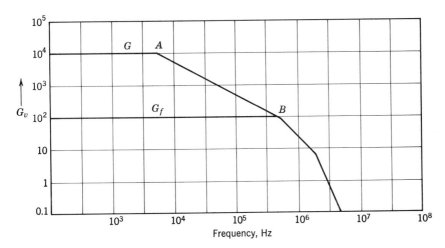

Fig. 13.21.   Illustration of stabilization by lowering the upper-cutoff frequency of one stage.

pensation is employed in the feedback network. Then, the closed-loop gain G_f can be small in comparison with the open-loop gain G over the mid-frequency range, or in other words, $K\beta$ can be large. In the amplifier of Fig. 13.21, $K\beta \simeq 100$. One convenient design technique is to determine the desired value of feedback factor β from some criterion such as distortion reduction or the fixing of the closed-loop gain. The closed-loop gain line is then drawn at the value $1/\beta$. The second break point B is then placed slightly below the closed-loop gain line G_f to allow adequate phase margin. This second break frequency is determined from either theoretical calculations or lab measurements, as was previously discussed. It was given as 5×10^6 Hz for the amplifier of Fig. 13.21. The desired open-loop gain curve G is then constructed by drawing a straight line from this desired second break point B, with a slope $- 20$ dB/decade, until it intersects the horizontal part of the open-loop gain curve. This intersection is shown as point A in Fig. 13.21, and the horizontal part of the open-loop gain curve represents the midfrequency gain of the open-loop amplifier. The upper-cutoff frequency of the stage that had the lowest frequency break point must then be lowered by adding capacitance to coincide with the frequency indicated at point A (5 kHz in Fig. 13.21).

The upper-cutoff frequency of an amplifier can be easily lowered by either increasing the resistance or shunt capacitance in the amplifier. One easy method is to add a capacitor in parallel with the load resistor as shown in Fig. 13.22a. You may recall that the upper-cutoff frequency of a stage is determined by the total shunt resistance and the effective shunt capacitance. The total shunt resistance R_{sh} in parallel with the capacitor C is the load resistor R_C in parallel with both the output admittance (approximately h_{oe}) of the preceding transistor and the input resistance (approximately h_{ie}) of the following transistor. The capacitor C is usually much larger than the other capacitances in the amplifier. Therefore, since $\omega_h = 1/R_{sh}C$,

$$C = \frac{1}{R_{sh}\omega_h} \tag{13.43}$$

where ω_h is the upper-cutoff frequency of the amplifier, determined as point A by the compensation technique illustrated in Fig. 13.21. For example, the lowest break frequency, point A, in Fig. 13.21 is 5×10^3 Hz or 3.14×10^4 rad/s. Thus, if $R_C = 10$ kΩ, $h_{oe1} = 20$ μmhos, and $h_{ie2} = 4$ kΩ, $R_{sh} \simeq 2.7$ kΩ, and $C = 1/2.7 \times 3.14 \times 10^7 = 0.012$ μF.

If a small resistance R_X is placed in series with the added capacitance C, as shown in Fig. 13.22b, the relative phase shift can be decreased at higher frequencies in the neighborhood of the second break frequency at point B, with a consequent increase of phase margin. The greatest phase correction occurs at the radian frequency $\omega_x = 1/R_X C$. Therefore, ω_x should be approximately equal to $\omega_B = 2\pi f_B$, where f_B is the second break frequency at point B, Fig. 13.21. If this relationship is used in the example of Fig. 13.21 where $\omega_B = 100\,\omega_A$, then $R_X = R_{sh}/100 = 27\,\Omega$.

PROBLEM 13.10 A given amplifier has an open-loop voltage gain $K_v = 5000$ in the midfrequency range. Its three stages have upper-cutoff frequencies of 10^5, 3×10^5, and 10^6 Hz.

(a) What maximum value of feedback factor β_v can be used, assuming a satisfactory phase margin is maintained and no compensation is used? What is the approximate upper-cutoff frequency of the amplifier with this feedback?   *Answer:* $\beta_v \simeq 6 \times 10^{-4}, f_h \simeq 3 \times 10^5$ Hz.

(b) What maximum value of feedback factor β_v can be used (with satisfactory phase margin) if phase-lead compensation is used in the feedback network? What is a good, or optimum, value of the $R_f C_f$ time constant? What voltage gain and bandwidth will the amplifier have with this feedback?
    Answer: $\beta_v \simeq 1.8 \times 10^{-3}, K_{vf} \simeq 500, \tau_f \simeq 1.8 \times 10^{-7}, f_h \simeq 9 \times 10^5$ Hz.

(c) It is desired that the voltage gain of the amplifier with feedback be 100. Assume that you will decrease the bandwidth of one stage sufficiently to obtain good stability with an uncompensated feedback network. What should be the upper-cutoff frequency of the modified stage? What value of shunt capacitance is needed if $h_{oe1} = 40\,\mu$mho, $R_C = 5.0\,$kΩ, and $h_{ie2} = 2.0\,$kΩ? (R_b, if any, can be neglected.)   *Answer:* $f_h = 6\,$kHz, $C \simeq 0.02\,\mu$F.

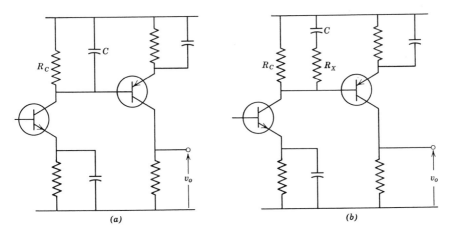

(a)   (b)

Fig. 13.22.   Typical methods of reducing the upper-cutoff frequency of an amplifier.

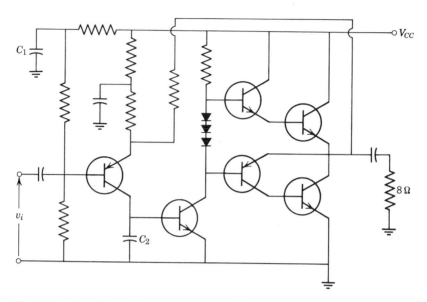

Fig. 13.23.   A quasi-complementary symmetry of amplifier with negative feedback.

(d) What will be the upper-cutoff frequency of the amplifier of part (c)? What value of resistance R_X (Fig. 13.22b) will give optimum phase margin, approximately?   *Answer:* $f_h \simeq 3 \times 10^5$ Hz, $R_X \simeq 27 \; \Omega$.

(e) What measures must be taken to provide stability and good transient response at low frequencies for all values of feedback?

PROBLEM 13.11   The amplifier of Fig. 13.23 has $V_{CC} = 40$ V and R_L is an 8-Ω loudspeaker. The amplifier should provide its full output power with a 1-V peak input signal V_i. Choose the transistor types and all components for the amplifier. Assume that the maximum distortion of the amplifier without feedback is 5 percent, and determine the maximum distortion with feedback. Also, determine the approximate maximum power output, with sinusoidal input signal, the approximate output impedance and input impedance of the amplifier. Determine the dc voltage gain of the amplifier. What does the dc voltage gain have to do with the thermal stability? The capacitor C_1 is a decoupling and filter capacitor. C_2 is a feedback compensating capacitor.

PROBLEM 13.12   Add an input stage, if necessary, and design a feedback circuit for the amplifier of Fig. 12.20. The amplifier should produce full output power with $V_i = 1$-V peak. Choose your own V_{CC} and R_L values and design the circuit.

CHAPTER **14**

INTEGRATED
CIRCUITS

One advantage of a transistor over a vacuum tube is the great size reduction possible for a given power output. Thus, it is quite surprising to open a conventional transistor container and discover the actual active portion of germanium or silicon is so much smaller than the total container. Most of the volume is used for mounting the leads and protecting the silicon or germanium chip from an unfavorable environment. To utilize some of this waste space, some semiconductor manufacturers began mounting several diodes in one container. Then two transistors in one case began to appear on the market. By making two internal connections, two transistors can be mounted in a Darlington configuration, as shown in Fig. 14.1. As noted in Chapter 10, the current gain of a Darlington configuration is approximately the product of the current gains of the individual transistors. Thus, very high current gains from a single three lead "transistor" case are possible.[1]

As manufacturing processes were refined, the complexity of circuitry on a single chip increased. Now, complex switching circuits (for use in digital computing circuits) and entire amplifiers are packaged as single units. Basically, two different approaches are used in constructing these *integrated circuits*. One procedure produces *thin film circuits* while another procedure produces *semiconductor monolithic circuits*.

[1]For example, Motorola advertises Uniblock Darlingtons with h_{FE} values from 5000 to 75,000.

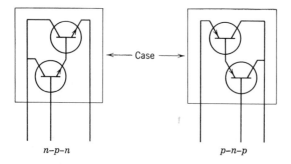

Fig. 14.1.   Two transistors internally connected to form a Darlington configuration.

14.1   THIN FILM CIRCUITS

The thin-film circuits are constructed on a flat plate of insulating material (for example, glass or ceramic), which is known as the substrate. Then, by silk screening, sputtering, vacuum evaporation, or other processes, thin layers of conducting, insulating, or resistive material are deposited to form the wiring and passive elements (resistors and capacitors) in the circuit. At the present time, it is not feasible to construct useful inductors in the limited space of an integrated circuit. Therefore, if inductors *must be used*, leads are provided so that external inductors may be connected to the integrated circuit. Figure 14.2 shows how the different elements may be constructed. Of course, the film thickness has been exaggerated in this diagram. The active elements (such as transistors and diodes) may be added as separate chips with connecting leads welded to the thin film. Some manufacturers are perfecting techniques that will permit them to deposit the active elements on the substrate much as the passive elements are deposited. In fact, Westinghouse[2] reports constructing thin film circuits, including transistors, on a flexible substrate! A typical thin-film circuit is shown in Fig. 14.3.

[2]Westinghouse R & D Letter 1968.

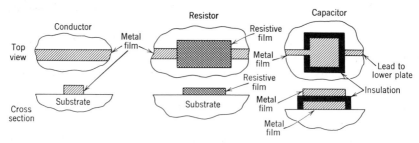

Fig. 14.2.   Typical thin-film construction.

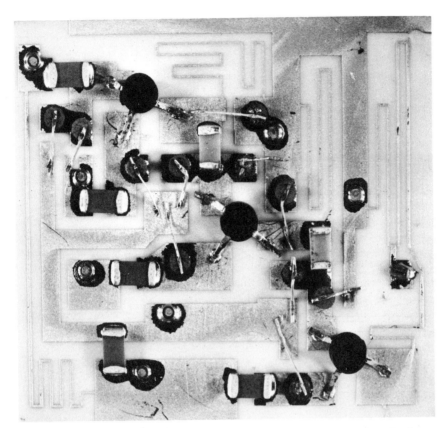

Fig. 14.3.   A typical thin-film circuit. (Courtesy of Motorola Semiconductor Products Phoenix, Ariz.)

14.2   MONOLITHIC CIRCUITS

Most of the integrated circuits on the market at the present time are semiconductor monolithic circuits. The substrate of these circuits is a silicon (at least at the present time) chip. The n and p regions are then diffused down into this chip. The procedure used in constructing a transistor from a chip will help clarify this process. As shown in Fig. 14.4a, a substrate of p-doped silicon is subjected to an oxidizing atmosphere (a small amount of water vapor speeds up the process considerably) at an elevated temperature. As a result of this step, a thin layer of silicon oxide is formed over the substrate (Fig. 14.4a). By the use of a photoengraving process, the layer of silicon oxide is removed over the required collector area, as shown in Fig. 14.4b. The silicon chip is then placed in an n-type

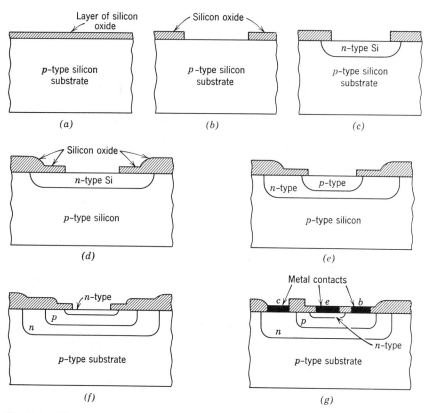

Fig. 14.4.   The process used in constructing an *n-p-n* transistor in a silicon substrate.

atmosphere at an elevated temperature. The *n*-type atoms diffuse into the silicon (but do not diffuse into the silicon oxide) creating a layer of *n*-type silicon, as shown in Fig. 14.4c. Again, a layer of silicon oxide is formed over the entire chip. (The old silicon oxide layer becomes a little thicker.) The photoengraving process is repeated to remove the silicon oxide over the required base region (Fig. 14.4d). The chip is then placed in a *p*-type atmosphere at an elevated temperature and a layer of *p*-type silicon is formed in the *n*-type collector region, as shown in Fig. 14.4e. Again, a layer of silicon oxide is formed over the entire chip. The photo engraving process is repeated to remove the silicon oxide over the required emitter area. Then the chip is subjected to an *n*-type atmosphere until a layer of *n*-silicon is formed in the base region *p*-silicon (Fig. 14.4f). A final silicon oxide layer is formed over the chip and the silicon oxide is again removed by the photoengraving process. This time, the silicon oxide is removed over the points where ohmic contact is to be made with the silicon (emitter, base, and collector lead junctions). By vacuum evaporation or sputtering, a layer of metal is deposited onto the chip (Fig. 14.4g). This metal is also deposited over the insulating silicon oxide to form connecting leads to the other circuit elements. If the junction between the substrate and the collector is maintained in a reverse bias condition, the leakage current will be small and the transistor is effectively insulated from the substrate. In Fig. 14.4, the substrate should be connected directly to the negative lead of the power supply to ensure proper bias of the substrate. The configuration shown in Fig. 14.4e could be used (with proper ohmic contacts) for a diode and the configuration shown in Fig. 14.4c could be used for a resistor. In the latter case, the n-material is the resistor, and an increase of length or decrease of width of this region would increase the resistance.

Of course, the techniques described above could be used to produce field-effect transistors (FET) or MOSFET transistors. Actually, fewer steps are required to construct MOSFET circuits than for bipolar circuits. Both types of circuits are presently being manufactured on a commercial basis.

A typical monolithic integrated circuit is shown in Fig. 14.5. In some instances, it is desirable to combine thin-film and monolithic techniques to produce *Hybrid integrated circuits.*

The design of integrated circuits is beyond the scope of this book. However, some interesting considerations that are used in the design of these integrated circuits will be noted. In the past, the active circuit elements (tubes, transistors, and the like) have been relatively expensive when compared to the passive circuit elements (resistors, capacitors, and inductors). Consequently, conventional designs usually incorporate as few active elements as possible. This design criteria is still used with thin film circuits that have discrete active elements. In contrast, a whole new concept is used in designing monolithic circuits or thin film circuits with deposited active elements. All the transistor collectors in the circuit can be formed at the same time and with the same process. Similarly, all the bases are formed simultaneously and all the emitters are formed in a given time period. Consequently, an integrated circuit with ten or fifteen transistors can be produced at a cost essentially the same as an integrated circuit with four or five transistors. In addition, the cost of the chip with the circuit in place is very small (10 percent or so) in comparison to the cost of testing and packaging the circuit. Thus, when designing monolithic circuits, *the cost of additional active elements is almost negligible.* In fact, since a transistor uses less space on the silicon chip than a resistor, it is desirable to replace resistors with transistors where possible! Also, it is possible to use the junction capacitance of a reverse-biased diode instead of constructing a capacitor in the circuit.

As shown in Fig. 14.4, a p-type substrate is used for n-p-n type mono-

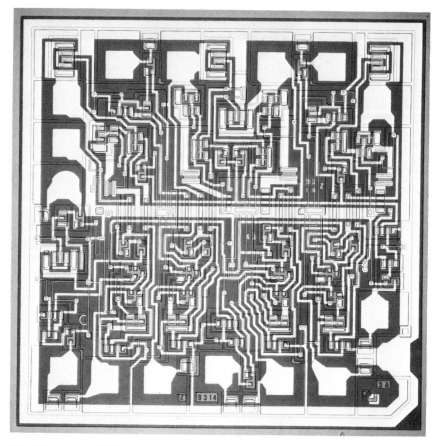

Fig. 14.5 A monolithic integrated circuit. (Photograph courtesy of Fairchild Semi-conductor.)

lithic circuits. Conversely, an n-type substrate is required for p-n-p type monolithic transistors. Therefore, it seems impossible to construct both n-p-n and p-n-p transistors on the same monolithic substrate, and thus take advantage of complementary symmetry and other convenient n-p-n–p-n-p arrangements. However, a *lateral p-n-p* transistor that uses a p-type substrate has been developed (Fig. 14.6), thus making possible the use of both n-p-n and p-n-p transistors on the same monolithic chip. The lateral p-n-p transistor has very low β (of the order of one) however, so a high gain n-p-n driver is always used with it to provide the needed current gain.

Thermally stable, high-resistance resistors were difficult to produce and space consuming until recently. Therefore, a transistor, acting as a current

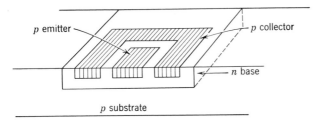

Fig. 14.6.   A lateral *p-n-p* transistor.

source, was often used as a high load resistance for another transistor. However, high-resistance, low-volume, thermally stable resistors can now be created by diffusing chromium into silicon.

14.3 BASIC LINEAR INTEGRATED CIRCUITS

The integrated-circuit manufacturer attempts to fill many of the needs of the user with a single type of integrated circuit, thereby creating large sales volume for each type. Since the differential amplifier, discussed in Chapter 10, can have excellent thermal stability and is useful for either dc or ac amplifiers in either balanced or single-ended operation, the differential amplifier is used in almost all linear integrated circuits.

A versatile amplifier should have high input impedance so that it will not load the commonly used driving sources; low output impedance so that it can drive a wide variety of load resistances; high gain, which can be controlled over wide limits with feedback; and broad bandwidth, which can also be controlled by feedback and compensation. The techniques used for accomplishing or controlling these characteristics will be discussed in the remainder of this chapter.

A simple, basic, linear integrated circuit (IC) is shown in Fig. 14.7. This circuit consists of a differential amplifier (T_1 and T_2) with a current source T_5 acting as a high emitter impedance, as discussed in Chapter 10, and a pair of emitter followers (T_3 and T_4) to provide low output impedance. This circuit may be ideally suited for some applications, but has the following disadvantages for other applications.

1. Insufficient voltage or current gain.
2. Only moderate input impedance.
3. Large offset voltage (volts) between both the input and output terminals with respect to ground, probably requiring the use of blocking capacitors.
4. Bias currents for transistors T_1 and T_2 must be provided from the external circuit.

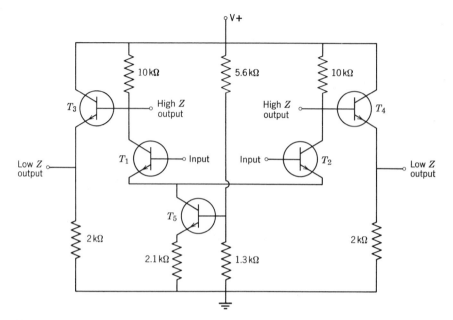

Fig. 14.7.   Schematic diagram of Fairchild μA 730 IC.

The integrated circuit shown in Fig. 14.8 eliminates most of the short-comings listed above. In this circuit the input differential amplifier (T_1 and T_2) is followed by a second differential amplifier (T_3 and T_4) to provide additional gain, and two emitter followers (T_5 and T_6) are connected in cascade to provide low impedance output. The resistor R_5 is placed in the emitter circuit of transistor T_5 to provide a dc voltage drop so that the output terminal will be at approximately dc ground potential. The transistors T_7, T_8, and T_9 are used to control the bias currents in the other transistors, as discussed later. Transistor T_9 also provides a *bootstrapping* function, which increases the voltage swing available at the output terminal and actually provides voltage gain between the base of transistor T_4 and the output of the amplifier. We shall illustrate this bootstrapping technique by assuming that the base of transistor T_5 is driven positive, therefore causing the emitter of T_5 and the base of T_6 to become more positive. If the bootstrapping circuit were missing, the potential at the base of T_6, and hence the output potential, would not rise as much as the emitter potential of T_5 because of the increased iR drop across R_5. However, the increased emitter current of the output transistor T_6 flows through resistor R_{11} in the emitter circuit of T_9 and thus tends to reverse bias T_9. But the reduced forward bias of T_9 reduces its collector current that flows through R_5. Therefore, the *total* current through R_5 actually

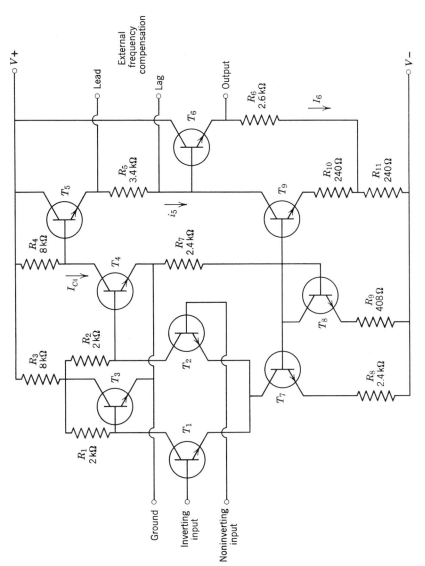

Fig. 14.8.   Schematic diagram of Fairchild μA 702 IC.

decreases and the base potential (and hence the emitter potential) of T_6 experiences a *greater* rise than either the base or emitter potential of T_5. Another way to look at the output circuit is to recognize that positive feedback is applied from the output through T_9 (as a common-base amplifier) to the input of T_6. The loop gain must be less than one in order that oscillation does not occur.

Transistors T_7, T_8, and T_9 provide stabilized bias to all the amplifiers in the IC, as illustrated in Fig. 14.9. This bias circuit is identical to its counterpart in Fig. 14.8 but is rearranged slightly for convenience. Transistor T_8 is connected as a diode so that the voltage drop across $R_7 + R_9$ is $V-$ minus V_{EB}. Thus the collector and emitter currents of T_8 are fixed almost completely by the supply voltage $V-$ if the transistor current gains are high so the base currents of T_9 and T_7 can be neglected. Also, since the transistors are essentially identical, their values of V_{BE} are very nearly equal and their emitter potentials are almost identical. Therefore the emitter, and hence collector, currents of the biasing transistors are controlled almost completely by their emitter circuit resistors.

EXAMPLE 14.1   The bias currents for the circuit of Figs. 14.8 and 14.9 will be determined as an example of the bias circuit design and operation. We assume that $V- = -8$ V and $V_{BE} = 0.5$ V for all transistors. Then for transistor T_8, $I_C \simeq |I_E| = (8 - 0.5)\text{V}/(2.4 \text{ k}\Omega + 480)\Omega = 2.6$ mA.   The voltage drop across the emitter resistor of T_8 is therefore approximately $(2.6 \text{ mA})(480 \,\Omega) = 1.25$ V. The voltage across the 2.4 kΩ resistor in the emitter circuit of T_7 must also be 1.25 V, so that the emitter current of

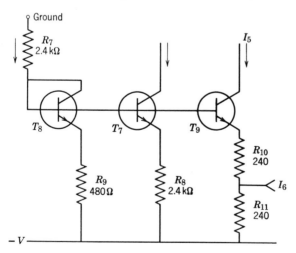

Fig. 14.9.   The stabilized bias circuit of Fig. 14.7.

transistor T_7 is 1.25 V/2.4 kΩ = 0.52 mA. The collector current of T_7 is essentially 0.52 mA, and this current divides equally between T_1 and T_2 whose collector currents must therefore each be about 0.26 mA and the voltage drop across R_1 and R_2 is (2 kΩ)(0.26 mA) = 0.52 V. The voltage drop across R_3 must therefore be $V+$ (8 V) minus (0.52 V + V_{BE}) = 6.98 V. The current through R_3 is then 6.98 V/8 kΩ = 0.87 mA and the collector current of transistor T_3 = (0.87 − 0.52) mA = 0.35 mA. The collector current of transistor T_4 must also be 0.35 mA since T_3 and T_4 form a balanced differential amplifier. The emitter current of transistor T_9 is not as easy to calculate because only current I_5 (approximately) flows through R_{10}, but both I_5 and I_6 flow through R_{11}. We *do* know that the sum of the voltage drops in the emitter circuit is 1.25 V. Therefore

$$I_5R_{10} + (I_5 + I_6)R_{11} = 1.25 \text{ V} \qquad (14.1)$$

we can find the value of I_5 required to provide zero dc output voltage, which is desired. If we neglect the base currents, Fig. 14.8 shows that

$$I_5R_5 + I_{C4}R_4 = V_+ - 2V_{BE} \qquad (14.2)$$

We have already found I_{C4} to be approximately 0.35 mA.

PROBLEM 14.1   Continue this example and determine both I_5 and I_6 assuming $V_{BE} = 0.5$ V and $V_+ = 8$ V. *Answer:* $I_5 \simeq 1.23$ mA, $I_6 = 2.7$ mA.

14.4 OPERATIONAL AMPLIFIERS

The term *operational amplifier* was coined by people in the analog computer field and was used to designate an amplifier that has very high open-loop gain, high input impedance, and low output impedance. This type of amplifier is very versatile because feedback can be used to control its characteristics, thus making it useful for a wide variety of applications. Operational amplifiers usually have values of open-loop voltage gain between 10,000 and 100,000. A triangular symbol is used to represent the entire amplifier (Fig. 14.10). The amplifier is normally an integrated circuit with differential input, as was previously discussed. The feedback circuit must always be connected to the *inverting* input for negative feedback, but the signal may be applied to either input terminal, depending on the required input impedance of the amplifier. When the signal is applied to the noninverting input, as shown in Fig. 14.10, the amplifier input impedance is very high at both input terminals because the feedback voltage is nearly equal to the input voltage and the two voltages

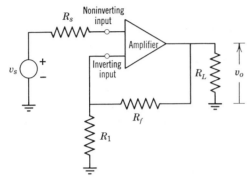

Fig. 14.10. An operational amplifier used as a linear amplifier with stable gain and very high input impedance.

appear as a common-mode input signal. This signal sees a very high input impedance because of the very high common-mode impedance in the emitter circuit. The difference between the signal source and feedback signals is the differential signal that is effective in driving the amplifier. In other words, the feedback arrangement of Fig. 14.10 is series feedback and the open-loop differential input impedance of the amplifier is multiplied by the factor $1+K\beta$ as a result of the feedback, as was discussed in Chapter 13. Since the effective amplifier impedance is so high, the feedback factor is

$$\beta = \frac{R_1}{R_1 + R_f} \qquad (14.3)$$

and the reference gain of the amplifier is

$$K_f = \frac{1}{\beta} = \frac{R_1 + R_f}{R_1} \qquad (14.4)$$

assuming the open-loop gain is much higher than the closed-loop gain so that $K\beta \gg 1$, as discussed in Chapter 13. The amplifier may require compensation if small values of closed-loop gain are used. However, some integrated circuits have built-in compensation, which provides stable operation for all values of closed-loop gain down to unity. Compensation techniques will be discussed in Section 14.5.

Sometimes it is desirable to have very small impedance and very small voltage at the amplifier input terminals. These characteristics are obtained when the input and feedback signals are both applied to the inverting input terminal of the amplifier, as shown in Fig. 14.11. This is the *series* input connection discussed in Chapter 13. The noninverting input terminal

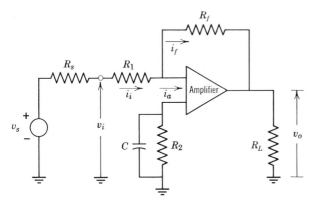

Fig. 14.11. An operational amplifier with the input signal applied to the inverting input terminal.

is maintained at signal ground potential by capacitor C. The amplifier input current i_a is very small, usually much smaller than one microampere, so that both i_i and i_f are very large in comparison with i_a. Then $i_i \simeq i_f$. Also, the input voltage of the amplifier is very small. Being the differential input voltage, it is equal to the output voltage divided by the voltage gain of the amplifier. Therefore, the voltage at the inverting input terminal is of the order of microvolts and can be assumed to be essentially zero in comparison with either the output voltage or the driving source voltage. Then

$$i_i \simeq \frac{v_s}{R_s + R_1} \tag{14.5}$$

and

$$i_f \simeq \frac{v_o}{R_f} \tag{14.6}$$

But since $i_i \simeq i_f$, by using Eqs. 14.5 and 14.6,

$$\frac{v_s}{R_s + R_1} = \frac{v_o}{R_f} \tag{14.7}$$

Therefore, the closed-loop voltage gain of the amplifier is

$$K_v = \frac{v_o}{v_s} = \frac{R_f}{R_s + R_1} \tag{14.8}$$

Sometimes the voltage gain from the output to the input terminals labeled v_i is wanted rather than the ratio v_o/v_s. This gain is

$$K_v' = \frac{v_o}{v_i} = \frac{R_f}{R_1} \tag{14.9}$$

The resistor R_1 can be made large compared to R_s. Then $K_v' \simeq K_v$.

EXAMPLE 14.2   An example will illustrate the use of the relationships above and demonstrate the small error introduced by the approximations that were used. We assume that an amplifier is needed that will provide an input impedance of $10 \text{ k}\Omega$ and a voltage gain $v_o/v_i = 100$. The driving source resistance is $1 \text{ k}\Omega$. An operational amplifier with open-loop voltage gain $= 5 \times 10^4$ and input resistance $= 100 \text{ k}\Omega$ will be used. The resistor $R_1 = 10 \text{ k}\Omega$ (Fig. 14.11) is used to provide the desired input resistance and $R_f = 100 \ R_1 = 1 \text{ M}\Omega$ will provide the desired gain. The voltage gain from voltage source to output is $v_o/v_s = 10^6/(1.1 \times 10^4) = 91$. We shall now determine the errors resulting from the assumption that $i_a \simeq 0$ and $v_{ia} \simeq 0$, assuming that the output signal is 5.0 V. The amplifier input voltage is then $v_{ia} = 5/(5 \times 10^4) = 10^{-4}$ V, or 100 μV and the signal input current to the amplifier is $10^{-4} \text{ V}/10^5 \ \Omega = 10^{-9}$ A or 1 nA. The feedback current $i_f \simeq 5 \text{ V}/10^6 \ \Omega = 5 \ \mu$A, which is 5000 times as large as the amplifier input current i_a. Since the current from the driving source flowing through R_1 can differ from i_f only by the amount i_a, this difference can be only one part in 5000 or 0.02 percent. Similarly, the driving source voltage is 5 V/91 = 0.055 V, which is 550 times as large as the $v_{ia} = 10^{-4}$ V that we neglected. Therefore, the approximations are extremely good and the gain of the amplifier is essentially as stable as the resistance values of R_1 and R_f.

The bias currents for the differential input stage of the operational amplifier normally flow through the external resistors connected to the input terminals. Therefore, the dc resistance from each of the two input terminals and ground should be approximately the same. Otherwise the $I_B R$ drop across the two resistors will be different and a dc-*offset* voltage will appear between the two input terminals. This offset voltage is multiplied by the closed-loop gain of the amplifier and appears in the output. Of course, a small offset voltage appears at the input due to imperfect balance in the input differential amplifier, but this offset can be minimized by proper balancing of the input circuit resistance. For example, the resistor R_2 in Fig. 14.11 should be equal to the parallel combination of R_f and $(R_1 + R_s)$, assuming there is no blocking capacitor in the driving source. If there is *no* dc path through the driving source, then $R_2 = R_f$.

The operational amplifier is commonly used as a *mixer* or *summer* (Fig. 14.12). Using the reasoning above, $i_1 + i_2 + i_3 + i_4 = i_f$. But $i_1 = v_1/R_1$, $i_2 = v_2/R_2$, $i_3 = v_3/R_3$, $i_4 = v_4/R_4$ and $i_f = v_o/R_f$. Then

$$\frac{v_o}{R_f} = \frac{v_1}{R_1} + \frac{v_2}{R_2} + \frac{v_3}{R_3} + \frac{v_4}{R_4} \tag{14.10}$$

and

$$v_o = \frac{R_f}{R_1} v_1 + \frac{R_f}{R_2} v_2 + \frac{R_f}{R_3} v_3 + \frac{R_f}{R_4} v_4 \tag{14.11}$$

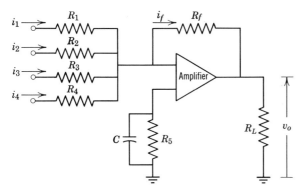

Fig. 14.12.   A summing amplifier.

Of course any desired number of inputs can be used. If a simple summing or adding operation is desired, then $R_1 = R_2 = R_3 = R_4 = R_f$. However, the voltage at any or all inputs can be amplified, as indicated by Eq. 14.11. The resistance of R_5 should be equal to the total dc (external) resistance from the inverting input terminal to ground, of course, for low off-set voltage.

PROBLEM 14.2   Use the circuit of Fig. 14.12 to add the signals from three microphones that have output voltages of 2 mV, 4 mV, and 10 mV, respectively, at normal voice levels of 1 microbar acoustic pressure. It is desired that each microphone produce 0.1 V at the output of the amplifier and that the load resistance for any of the microphones should be 5 kΩ or higher. Design the mixer-amplifier circuit using an operational amplifier with an open-loop gain of 5×10^4 with $R_L \geqslant 2\,\text{k}\Omega$. Assume $R_s \ll 5\,\text{k}\Omega$.
Answer: $R_1 = 5\,\text{k}\Omega$, $R_2 = 12.5\,\text{k}\Omega$, $R_3 = 25\,\text{k}\Omega$, $R_f = 250\,\text{k}\Omega$, R (non-inverting input) = 3.1 kΩ.

14.5   AMPLIFIER COMPENSATION

Some compensation techniques that will provide stable operation and acceptable transient response of an amplifier with feedback were discussed in Chapter 13. The ideas developed there are very briefly reviewed here with the aid of Fig. 14.13 where curve a is the straight-line approximation of the open-loop frequency response curve of a representative operational amplifier with 30,000 or 89 dB reference gain and break frequencies at 150 kHz, 1.5 MHz, and 15 MHz. You may recall from Chapter 13 that approximately 45° or more phase margin will be maintained if the straight-line approximation of the closed-loop response curve intersects the open-loop response curve at a point where the difference in

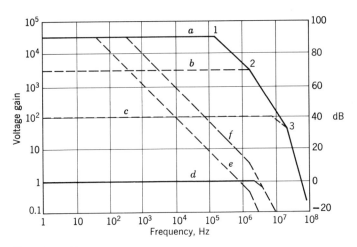

Fig. 14.13.   Frequency response curves.

their slopes is only 6 dB/octave or 20 dB/decade. Therefore, no compensation is required in either the amplifier or the feedback network if the closed-loop reference gain is 3000 or higher, as shown by curve b. This gain corresponds to 20 dB of feedback $(1 + K\beta = 10)$. Curve c shows that the closed-loop gain can be reduced to approximately 100 if phase-lead compensation is used in the feedback network with $R_f C_f \simeq 1/2\pi \times 10^7$ $(1 + K\beta = 300$, or 45 dB of feedback). Sometimes a lower closed-loop gain such as 10 is desired, or even a gain of 1, perhaps, which is common for a summing amplifier. Then the *amplifier* must be compensated to achieve adequate stability. A simple type of compensation is achieved by adding capacitance in parallel with the output of one of the amplifier stages, as was discussed in Chapter 13. If wide bandwidth is not important, no compensation is required in the feedback network if the compensating capacitor produces a sufficiently low break frequency so that the 6 dB/octave slope of the open-loop gain curve of the compensated amplifier intersects the closed-loop gain curve before the second break frequency is encountered. For example, in Fig. 14.13 the desired closed-loop gain curve d is drawn at unity gain. The compensating capacitor is added in the output of the amplifier stage that had the break frequency at 150 kHz, which is reduced to about 40 Hz by the compensation. The open-loop gain curve e then decreases at 20 dB/decade from 30,000 gain at 40 Hz to unity gain at 0.9 MHz, which is less than the second break frequency at 1.5 MHz. Thus, the amplifier is stable with an uncompensated feedback network. Some operational amplifiers have *built-in* compensation of this type and are therefore stable for any amount of uncompensated feedback to unity gain.

Bandwidth is sacrificed in this type of amplifier however, because additional bandwidth could be obtained by using phase-lead compensation in the feedback network, as shown by curve f, where the point of intersection of f with the closed-loop gain curve can occur at a frequency higher than the second break frequency. You may recall that the upper-cutoff frequency of the amplifier occurs approximately at the first break frequency in the closed-loop response. This break frequency is located at the intersection with the open-loop response curve when no phase-lead compensation is used in the feedback network, or it is the break frequency produced by the compensating network when phase-lead compensation is used.

Observe from Fig. 14.13 the large amount of bandwidth that is sacrificed when an operational amplifier with built-in compensation is used with small amounts of feedback that produce high closed-loop gains.

PROBLEM 14.3   Determine the approximate bandwidth of the amplifier of Fig. 14.13 if the closed-loop gain is 3000 and the amplifier is (a) internally compensated, (b) uncompensated.   *Answer:* (a) 400 Hz, (b) 1.5 MHz.

PROBLEM 14.4   How much capacitance is needed to compensate the amplifier of Fig. 14.13 if curve e is the desired open-loop response and the total effective resistance between the two collector terminals to which the compensating capacitor is attached is 8 kΩ.   *Answer:* 0.5 μF.

14.6   COMPENSATION FOR OPTIMUM RISE TIME AND SLEW RATE

The *slew rate* of an amplifier is the rate at which the output voltage of the amplifier rises toward the supply voltage when some or all of the stages of the amplifier are either cut off or in saturation. In other words, the amplifier is not operating in its normal linear mode. Slew rate is usually expressed in volts per microsecond. This slew rate becomes important when fast-rising rectangular pulses are applied to the input of an amplifier that has high open-loop gain, such as an operational amplifier, but has a large feedback factor that reduces the gain to a small value. Although the input pulse may seemingly be too small to overload the amplifier, the delay time and rise time of the amplifier prevent the feedback pulse from providing adequate cancellation of the leading edge of the input pulse. Thus a large, sharp, disabling spike is applied as a differential input signal to the amplifier, as shown in Fig. 14.14. The output voltage does not then rise at the rate predicted on the basis of linear amplifier theory because the amplifier is not entirely operative, but rises at the slower slew rate instead. Thus slew rate becomes an important parameter

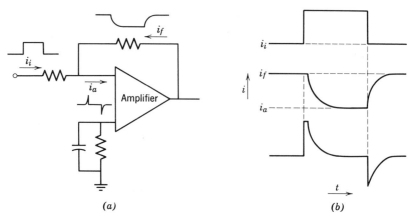

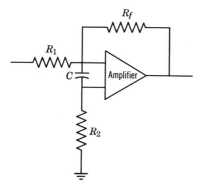

Fig. 14.14.   Illustration of the large input spikes that result from fast-rising pulses, large amounts of feedback, and amplifier delay and rise time.

in amplifier design. Compensating capacitors can greatly decrease or slow the slew rate when they are placed near the output of an amplifier because the lower level stages then cut off, and the output rises slowly due to the large time constants caused by the large compensating capacitance. On the other hand, a compensating capacitance near the input of the amplifier may reduce the magnitude of the input spike and help maintain linear operation of the amplifier, as illustrated in Fig. 14.15, where the compensating capacitor C is placed at the input of the amplifier. This capacitor slows the rate of rise of the input pulse and thus greatly reduces the size of the input spike to the amplifier, thereby permitting linear amplifier operation over a wide range of magnitude of input signals. This single compensating capacitor may not provide optimum rise time for the amplifier, however, because a compensating capacitance usually needs to either

Fig. 14.15.   Input compensation that increases rise time of the circuit.

absorb or remove a low-frequency break point in the open-loop gain curve in order to obtain wide bandwidth, and the input capacitor C may not accomplish this function. However, input compensation may be used in conjunction with another compensating scheme to provide maximum bandwidth and hence minimum rise time, as well as immunity from slow slew rates. A compensating scheme that has these characteristics will be illustrated by the following example.

EXAMPLE 14.3   We assume that a high-speed summing amplifier is needed to provide less than a 100-ns rise time and less than a 5-percent overshoot at unity gain. The desired input resistance is 12 kΩ. A μA715C integrated circuit is chosen as the operational amplifier because of its published frequency characteristics. An equivalent circuit of the μA715C is given in Fig. 14.16. The input stage of this amplifier is a Darlington-connected cascode differential amplifier utilizing transistors Q_1, Q_3, and Q_{16} on one side and Q_2, Q_4, and Q_{17} on the other side of the differential pair. The second stage is a differential amplifier consisting of Q_{18} and Q_{19}. Compensation terminals are brought out at both the input and the output of this stage. A Darlington-connected emitter follower (Q_{21} and Q_{22}), which prevents loading and provides dc level adjustment in the output of the second stage, is followed by a Darlington-connected common-emitter stage (Q_{10} and Q_{11}), which drives the complementary symmetry emitter follower output (Q_{14} and Q_{15}). All the other transistors in the circuit are used to provide the proper bias currents for the amplifying circuits, as was previously discussed.

The straight-line approximation of the frequency-response curve of the μA715C is given in Fig. 14.17. The open-loop response curve (without compensation) shows that there are three break frequencies at about 110 kHz, 1 MHz, and 20 MHz, respectively. The desired closed-loop gain (with feedback) is unity. Therefore, the desired open-loop response with compensation in the amplifier intersects the closed-loop response at a frequency somewhat lower than the third break frequency. This point of intersection was chosen at 10 MHz, which provides a theoretical rise time of $2.2/(2\pi \times 10^7) = 35$ ns. A compensation scheme can hopefully be found to cause the open-loop voltage gain to rise uniformly at the rate of 6 dB/octave or 20 dB/decade as the frequency is decreased until the low-frequency open-loop gain of 30,000 is reached. The single break frequency to the left of the intersection of the open-loop and closed-loop gain curves then occurs at $10^7/(3 \times 10^4) = 330$ Hz.

Part of the compensation is accomplished at the amplifier input to relieve the slew-rate problem (Fig. 14.18). The 1.3-kΩ resistor is connected in series with the compensating capacitor C to eliminate the phase shift of this capaci-

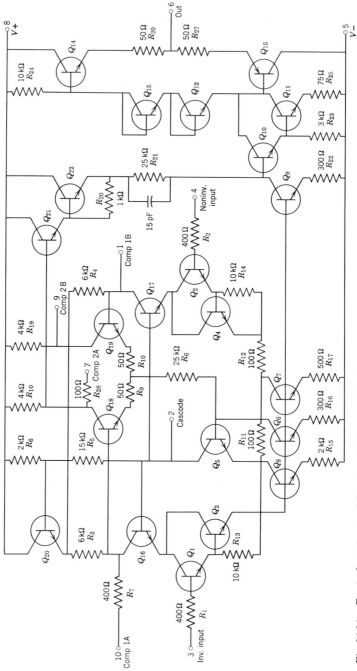

Fig. 14.16. Equivalent circuit of the μA715C. Courtesy Fairchild Semiconductor Company.

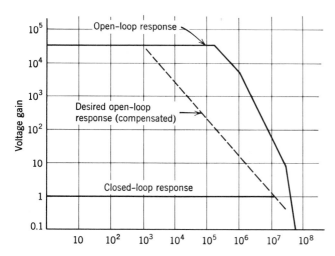

Fig. 14.17.   Frequency response curves for the μA715C.

tor at higher frequencies, thus permitting the use of additional compensation at higher frequencies. The frequency and phase response of the equivalent open-loop input circuit (feedback circuit grounded at output end) is given in Fig. 14.19. The input resistance of the μA715C is assumed to be so high that it can be neglected. The voltage v_a is the voltage at the input terminals of the amplifier. The gain and phase response shown can be easily verified qualitatively by recognizing that essentially all the input voltage is applied to the amplifier input terminals at low frequencies where $\chi_c \gg$ 13.3 kΩ and the phase shift is essentially zero because the capacitor behaves like an open circuit. On the other hand, the capacitor behaves like a

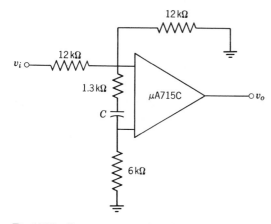

Fig. 14.18.   Input compensation circuit.

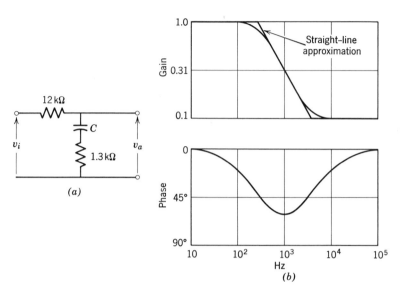

Fig. 14.19.   The equivalent compensated input circuit of Fig. 14.16 and its relative gain and phase. (a) Equivalent circuit and (b) gain and phase.

short circuit at frequencies for which $\chi_c \ll 1.3\ k\Omega$ and the circuit appears as a resistive voltage divider with 0.1 gain and zero phase shift. The lower break frequency in the straight-line approximation is often called a *pole* and the response is said to *roll off* at 6 dB/octave (or 20 dB/decade) above this frequency. The value of capacitance required to produce this desired pole at 330 Hz is $C = 1/(2\pi \times 330 \times 13.3 \times 10^3) = 0.036\ \mu F$. The higher break frequency in the straight-line approximation where the roll-off ceases, occurs where $\chi_c = 1.3\ k\Omega$ and is often called a zero. This frequency occurs at about 3.3 kHz.

The next part of the compensation strategy is to compensate the second differential amplifier stage in the $\mu A715C$ so its lower break frequency, or pole will occur at 3.3 kHz and thus continue the 6 dB/octave roll-off initiated by the input compensation. This compensation can be accomplished with minimum capacitance, and hence maximum slew rate, if the compensating capacitance is included in a feedback circuit as shown in Fig. 14.20. The 500-Ω resistor is included internally in series with compensation terminals 7 and 10. A 500-Ω resistance must be added externally in series with C between 1 and 9 to keep the differential amplifier balanced. Although balance is not necessary for satisfactory operation, it does simplify the theoretical compensation calculations. With the amplifier balanced we can look at only one side of it, as shown in Fig. 14.21a. The open-loop gain characteristics were deduced by observing that the emitter

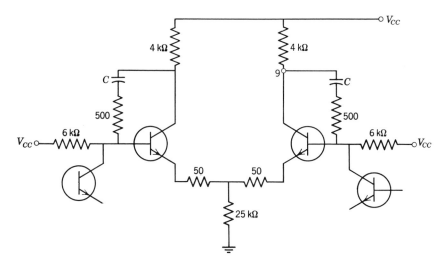

Fig. 14.20.   Circuit showing the compensating capacitor C in a feedback loop around the second differential amplifier.

circuit of this stage has a 25-kΩ resistor in it, across which the dc voltage drop may not exceed about 25 V if $+15$ and -15 V power supplies are used, as suggested by the manufacturer. Thus the q-point collector currents of each of the two transistors are about 0.5 mA each. Then the internal emitter resistance is approximately $r_e = 25/0.5 = 50\ \Omega$. If the transistor current gain is high, the input voltage is about the same as the voltage drop

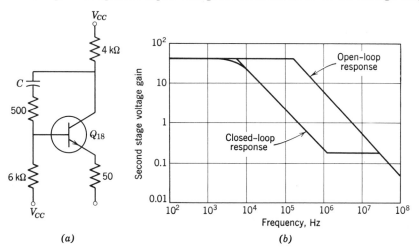

Fig. 14.21.   The equivalent circuit and frequency response of the second differential stage.

across the total emitter circuit resistance so that the approximate voltage gain K_v is $R_C/(r_e + R_E)$, as was discussed in Chapter 7. Therefore, $K_v \simeq$ 4 kΩ/100 = 40. The break frequency, or pole, at 110 kHz was assumed to be the upper-cutoff frequency of this stage, since the input stage is cascode and the other common-emitter stage, which drives the output, is preceded by an emitter follower that increases its bandwidth. With these two assumptions, the open-loop frequency response curve given in Fig. 14.21b was drawn. However, the desired compensated response has a break frequency, or pole, at 3.3 kHz to continue the 6 dB/octave roll-off initiated by the input compensation.

Since the compensating capacitor C (Fig. 14.21a) is in a feedback loop, the feedback equation must be used to calculate the required capacitance to produce a break frequency, or pole, at 3.3 kHz. This equation, developed in Chapter 13, is

$$G_f = \frac{G}{1 + G\beta(\omega)} \qquad (14.12)$$

where G_f is the gain with feedback, G is the forward open-loop gain, and $\beta(\omega)$ is the feedback factor that is a function of frequency because of the capacitor C. The pole, or break frequency, in the closed-loop gain G_f occurs when $G\beta(\omega)$ has a magnitude of one and a phase angle of 90°. At the break frequency, χ_c is much larger than the resistances in the circuit, as seen later, so the phase angle is near 90°. Then, at the break frequency, f_b

$$|G\beta(\omega_b)| = 1 \qquad (14.13)$$

But the magnitude of G (or K) has already been estimated at 40 and inspection of Fig. 14.21 shows that

$$\beta(\omega) = \frac{6k \| r_{in}}{(6k \| r_{in}) + 500 + (1/j\omega C)} \qquad (14.14)$$

Since $r_{in} = r_b + (r_e + R_E)h_{fe} \simeq 100h_{fe}$, and since a typical value for h_{fe} is 100, r_{in} is of the order of 10 kΩ, and 6 kΩ in parallel with 10 kΩ is about 4 kΩ. Then

$$\beta(\omega) = \frac{4 \times 10^3}{4.5 \times 10^3 + (1/j\omega C)} \qquad (14.15)$$

Rationalizing Eq. 14.15,

$$\beta(\omega) = \frac{4 \times 10^3 (j\omega C)}{1 + 4.5 \times 10^3 (j\omega C)} \qquad (14.16)$$

Substituting this expression for $\beta(\omega)$ and $K = 40$ into Eq. 14.13

$$\left| \frac{40 \times 4 \times 10^3 (j\omega_b C)}{1 + 4.5 \times 10^3 (j\omega_b C)} \right| = 1 \tag{14.17}$$

Since the magnitude of the denominator of the fraction on the left of Eq. 14.17 must be equal to the magnitude of the numerator, but the j term in the numerator is nearly 40 times as large as the j term in the denominator, this denominator j must be negligible in comparison with one. Then

$$|j1.6 \times 10^5 \omega_b C| = 1 \tag{14.18}$$

The j in the left side of Eq. 14.18 can be dropped because of the magnitude sign. Its presence indicates the needed $90°$ relationship previously discussed. Then

$$C = \frac{1}{R_b' K \omega_b} = \frac{1}{1.6 \times 10^5 \times 2.1 \times 10^4} = 280 \text{ pF} \tag{14.19}$$

where R_b' is the total shunt resistance between the base of Q_{18} and ground. The nearest stock size capacitor is 270 pF. The 500 Ω resistor in series with C causes a zero in the response of the second stage at $\omega = 1/500 \times 2.7 \times 10^{-10} = 7.4 \times 10^6$ or $f = 1.2 \times 10^6$. Therefore the 6 dB/octave roll-off in response stops at this frequency as shown by the closed-loop response curve of Fig. 14.21b. However, one of the other stages in the μA715C has a break frequency or pole that occurs at about 1.0 MHz (Fig. 14.18). The resistance in series with C in the feedback network of the second stage can be increased slightly to move the zero in the second stage back to 1.0 MHz, and then the pole at 1.0 MHz will neutralize the effect of the zero and the 6-dB/octave roll-off will continue until the compensated open-loop gain of the entire μA715C crosses the closed-loop gain curve at unity gain as desired. The resistance required to place the zero in the second stage at 1.0 MHz is $R = 1/\omega C = 1/6.28 \times 10^6 \times 2.7 \times 10^{-10} \simeq 600 \ \Omega$.

The phase margin of the operational amplifier can be increased by adding a phase-lead capacitor in parallel with the 12 kΩ feedback resistor. The break frequency of the phase-lead network should be near the intersection of the open-loop and closed-loop gain curves at 10 MHz. This criterion gives $C_f = 1/6.28 \times 10^7 \times 1.2 \times 10^4 = 1.3$ pF. However, the amplifier *without* the phase-lead capacitor, as shown in Fig. 14.22a, gave ideal time response to a rectangular input pulse, as shown in Fig. 14.22b.

PROBLEM 14.5 A given IC-operational amplifier has the same equivalent circuit as the μA715C except the 25-kΩ resistance in the second

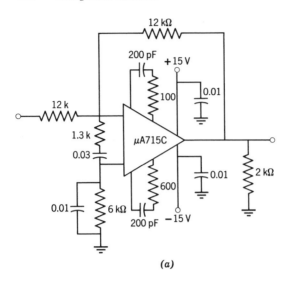

(a)

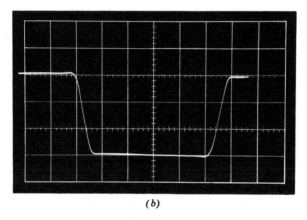

(b)

Fig. 14.22.   (a) Circuit diagram and (b) oscillogram of the time response of the unity gain amplifier. Vertical scale is 1 V/cm and horizontal scale is 50 ns/cm.

differential stage is reduced to 12 kΩ, the open-loop voltage gain is 5×10^4, the three break frequencies in the open-loop frequency-response curve occur at 150 kHz, 2.0 MHz, and 20 MHz, and there is no built-in resistance in series with the compensation terminals. Design a compensation circuit similar to the one in Example 14.3 using $R_f = R_1 = 12$ kΩ for unity gain.

Typical answer: Same input compensation. Second stage $C = 200$ pF, $R = 400$ Ω.

14.7 ADDITIONAL TYPES OF INTEGRATED CIRCUITS

Many types of linear integrated circuits are available to meet special needs. Their circuits, performance characteristics, and applications are available from the various manufacturers. Also selection guides are occasionally published in the electronics periodicals. The electronics technician or engineer should become familiar with these sources of information. Only a few types of special linear IC's can be presented here.

One special type of linear IC is the wideband video amplifier. The circuit diagram of a typical video amplifier is given in Fig. 14.23a and its frequency response curves for $R_s = 50\ \Omega$ are given in 14.23b. The amplifier consists of two differential stages with negative feedback around the second stage, and a balanced emitter-follower output. The gain of the amplifier is controlled by the degenerative resistance in the emitter circuit of the first differential amplifier. The maximum voltage gain of 400 is obtained by connecting terminal G_{1A} to G_{1B}. The intermediate gain of 100 is obtained by connecting terminal G_{2A} to G_{2B}. The low gain (10) is obtained without any connections of the gain-select terminals. Continuously variable gain from about 10 to 400 may be obtained by connecting a 10 $k\Omega$ potentiometer between terminals G_{1A} and G_{1B}. No external feedback or compensation is needed for any of these gain values. The amplifier may be used either single-ended input or single-ended output, or both. In fact, the voltage gains given in Fig. 14.23b are for single-ended output and are therefore one-half or 6 dB lower than the gain values given above.

Another special type of integrated circuit is the RF/IF amplifier. These amplifiers differ from the operational or video amplifiers in that the amplifier output impedance needs to be high, rather than low, to permit proper bandwidth in the tuned coupling circuit, as discussed in Chapter 9. Some amplifiers bring out high-impedance terminals to accommodate tuned as well as untuned loads. Provision for automatic gain control is also convenient in an RF/IF amplifier because they are often used in radio receivers where automatic gain control is needed to maintain relatively uniform output voltage in spite of the wide variety of input signal levels.

The major part of a radio receiver is available in a single IC, as shown in Fig. 14.24. This μA719 amplifier has three differential, or emitter-coupled, stages between the input and output of section 1. The input, usually from a tuned circuit, is applied between the high input 1 and the low input 1 terminals. The output of this section is obtained at high impedance between the collector of Q_s, which is emitter coupled to Q_6, and $V+$. A tuned circuit would be used in the output if section 1 is used only as an RF or IF amplifier. However, either AF or video can be obtained across a

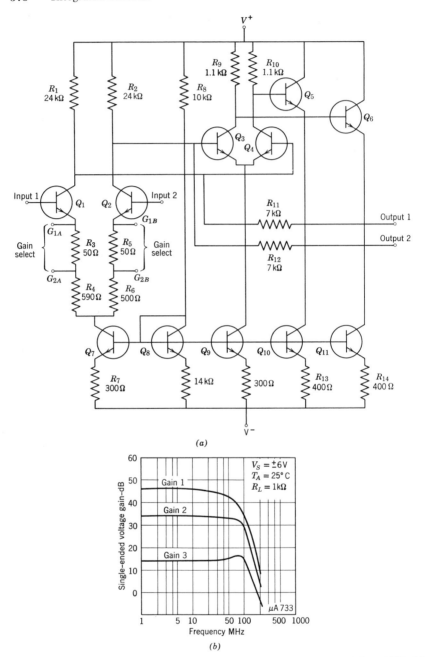

Fig. 14.23. (a) Circuit diagram and (b) frequency response curves of the μA733 video amplifier. Courtesy Fairchild Semiconductor Company.

resistance (about 25 kΩ with RF bypass) if a parallel tuned circuit is connected between the QUAD terminal and the low INPUT 1 terminal, so the
transistors Q_6, Q_7, and Q_8 become a quadrature FM detector. Section 2
has about a 50-MHz bandwidth and will serve as an RF, video, or AF
amplifier with about a 35-mmho transconductance if the DECOUPLE
terminal is bypassed to ground or about an 8-mmho transconductance if
the DECOUPLE terminal remains free.

The final special type integrated circuit to be considered is the voltage
comparator. The comparator is normally used to indicate whether or not
the noninverting input is positive with respect to the inverting input. The
indication, of course, is the output voltage that changes from negative to
positive very quickly and distinctly as the relative polarity of the input
changes. Therefore a high-gain, nonsaturating amplifier is needed. The
nonsaturating feature is needed because of the time required to pull a
transistor out of saturation. This time is known as *storage time* and is
discussed in Chapter 16.

The schematic diagram and the transfer characteristics of a typical
comparator are given in Fig. 14.25. The differential input stage Q_1 and Q_2
is followed by a second differential stage Q_3 and Q_4, which in turn drives a
single-emitter follower Q_7. As the noninverting input goes positive with
respect to the inverting input, the base and emitter of Q_7 go positive until
the base of the diode-connected transistor Q_6 becomes positive with respect
to its emitter. Then resistors R_4 and R_5 are clamped at almost the same
potential at their lower ends, and the collector currents of both Q_3 and Q_4
flow through these two resistors. But since Q_3 and Q_4 are the two halves of a
differential amplifier, the sum of their collector currents is constant and
the drops across R_4 and R_5 are constant. Thus the output voltage cannot
rise above about 3 V, as shown by the transfer characteristics of Fig. 14.25,
which is $V+$ minus the sum of the drop across R_5, plus V_{BE7}, plus the 6.2 V
reference voltage. When the noninverting input goes negative with respect
to the inverting input, the transistor Q_4 is driven to saturation, which also
places Q_8 approximately in saturation because the output voltage must be
$(6.2\text{ V} + V_{BE7})$ negative with respect to the collector of Q_4, which is only
$(6.2 + V_{CE\,sat})$ positive with respect to ground. Therefore, the output voltage
is clamped at approximately -0.5 V which is $(V_{BE7} - V_{CE\,sat})$ while the
input is negative, as shown by the voltage transfer characteristic in Fig.
14.25. Note that approximately a 1-mV input signal is required to drive the
output to its clamped voltage in either direction. The biasing circuits control
the currents so that the transistors are not driven appreciably into saturation during normal operation.

The voltage comparator is used in digital voltmeters or analog-to-digital
converters where a stable, precise reference voltage is applied to the

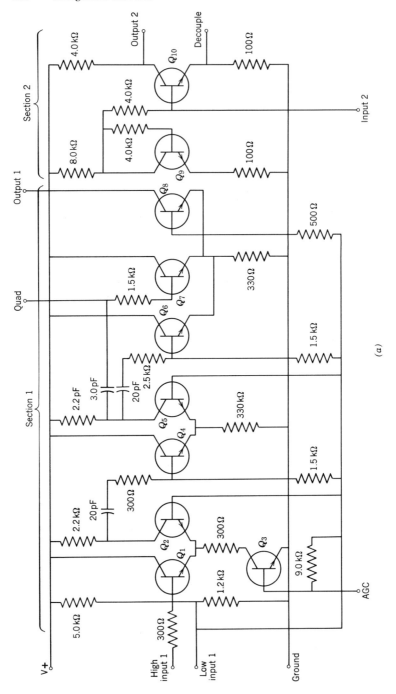

(a)

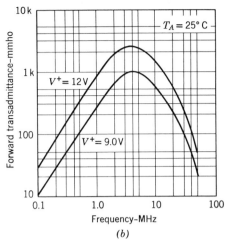

Fig. 14.24.   A high-gain RF/IF amplifier, FM detector and AF/video amplifier μA719. Courtesy Fairchild Semiconductor Company. (a) Schematic diagram and (b) section 1, forward transadmittance as a function of frequency.

inverting input and the output voltage operates some type of read out when the reference voltage is exceeded. The comparator is also convenient for changing sine waves to square waves, cleaning up a noisy binary signal, or performing many other functions described in the literature.

A careful study of the manufacturers data and literature is almost essential prior to the choosing and application of a linear IC to a specific circuit problem. The study of this chapter will hopefully help you to understand the literature and data sheets.

PROBLEM 14.6   The comparator of Fig. 14.25 is used to produce a square wave from a 1.0-V peak, 60-Hz sine wave. What will be the approximate peak-to-peak magnitude and rise time of the square wave? Hint: Use $\sin \theta \simeq \theta$ (in radians) for small angles.   *Answer:* 3.5 V, 5 μs.

PROBLEM 14.7   For the amplifier of Fig. 14.24, determine the voltage gain of section 1 when $V_+ = 12$ V, the signal is a 10.7 MHz IF, and the impedance of the tuned circuit in the output is 5 kΩ.   *Answer:* 6000.

PROBLEM 14.8   What is the gain-bandwidth product and the rise time of the amplifier of Fig. 14.23 with single ended (a) gain 1, (b) gain 2?
          Answer: (a) 8000 MHz, 9 ns, (b) 5000 MHz, 3.5 ns.

PROBLEM 14.9   A μA715C amplifier is to be used to provide voltage

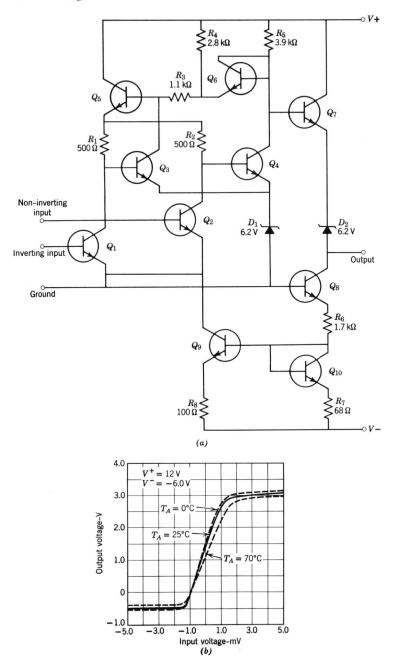

Fig. 14.25.   (a) The schematic diagram and (b) voltage transfer characteristics of a typical voltage comparator, μA710. Courtesy Fairchild Semiconductor Company.

gain $= 5000$. The driving source resistance is $10\,k\Omega$ and the amplifier should present a very high impedance to the source. Devise a feedback network and compensation, if necessary, to give the desired characteristics. What is the expected bandwidth and rise time of your amplifier?

Answer: $R_1 = 10\,k\Omega$, $R_f = 5$ meg with $10:1$ voltage divider in the output.
$$B = 1\,\text{MHz}, t_r = 0.35\,\mu\text{s}.$$

PROBLEM 14.10   Repeat Prob. 14.9 for voltage gain $= 1000$.

Answer: $R_f = 90\,k\Omega$, $C_f = 1$ pF with $100:1$ voltage divider in the output.
$$B \simeq 2\,\text{MHz}, t_r \simeq 170\,\text{ns}.$$

PROBLEM 14.11   A μA715C is to be used as a summing amplifier with a 10-$k\Omega$ input resistance, voltage gain $= 10$, and rise time $= 40$ ns or less. Design a feedback and compensation system for the amplifier.

CHAPTER 15

POWER SUPPLIES

Essentially all amplifiers require a source of dc power for their operation. In the preceding work we have assumed that a battery of the desired voltage and current capacity is always available. Although batteries are frequently used as the power source, cost and convenience often dictate the use of rectifiers and filters that convert ac power to dc power. These ac to dc converters are commonly known as *ac power supplies*. Motor generator sets, thermocouples, solar panels, and fuel cells are occasionally used as dc power sources, but these devices will not be discussed in this book.

15.1 RECTIFIERS WITH CAPACITOR-INPUT FILTERS

Half-wave and full-wave rectifiers with capacitors as smoothing or filtering elements were discussed briefly in Chapter 3. Full-wave rectifiers with capacitor smoothing elements are used in almost all ac power supplies designed for semiconductor circuits. Therefore, they warrant further discussion here. Inductors or *chokes* were frequently used in power supplies designed for electron-tube circuits and are discussed in the literature.[1]

Figure 15.1 shows a full wave rectifier circuit which incorporates a center-tapped transformer and a smoothing capacitor. The input power

---

[1]See, for example, *Electronic Engineering*, C. L. Alley and K. W. Atwood, Second Edition, John Wiley and Sons, Inc., New York, N. Y.

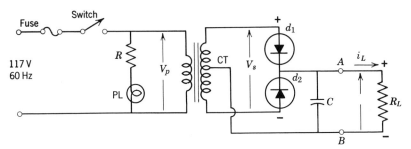

Fig. 15.1.   A full-wave rectifier and filter using a center-tapped transformer.

is assumed to be 117 V, 60 Hz. A typical fuse, switch, and pilot light arrangement is shown. If the pilot light is across the transformer primary, as shown, a neon light is usually used with a series current-limiting resistor R of about 50 kΩ. The load resistor R_L represents the power consumed by the electronic device (such as an amplifier).

When the upper terminal of the transformer (Fig. 15.1) is positive with respect to the center tap (CT), the upper diode d_1 conducts and current flows through the load in the direction shown. In addition, the capacitor charges. The voltage across both R_L and C is equal to the voltage across *one-half* of the secondary, or $V_s/2$, minus the drop across the diode (about 1 V) during the time the diode conducts. Therefore, the capacitor charges to almost the peak value of the half-secondary voltage during the peak of the input voltage. The capacitor cannot discharge through the diodes, however, and therefore discharges through the load and provides the power to the load during the times when the diodes are not conducting. During the next half cycle, the lower diode d_2 becomes forward biased and the capacitor is again charged to near the peak of the half-secondary voltage.

Equations that express the approximate relationships between peak input voltage, average output voltage, peak-to-peak ripple ratio, load resistance, and filter capacitance were developed in Chapter 3. However, those equations were based on the following assumptions.

1. The peak-to-peak ripple ratio is small (about 10 percent or less).
2. The resistance of the transformer windings is negligible.
3. The ohmic, or bulk, resistance of the rectifier diodes is negligible.

Unfortunately, these constraints are not always met in actual practice. In fact, the iR drops in the transformer and the rectifiers may be significant, even though the resistances of these elements are small, because the peak currents may be very high in comparison with the average current in the load. Therefore, J. Phillip Stringham, a Systems Design Engineer of Ball Brothers Research Corporation, developed a computer program and

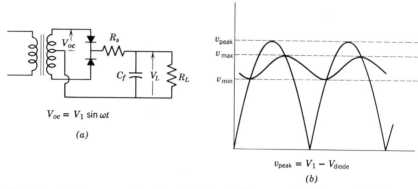

Fig. 15.2.   A full-wave rectifier circuit. (*a*) Circuit diagram and (*b*) voltages defined.

obtained a solution, which is *not* based on the above assumptions, for the full-wave rectifier with a capacitive filter. Stringham prepared a nomograph from his solution, which provides the needed relationships among the parameters with accuracy limited only by the tolerances of the components and the preciseness of reading the nomograph. The nomograph parameters are illustrated in Fig. 15.2 and the nomograph is given in Fig. 15.3*a*. A second nomograph prepared by Stringham, which can be used to determine the peak diode currents is given in Fig. 15.3*b*.

15.2   CIRCUIT AND VOLTAGES FOR THE FULL-WAVE RECTIFIER

Figure 15.2*a* shows the circuit diagram of the full-wave rectifier with the open circuit (no load) voltage V_{oc} across one-half of the secondary. The resistance R_s includes the dynamic resistance of the diode plus the ac resistance looking into one-half of the secondary of the transformer. The value of R_s could be experimentally determined by disconnecting the filter capacitor and plotting the voltage V_L across R_L as a function of $I_L = V_L/R_L$. Then R_s is the slope of the curve, $\Delta V_L/\Delta I_L$. Typical values of R_s for some representative power supply ratings are given in Table 15.1. This source resistance is included in the nomograph of Fig. 15.3*a* by defining the parameter R_{eq} as

$$R_{eq} = R_L/R_s \tag{15.1}$$

Also, this nomograph can be used for any primary power frequency f by defining C_{eq} as

$$C_{eq} = R_s C_f (f/60) \tag{15.2}$$

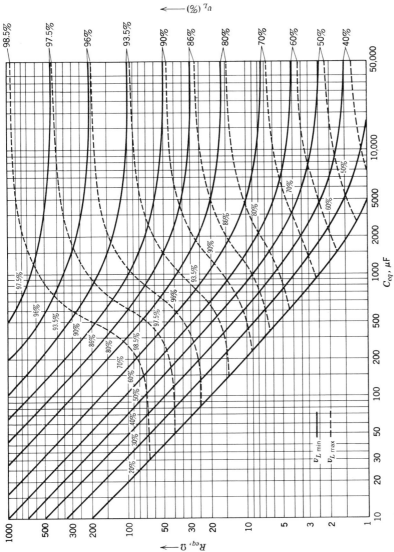

Fig. 15.3 (*a*) Nomograph to determine the circuit parameters of Fig. 15.2.

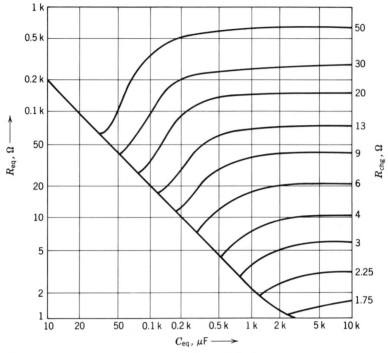

Fig. 15.3. (*b*) Nomograph to determine the peak diode currents.

The nomograph gives v_{Lmax} and v_{Lmin} as percentages of V_{Lpeak}, where V_{Lpeak} is the peak voltage across the load when the load resistance is very high, or when R_L approaches infinity. Thus V_{Lpeak} can be obtained from the relationship

$$V_{Lpeak} = V_1 - V_{diode} \tag{15.3}$$

where V_{diode} is the forward voltage drop across the diode when the diode current is small (perhaps 10 percent of the average value).

Table 15.1 Typical R_s as a Function of Secondary Voltage and Current Ratings

	10 V	25 V	50 V	100 V	200 V
0.02 A	70 Ω	120 Ω	220 Ω	420 Ω	820 Ω
0.2 A	7.0 Ω	12.0 Ω	22 Ω	42.0 Ω	82 Ω
1 A	1.4 Ω	2.4 Ω	4.4 Ω	8.4 Ω	16.4 Ω
2 A	0.7 Ω	1.2 Ω	2.2 Ω	4.2 Ω	8.2 Ω
5 A	0.3 Ω	0.5 Ω	0.9 Ω	1.7 Ω	3.3 Ω

Finally, after the parameters R_{eq} and C_{eq} have been determined (Eqs. 15.1 and 15.2), a parameter known as R_{chg} can be found from the nomograph given in Fig. 15.3b. Then the peak diode current I_{peak} can be determined from the following equation:

$$I_{peak} = \frac{V_{Lpeak}}{R_s R_{chg}} \qquad (15.4)$$

The proper use of the nomographs is illustrated by the following examples.

EXAMPLE 15.1   We want to determine the minimum output voltage v_L, the peak-to-peak ripple voltage, and the peak diode current of a full-wave circuit such as Fig. 15.2a when V_{oc} (1/2 secondary) $= 35$ V, $R_L = 10\ \Omega$, $R_s = 0.35\ \Omega$, $C_f = 5000\ \mu\text{F}$, $f = 60$ Hz, and $V_{diode} = 0.5$ V.

1. $R_{eq} = R_L/R_s = 10/0.35 = 28.6$ (from Eq. 15.1).
   $C_{eq} = R_s C_f f/60 = 0.35 \times 5000 \times 60/60 = 1750\ \mu\text{F}$ (Eq. 15.2).
2. Using these values of R_{eq} and C_{eq}, read $v_{Lmax}(\%)$ and $v_{Lmin}(\%)$ from Fig. 15.3a, then compute the V_{peak} and the minimum and maximum load voltages.

$$v_{Lmax}(\%) \simeq 90\% \text{ (dashed line)}$$
$$v_{Lmin}(\%) \simeq 80.5\% \text{ (solid line)}$$
$$V_{peak} = (1.414 \times 35 - 0.5) = 49 \text{ V}$$
$$v_{Lmax} = 49 \times 0.90 = 44.1 \text{ V}$$
$$v_{Lmin} = 49 \times 0.805 = 39.4 \text{ V}$$
$$\text{Peak-to-peak ripple voltage} = 44.1 - 39.4 = 4.7 \text{ V}$$

3. The parameters R_{chg} can be obtained from Fig. 15.3b.

$$R_{chg} \simeq 8.5\ \Omega$$
$$I_{peak} = V_{peak}/R_s R_{chg} = 49/0.35 \times 8.5 = 16.5 \text{ A}$$

EXAMPLE 15.2   The transformer and diodes of Example 15.1 are used in the circuit of Fig. 15.2a, but the frequency is increased to 400 Hz and the load resistance is increased to 12 Ω. A voltage regulator will follow the rectifier and filter. The voltage input to the regulator must not fall below 41 V. We need to determine the required filter capacitance C_f and the peak diode currents I_{peak}.

1. $R_{eq} = R_L/R_s = 12/0.35 = 34\ \Omega$
   $v_{Lmin}(\%) = 41 \times 100/49 = 83.6\%$
   $C_{eq} \simeq 2500\ \mu\text{F}$ (Fig. 15.3a)
   $C_f = C_{eq} \times 60/R_s f = 2500 \times 60/0.35 \times 400 = 1070\ \mu\text{F}$ (Eq. 15.2)

2. Using Fig. 15.3b,

$$R_{chg} \simeq 8 \; \Omega$$
$$I_{peak} = V_{Lpeak}/R_s R_{chg} = 49/0.35 \times 8 = 17.5 \text{ A (Eq. 15.4)}$$

The nomographs can be used for the bridge rectifier but R_s includes the dynamic resistance of two diodes in series and $V_{peak} = V_1 - 2V_{diode}$.

A commonly encountered term in power supply design is the *peak-to-peak ripple ratio* (pprr) which is defined as $(v_{omax} - v_{omin})/V_{oave}$. The power supply discussed above has rather poor regulation unless $R_L C \gg T$ because the output voltage is a strong function of the rate of discharge of the filter capacitor through the load. This drop adds to the diode drop and the voltage drop in the transformer windings. A sketch of V_o as a function of I_L is given in Fig. 15.4 for two different values of capacitance in a typical rectifier circuit.

Power supply *output resistance* is a common and useful term to express the relationship between output voltage and load current. This dynamic output resistance is defined as

$$r_o = \frac{\Delta V_o}{\Delta I_L} \tag{15.5}$$

For example, the power supply that has the output characteristics given in Fig. 15.4 has an average value of output resistance $r_o = 10 \text{ V}/1 \text{ A} = 10 \; \Omega$ when the filter capacitance is 1000 μF.

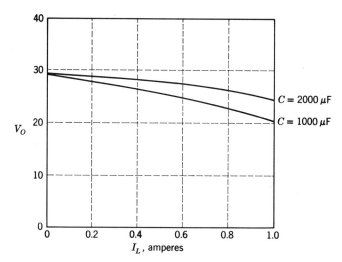

Fig. 15.4. Output voltage as a function of output current for two different values of filter capacitance.

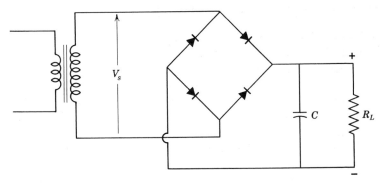

Fig. 15.5.   A full-wave bridge rectifier circuit.

A full-wave *bridge* rectifier was discussed in Chapter 3 and is shown in Fig. 15.5 for convenience. Observe that this rectifier does not require a center-tapped transformer. Therefore the transformer is required only for voltage transformation and may be eliminated if this function is not required. Note that the capacitor C charges to the maximum value of the *full secondary* voltage minus the drop across two forward biased diodes. Therefore, the bridge rectifier provides twice as much dc output voltage, for the same full secondary voltage, as the circuit of Fig. 15.1. The peak inverse voltage across the diodes is the same in both circuits, however, so the ratio of output voltage to peak-inverse voltage is twice as high in the bridge circuit. The bridge rectifier has the disadvantage of requiring four diodes, and in very low-voltage circuits the rectifier efficiency may be seriously reduced because two diodes are in series with the load.

PROBLEM 15.1   Determine the average value of output resistance for the power supply with characteristics shown in Fig. 15.4 with $C = 2000\ \mu F$.
Answer: $6\ \Omega$.

PROBLEM 15.2   Design a power supply to provide 25 V and 1.5-A dc output with a 3.75 V peak-to-peak ripple voltage. Use a center-tapped transformer with 120-V, 60-Hz primary power. List component ratings.
Answer: $C = 3300\ \mu F$, $v_s = 43$ V, peak inverse = 56 V.

PROBLEM 15.3   Design a power supply that will provide 40 V at 2-A dc output, using a bridge rectifier and 120-V, 60-Hz primary power. Use pprr = 0.1.   *Answer:* $C = 4100\ \mu F$, peak inverse voltage = 44 V.

15.2 EMITTER-FOLLOWER REGULATORS

Reference diodes (Zener diodes) were discussed in Chapter 3 and a simple circuit, that will maintain a fairly constant voltage across a load resistor, was analyzed. A circuit of this type can be used to improve the voltage regulation and reduce the ripple in a power supply. However, the load currents usually encountered require high-power dissipation capability in the reference diode and the power supply efficiency is low because of the power loss in this diode. One or more transistors can be used in conjunction with the reference diode, however, to greatly increase the efficiency of the regulator by reducing the current through the reference diode. A typical circuit of this type is shown in Fig. 15.6. This circuit is known as an *emitter-follower regulator* because the output voltage follows (is nearly equal to) the reference voltage; the difference being the base-emitter voltage V_{BE}.

The emitter-follower regulator operates in the same manner as the simple Zener regulator except that the currents through the reference diode are reduced by the factor $(\beta+1)$ because of the transistor. To illustrate this behavior, let us assume both the input voltage and the reference-diode voltage to be constant, then the current I through R_B is constant. A reduction in load current I_L will reduce the base current I_B since $I_B = I_L/(h_{FE}+1)$, but the reference diode current I_D will increase because the base voltage will tend to rise. Thus the sum of the currents $(I_B + I_D)$ is essentially constant. The diode voltage, and hence the base voltage, does change by the amount $\Delta V_B = \Delta I_D r_d$ where r_d is the dynamic resistance of the diode, discussed in Chapter 3. The load voltage V_O increases more than the base voltage because of the reduced V_{BE}. If the change in load current is small so that r_d and r_e are essentially constant over the current range, the change in output voltage is

$$\Delta V_O = \Delta I_L r_e + \Delta I_D r_d \tag{15.6}$$

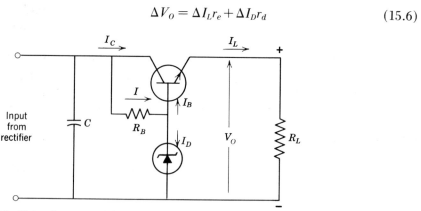

Fig. 15.6.   An emitter-follower regulator.

But $\Delta I_D \simeq \Delta I_B = \Delta I_L/(h_{FE}+1)$, so

$$\Delta V_0 \simeq \Delta I_L \left(r_e + \frac{r_d}{h_{FE}+1} \right) \tag{15.7}$$

and the output resistance of the regulator is

$$r_o = \frac{\Delta V_O}{\Delta I_L} \simeq r_e + \frac{r_d}{h_{FE}+1} \tag{15.8}$$

Of course, since the input voltage does not remain constant, the current I through R_B changes with input voltage and this current change must be absorbed by the reference diode.

EXAMPLE 15.3 An example will illustrate the design of an emitter-follower regulator. Let us assume that the required output voltage is 20 V and the maximum load current is 1.0 A. Let us also assume that the chosen transistor is silicon with $h_{FE} = 50$ and $V_{BE} = 0.6$ V at $I_E = 1.0$ A. The reference diode current is minimum when the load current is maximum and we shall choose this minimum diode current to be 5 mA. Then, the reference diode voltage should be $(V_O + V_{BE}) = 20.6$ V at $I_D = 5$ mA. We must choose the minimum voltage that can be tolerated across R_B, keeping in mind that the higher this voltage is, the higher the required input voltage and hence the lower the efficiency of the power supply. However, we shall see shortly that low values of R_B, which result from choosing a low minimum voltage, result in high diode dissipation and inferior regulator characteristics. In this example, we choose 2 V as the minimum voltage, $V_{R\min}$, across R_B. Now the value of R_B can be calculated from the following relationship.

$$R_B = \frac{V_{R\min}}{I_{D\min} + (I_{L\max}/h_{FE}+1)} \tag{15.9}$$

For this example, $R_B = 2/(5 \text{ mA} + 20 \text{ mA}) = 80 \ \Omega$.

The minimum voltage input to the regulator is $20.6 + 2 = 22.6$ V. This is the voltage $v_{0\min}$ shown in Fig. 15.3.

The required value of filter capacitance and the voltage rating of the transformer can be determined from Fig. 15.3a after we have chosen the current rating of the transformer secondary so we can determine R_{eq}. If the load on the power supply is expected to draw near 1.0 A dc for long periods of time, then the transformer secondary must have a current rating greater than 1.0 A because the high peak currents that flow in the transformer windings cause higher power dissipation than sinusoidal currents of the same average value. Therefore, a generous safety factor (perhaps 30 percent) must be allowed for the current rating of the transformer. On the other hand, a 1.0 A rating is adequate if the load draws 1.0 A only occasionally. Audio-frequency amplifiers have this latter characteristic and we shall assume our load to be of that type. Then, from Table 15.1, $R_s \simeq 2.4\,\Omega$, and from Eq. 15.1, $R_{eq} = 20\,\Omega/2.4\,\Omega \simeq 8\,\Omega$. We can now move to the right along the $R_{eq} = 8\,\Omega$ line on Fig. 15.3a. Observe that as we move to the right, the ripple decreases and the capacitance C_{eq} increases. Let us assume that a 10 percent difference between $v_{L\max}$ and $v_{L\min}$ is satisfactory. This difference is obtained approximately at $C_{eq} = 5,000\,\mu F$ where $v_{L\min} = 67$ percent and $v_{L\max} = 76$ percent, and the actual difference is 9 percent. Thus, using Eq. 15.2, the actual filter capacitance $C_f = C_{eq}/R_s = 5000/2.4 \simeq 2,000\,\mu F$. The maximum *open circuit* voltage V_{peak} out of the rectifier is therefore $v_{L\min}/0.67 = 22.6/0.67 = 33.7$ V, and we must add approximately 1.0 V for the drop across the two rectifiers, giving 34.7 as the peak open circuit voltage across the transformer secondary. However, transformers are normally rated at full load voltage, which is about 10 percent less than the open circuit voltage, but the open circuit voltage we have determined is at *minimum* line voltage, which is about 10 percent lower than the nominal line voltage. Therefore 34.7 V is also the peak secondary voltage at full load and nominal input voltage. The secondary *rms* voltage rating is, therefore, $34.7 \times 0.707 = 24.5$ V. Either a 24 V or a 25 V transformer would be suitable.

The maximum power dissipation of the reference diode must yet be determined. This dissipation will occur when the primary voltage is maximum and the load current is minimum. The maximum full load rms secondary voltage of 27 V occurs when the primary voltage is 130 V. The corresponding no load voltage is 10 percent above this value or about 30 V. The peak voltage into the rectifiers is therefore $(2)^{1/2}(30) = 42$ V and the maximum voltage applied to the regulator is approximately $42 - 1 = 41$ V, allowing 0.5 V for each rectifier at minimum current. Let us assume that the minimum load current is zero. Then the ripple is approximately zero and the dc voltage applied to the regulator is nearly equal to the 41 V maximum value. The current through R_B is $I = (41 - 20.6)\text{ V}/80\,\Omega = 255$ mA, which all flows through the reference diode since $I_B = 0$ when $I_L = 0$. Thus, the maximum power dissipation of the reference diode is approximately 20.6 V (0.255 A) = 5.25 W. A 10-W diode would probably be purchased.

Notice that the high dissipation requirement of the reference diode in the preceding circuit resulted from the low value of R_B, which in turn resulted from the low value of minimum voltage (2 V) that we allowed across R_B. The large change in reference-diode current also causes a wider voltage variation than may be desired. On the other hand, an increase of the minimum voltage across R_B requires higher input voltages and lower efficiency. Of course, R_B is larger and ΔI_D is smaller if the maximum load current is smaller.

The characteristics of the emitter-follower regulator can be improved and the reference diode dissipation can be reduced greatly in high-current regulators if a Darlington-connected amplifier is used in the circuit (Fig. 15.7). The maximum base current I_{B1} is then $I_L/(h_{FE2}+1)(h_{FE1}+1)$, which is usually less than 1 mA. Then a low-power reference diode with minimum currents of the order of 1 mA can be used. In this circuit, R_B can be much larger than in Example 15.3. For example, let us add a silicon transistor T_1 to the circuit of Example 15.3 and assume T_1 has $h_{FE1} = 100$ and $V_{BE} = 0.6$ V at $I_E = 20$ mA. Then $I_{B1max} \simeq 20/100 = 0.2$ mA. Then, if $I_{dmin} = 1.0$ mA and we allow a 2.0-V minimum across R_B, the value of R_B is 2.0 V/1.2 mA $= 1.67$ kΩ. Any standard value between 1.5 kΩ and 1.8 kΩ would be suitable. The design would then proceed as before, except the additional 0.6 V must be added to 20.6 V to yield a 21.2 reference-diode voltage. The voltages all along the line back to the primary voltage must therefore be appropriately increased. The important result is that $I_{Dmax} = 19.8$ V/1.67 kΩ $= 12$ mA and the maximum power dissipation of the reference diode is $(21.2 \text{ V})(12 \text{ mA}) = 254$ mW, approximately. In addition, the change of I_D is $\Delta I_D = 11$ mA. The maximum dissipation of the power transistor (T_2) must also be considered. This dissipation occurs at maximum line voltage and maximum load current. The minimum input voltage to the Darlington regulator is $22.6 + 0.6 = 23.2$ V at full load and 105 V input to the transformer primary. Then, if the input to the transformer rises to

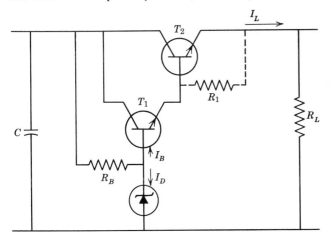

Fig. 15.7.   An emitter-follower regulator using a Darlington-connected amplifier.

130 V, $v_{min} = 23.2 \times 130/105 = 28.8$ V. Since the ripple is about 10 percent, the average input voltage to the regulator is about 5 percent above 28.8 V or 30.3 V. The average voltage across the regulating transistor under these conditions is $30.3 - 20 = 10.3$ V and the maximum power dissipation is 10.3 V $\times 1$ A $= 10.3$ W. If germanium transistors are used the dashed resistor R_1 (Fig. 15.7) may be needed to reduce the thermal currents, as was discussed in Chapter 10.

The reduction of the ripple voltage by the emitter-follower regulator can be determined from the equivalent circuit of Fig. 15.8. This equivalent circuit was initially discussed in Chapter 3. The ripple voltage is the ac component in either the input or output of the regulator. The ripple voltage in the output is essentially the same as the ripple voltage, v_r, across r_d. Thus,

$$v_r = \frac{v_r' r_d}{R_B + r_d} \tag{15.10}$$

where v_r' is the ripple voltage across the filter capacitor. The ripple voltage is maximum at full load and should be calculated for this condition. For example, if $r_d = 30\ \Omega$ (a typical value) at $I_D = 1$ mA and the peak-to-peak ripple voltage across the capacitor is about 3.5 V, as considered in the preceding examples, the peak-to-peak ripple voltage in the output of the Darlington connected circuit example is $v_r = 3.5 \times 30/1.7\ k\Omega = 0.06$ V.

Sometimes the good regulation of a regulated supply is not needed but low ripple voltage is required. Then the reference diode in the circuit of Fig. 15.7 may be replaced by a capacitor as shown in Fig. 15.9. This circuit can be designed in the same manner as the emitter-follower regulator except there is no reference diode current and jX_C replaces r_d in determining the output ripple from Eq. 15.10.

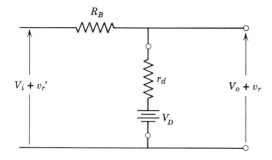

Fig. 15.8.   An equivalent circuit used to determine the ripple voltage at the regulator output.

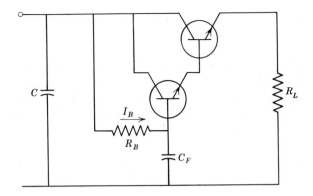

Fig. 15.9.   An active filter for reducing ripple.

EXAMPLE 15.4   Let us assume that the active filter circuit of Fig. 15.9 will be used to replace the Darlington connected emitter-follower regulator of Fig. 15.7, the power supply requirements and transistors being the same as in the preceding example. Thus, $V_{0\min} = 20\,V$, $I_{L\max} = 1$ A, $h_{fe2} = 50$, $h_{FE1} = 100$, $C = 2000\,\mu$F. Now $I_{B\max}$ is $\simeq 1$ A/(50)(100) or 0.2 mA. If we allow a 2-V maximum drop across R_B, the value of R_B is 2 V/0.2 mA = 10 kΩ. The value of v_r can be calculated from Eq. 15.10 for a given value of C_F. However, a preferable procedure is to specify the desired value of v_r and calculate the required value of C_F. If jX_C is used to replace r_d in Eq. 15.10, this equation becomes $v_r = jv'_r X_C/(R_B + jX_C)$. However, if $R_B \gg X_C$, the denominator of this relationship is $\simeq R_B$ and the magnitude of v_r is

$$|v_r| = \frac{v'_r X_{CF}}{R_B} \tag{15.11}$$

Solving for X_{CF},

$$X_{CF} = \frac{1}{2\omega C_F} = \frac{|v_r|R_B}{v'_r} \tag{15.12}$$

Solving for C_F,

$$C_F = \frac{v'_r}{2\omega R_B |v_r|} \tag{15.13}$$

where ω is the angular frequency ($2\pi f$) of the primary power frequency. The factor 2 appears because of the full-wave rectification. In this example, let us specify the peak-to-peak ripple voltage in the output to be 0.01 V. Then with $v'_r = 3.5$ V and $f = 60$ Hz, $C_F = 3.5/(4\pi \times 60 \times 0.01 \times 10^4) = 47$ μF.

The filter capacitor C_F can also be used in parallel with a reference diode to further reduce the ripple voltage in the output of an emitter-follower regulator.

PROBLEM 15.4 Design an emitter-follower regulator that will provide 25 V at 2-A maximum to a load. Use $h_{FE2} = 50$ and $h_{FE1} = 100$, and use the circuit of Fig. 15.7. Allow for a primary voltage range of 105 V to 130 V (60 Hz). Specify component ratings and determine the ripple voltage in the output, assuming $r_d = 25\,\Omega$ and $I_{Dmin} = 1$ mA. Use pprr = 0.15 approximately.

Answer: For $V_{Rmin} = 2$ V, $R_B = 1.4\,k\Omega$, Zener $= 26.2$ V at 1 mA, 210 mW actual max diss; $C = 2500\,\mu F$, Trans. sec $= 33$ V.

PROBLEM 15.5 Use an active filter instead of the regulator in the power supply of Prob. 15.4 and determine the value of capacitance C_F that will provide a 5-mV peak-to-peak maximum ripple in the output.

Answer: $C_F = 220\,\mu F$.

15.3 CLOSED-LOOP REGULATORS

Although the emitter-follower regulators provide satisfactory performance for many applications, their output resistance cannot be reduced below the value given by Eq. 15.8. Also, large values of C_F (Fig. 15.9) are required to provide very low values of ripple. On the other hand, regulators that employ the principle of negative feedback can provide almost any desired value of output resistance and ripple quite easily. The basic philosophy of the closed-loop regulator is illustrated by the block diagram of Fig. 15.10. A fraction of the output voltage ηV_0 is compared with a reference voltage V_{REF}, and their difference is amplified and used to

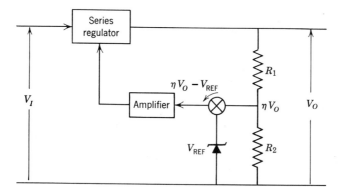

Fig. 15.10. The block diagram of a closed-loop regulator.

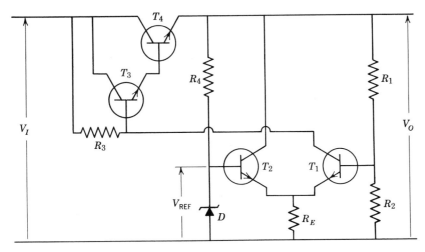

Fig. 15.11.   A typical closed-loop regulator circuit.

control the series regulator, which in turn controls the output voltage. A typical circuit diagram that will perform this basic function is shown in Fig. 15.11. In low-current regulators, the series regulator may be a single transistor, but a Darlington-connected amplifier is usually used as shown in Fig. 15.11 and as previously used in the emitter-follower regulator (Fig. 15.7). The differential amplifier consisting of transistors T_1 and T_2 provides both the voltage comparison and the amplification functions. The resistors R_1 and R_2 chosen so that their ratio provides the desired ratio between V_{REF} and V_O while their sum provides a bleeder current through the resistors that is large compared to the base current of transistor T_1. If these conditions are met,

$$V_{\text{REF}} \simeq \frac{R_2}{R_1+R_2} V_O \tag{15.14}$$

and the current I_R through R_1 and R_2 is

$$I_R = \frac{V_O}{R_1+R_2} \tag{15.15}$$

These equations can be solved simultaneously to find R_1 and R_2. By substituting Eq. 15.15 into Eq. 15.14

$$V_{\text{REF}} = R_2 I_R \tag{15.16}$$

and

$$R_2 = \frac{V_{\text{REF}}}{I_R} \tag{15.17}$$

Substituting this expression for R_2 into Eq. 15.15 and solving for R_1

$$R_1 = \frac{V_0}{I_R} - R_2 \qquad (15.18)$$

The resistor R_4 is chosen so that the current through the reference diode D is large in comparison with the base current of transistor T_2. Then, the current through D is essentially constant and hence the reference voltage V_{REF} is very constant. The diode D should have essentially zero temperature coefficient if V_{REF} and hence V_0 are to be independent of temperature.

The operating principles of the closed-loop regulator will be illustrated by assuming that the output voltage V_0 becomes more positive because of either a reduction in load current or an increase of input voltage or both. Then, the base of transistor T_1 becomes more positive than V_{REF} and the current through T_1 increases. The increased current through T_1 causes the drop across resistor R_3 to increase. Then, the forward bias of the series regulator transistors T_3 and T_4 is also decreased. This decreased forward bias increases V_{CE} and therefore reduces the output voltage, or in other words, tends to cancel the assumed rise in output voltage. Note that a polarity reversal must be provided in the amplifier so that negative feedback is obtained by the closed loop.

The effectiveness of the amplifier and feedback system in improving the characteristics of the power supply will now be investigated. A semi-block-diagram similar to the one in Fig. 15.10 is drawn in Fig. 15.12, as an aid in the investigation. Observe that the output voltage v_{CE} of the series regulator transistor T_4 is the difference between the output voltage V_0 and the input voltage V_I or $(V_I - V_0)$. Let K_v be the voltage amplification between the input of the amplifier and the output (V_{CE}) of the series regulator (Transistor T_4). Then

$$V_I - V_0 = (\eta V_0 - V_{REF})K_v \qquad (15.19)$$

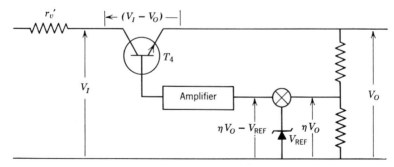

Fig. 15.12.   A block diagram of a closed-loop regulator.

But we are primarily interested in the ratio of the change in output voltage to the change in the input voltage in order to determine the reduction in ripple or input voltage variations. Therefore, we write Eq. 15.19 in terms of voltage variations instead of total voltages.

$$\Delta(V_I - V_O) = \Delta(\eta V_O - V_{REF})K_v \qquad (15.20)$$

But $\Delta V_{REF} = 0$ if the reference voltage is constant. Then

$$\Delta V_I - \Delta V_O = \Delta \eta V_O K_v \qquad (15.21)$$

Solving for ΔV_O, we have

$$\Delta V_O(1 + \eta K_v) = \Delta V_I \qquad (15.22)$$

or

$$\Delta V_O = \frac{\Delta V_I}{1 + \eta K_v} \qquad (15.23)$$

Observe from Eq. 15.23 that variations in the output voltage are reduced by the factor $(1 + \eta K_v)$ compared with variations of V_I that are caused by ripple, primary voltage variations, and so forth. Note Eq. 15.23 is a standard feedback equation with β replaced by η.

If both sides of Eq. 15.23 are divided by the change in load current ΔI_L, the effectiveness of the regulator in reducing the power supply output resistance may be obtained. Thus

$$\frac{\Delta V_O}{\Delta I_L} = \frac{\Delta V_I / \Delta I_L}{1 + \eta K_v} \qquad (15.24)$$

But $\Delta V_O/\Delta I_L$ is the output resistance of the power supply after regulation and $\Delta V_I/\Delta I_L$ is the output resistance r_o' of the filter before regulation, where we should think of the change in voltage ΔV_I as being caused by the change of current ΔI_L. Then

$$r_o = \frac{r_o'}{1 + \eta K_v} \qquad (15.25)$$

where r_o is the output resistance of the regulated supply. Observe that, as in all feedback circuits, the output resistance is reduced by the factor $(1 + \eta K_v)$. As noted in Chapter 11, low-power supply impedance is helpful in maintaining stability in a multistage amplifier.

EXAMPLE 15.5 Let us consider the design of the type regulator shown in Fig. 15.11 with the following data and specifications given: $V_O = 20$ V,

$I_{Lmax} = 1.0$ A, $h_{FE4} = 80$, $h_{FE3} = h_{FE2} = h_{FE1} = 100$, pprr $= 0.15$. Notice in Example 15.3 that $0.10 V_{peak}$ is about equal to $0.15 v_{AVE}$ since $v_{AVE} = 0.7$ V_{peak}. All transistors are silicon. R_3 (which was R_B in the preceding examples) can be determined as before. At maximum load $I_{B3} = 1$ A/80 $\times$ $100 = 0.12$ mA. If we assume that V_{Imin} occurs at I_{Lmax}, this is the worst case. Under these conditions, the minimum collector current flows in T_1. We shall let this minimum current be 0.2 mA. Then, if we allow 2.0 V minimum across R_3, $R_3 = 2/0.32$ mA $= 6$ kΩ. The rectifier circuit and filter capacitor could now be determined as in Example 15.3. We shall select $V_{REF} = 6.8$ V since diodes in this voltage area have minimum dynamic resistance and minimum thermal coefficient. It is preferable to calculate the remaining resistance values in the circuit using nominal input voltage, rather than minimum. This minimum V_I is approximately $21.2 + 2 = 23.2$ V. The peak-to-peak ripple at full load is about $24 \times 0.15 =$ 3.6 V so that the average V_I at 105 primary volts (assuming 117-V, 60-Hz nominal input power) is $23.2 + 3.6/2 = 25$ V. (We guessed 24 V in calculating this average above.) Then the *nominal* average $V_I = 25 \times 117/105 =$ 27.8 V. Under this condition, the average current through $R_3 = (27.8 -$ $21.2/6$ kΩ $= 1.1$ mA. Since $I_{B3max} = 0.12$ mA, the nominal collector current in transistor T_1 is $1.1 - 0.12 \simeq 1.0$ mA. Thus $I_{B1} = 1.0$ mA/100 $=$ 10μA. Since the current through the voltage divider circuit should be very large in comparison with I_{B1}, we choose this bleeder current $I_R = 10$ mA, which is only 1 percent of the full load current. Then, by using Eqs. 15.21 and 15.22, $R_2 = 6.8$ V/10 mA $= 680$ Ω, $R_1 = 2$ kΩ $- 680 = 1.32$ kΩ. An adjustable resistance that will span the 1.32 kΩ value is usually used for R_2 so that the output voltage can be adjusted.

At nominal conditions the base current $I_{B2} = I_{B1} = 10 \mu$A. The current through R_4 should be large compared with this value so that the current through D varies only a few percent. We choose this current to be approximately 2 mA. Then $R_4 = (20 - 6.8)$ V/2 mA $= 6.6$ kΩ. A standard 6.8 kΩ resistor will suffice.

The amplifier voltage gain K_v must be calculated to determine the characteristics of the regulated supply. The easiest method of calculating voltage gain is probably using the relationship $K_v = K_i R_L/R_{in}$. The following approximations will be used; $K_i = h_{FE}$, $R_{in} = 2h_{ie}$ for the differential amplifier, $R_{in} = (h_{ie3} + h_{FE3}h_{ie4})$ for the Darlington amplifier, and $h_{ie} \simeq \beta_o/g_m$. Then $h_{ie} \simeq 100/0.04 = 2.5$ kΩ and $R_{in} \simeq 5.0$ kΩ for the differential amplifier. In addition, $h_{ie} \simeq 80/40 = 2.0$ Ω for transistor T_4 at $I_L = 1$ A and $h_{ie} =$ $100/0.48 = 220$ Ω for transistor T_3. Then, $R_{in} = 220 + 200 = 420$ Ω for the Darlington connection. This resistance, in parallel with R_3, is the load resistance for the differential amplifier. This fact points up the need for the Darlington amplifier to provide a suitable (400 Ω) load resistance for the differential amplifier. The load resistance for the series regulator is the output resistance r_o' as seen in Fig. 15.12. Then, for the differential amplifier, $K_v \simeq 100(400)/5\ \mathrm{k\Omega} = 8$. The value of r_o' can be accurately obtained from a plot of V_I as a function of I_L, as given in Fig. 15.4. The value of r_o' can also be obtained by calculating the increase of the average value of V_I as I_L decreases from its full load value to essentially zero and then taking $\Delta V_I/\Delta I_L$. This voltage rise is approximately the peak ripple voltage plus about a 2-V reduced drop in the rectifier diodes and the transformer windings. The total voltage rise in this example is, therefore, about $(0.15 \times 25/2) + 2 \simeq 4\ \mathrm{V}$ and since ΔI_L is 1 A, $r_o' \simeq 4\ \Omega$. Then the voltage gain of the series regulator is $K_i r_o'/\mathrm{Rin} = 100 \times 80 \times 4/420 = 76$. Thus the total gain $K_v = 8 \times 76 = 608$ and $\eta = V_{\mathrm{REF}}/V_0 = 6.8/20 = 0.34$, so $\eta K_v = 207$ and $(1 + \eta K_v) = 208$. Therefore, the peak-to-peak ripple voltage is $3.7/208 = 0.018\ \mathrm{V}$ and $r_o = 4/208 = 0.02\ \Omega$ for the regulated power supply.

Integrated-circuit operational amplifiers can be conveniently used to provide excellent characteristics in a regulated power supply. These amplifiers usually provide open-loop gain values of several thousand. The operational amplifier can replace the differential amplifier directly in the circuit of Fig. 15.11 providing the output voltage V_0 and V_{REF} do not exceed the permissible output voltage range of the operational amplifier. Many different regulator configurations can be devised, but the circuit of Fig. 15.13 is one arrangement that allows the regulated output voltage to be used as the power source for the operational amplifier. The reference voltage V_{REF} should be near $V_0/2$. The transistor T_1 is used as a common-emitter driver, instead of being Darlington connected, to provide a base

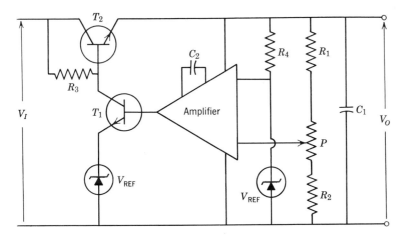

Fig. 15.13. A regulator that incorporates an operational amplifier.

potential that is the desired output potential for the operational amplifier. The voltage gain of T_1 is low because its collector load resistance is R_3 in parallel with h_{ie2}, both of which are small. The capacitor C_2 is included to indicate that the operational amplifier must be properly compensated (Chapter 14) to insure stable operation. The potentiometer P is included to provide adjustable output voltage and the capacitor C_1 maintains low output impedance at high frequencies where the gain of the regulating amplifier is low and therefore cannot provide very low output impedance. Capacitor C_1 and the potentiometer are included in most regulated supplies.

PROBLEM 15.6   Design a closed-loop regulator similar to the circuit of Fig. 15.11 that will deliver a 0.5-A maximum at 25 V to a load. Assume that the rectifier and filter provide voltage V_I with 0.1 pprr. All transistors have $h_{FE} \simeq \beta = 100$. Using $V_{\text{REF}} = 7 \text{ V}$, calculate suitable component values and estimate the peak-to-peak ripple voltage and the output resistance for the regulated supply.

Typical   Answer: $R_1 = 7 \text{ k}\Omega$, $R_2 = 18 \text{ k}\Omega$, $R_3 = 3.9 \text{ k}\Omega$, $R_4 = 18 \text{ k}\Omega$,
peak-to-peak ripple $= 3.25/470 = 7 \text{ mV}$, $r_o = 10.5/470 = 0.022 \ \Omega$.

PROBLEM 15.7   The circuit of Fig. 15.13 is used as a voltage regulator to provide 1.0 A at 15 V to the load. Transistor T_2 has $h_{FE} = \beta = 80$ and transistor T_1 has $\beta = 100$. The operational amplifier has open-loop voltage gain $= 4000$ for loads of 100 Ω or greater. If $V_{\text{REF}} = 7.0 \text{ V}$ and pprr $= 0.15$, determine suitable values for all the resistors and the potentiometer resistance if the output voltage is to be adjustable from 14 V to 16 V. Determine the peak-to-peak ripple in the output and the output resistance r_o if $r'_o = 4 \ \Omega$, and $r_{dref} = 10 \ \Omega$. Use bleeder current $= 5 \text{ mA}$ at $V_O = 16 \text{ V}$.

Answer: $R_1 = 1.6 \text{ k}\Omega$, $R_2 = 1.4 \text{ k}\Omega$, $P = 200 \ \Omega$, $R_4 = 1.8 \text{ k}\Omega$, $R_3 \simeq 120 \ \Omega$,
ripple 40 μVp-p, $r_o \simeq 4 \times 10^{-5} \ \Omega$.

15.4   CURRENT LIMITERS

One problem with transistor power supplies is that the ordinary fuse acts too slowly to protect the semiconductors in either the power supply or the load. This problem is usually solved by some technique that limits the maximum output current from the power supply to a value that will protect the semiconductors in either the power supply or the load, or both. Numerous current-limiting circuits have been devised but only three representative circuits will be discussed here. We shall first consider the circuit shown in Fig. 15.14. This is the regulator circuit of Fig. 15.13 with the resistor R_E and the diodes D_1 and D_2 added to perform the current limiting function. The load current flows through R_E and tends to forward

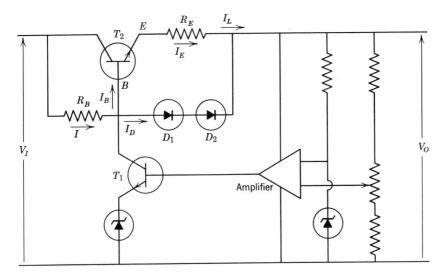

Fig. 15.14. A current-limiting circuit.

bias the silicon diodes D_1 and D_2. However, this forward bias is insufficient to cause appreciable current flow through the diodes until the limiting value of current is reached. Then, the diodes current rises sharply with a small increase of emitter current through R_E. Since the base current is $I_B = I - I_D$, the increasing diode current tends to hold the base current constant, which in turn maintains the emitter and hence the load current almost constant. The regulator tries desperately to hold the output voltage V_O constant, but succeeds only in cutting off the collector current in transistor T_1, and the voltage V_O decreases sharply toward zero.

A more quantitative view of the current limiter can be gained from Fig. 15.15, which is used in conjunction with Example 15.6.

EXAMPLE 15.6   The load current I_L in the circuit of Fig. 15.14 is to be limited to approximately 1 A. Determine the value of R_E if $h_{FE2} = 100$, and determine the approximate short-circuit current of the power supply. The characteristics of the two diodes in series and v_{BE} versus I_E are given in Fig. 15.15a, where $v_{BE} = 0.6$ V at $I_E = 1$ A. Also, the value of collector current of transistor T_1 was assumed to be 5 mA at full load current. Therefore current limiting will begin and the output voltage will begin to drop sharply when the diode current $I_D = 5$ mA. Under these conditions, the value of I_{C1} is zero. At $I_D = 5$ mA, the voltage across the two diodes in series is (from the characteristics of Fig. 15.15) 1.1 V. Therefore, $v_{RE} = (1.1 - 0.6)$ V $= 0.5$ V and $R_E = 0.5$ V/1 A $= 0.5$ Ω, with power dissipation

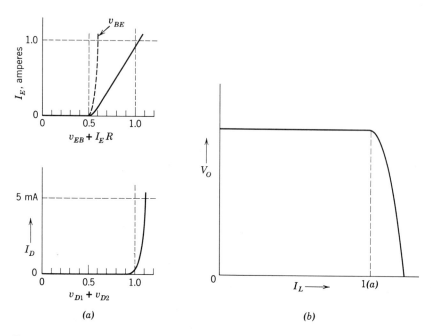

Fig. 15.15.   Current and voltage relationships in the current-limiting circuit.

$= I^2 R_E = 0.5$ W. Let us assume that $v_O = 20$ V and $v_I = 26.1$ V at $I_L = 1$ A. Then $IR_B = 26.1 - 21.1 = 5$ V and $R_B = 5$ V$/(10$ mA $+ 5$ mA$) = 330\ \Omega$. When the power supply is shorted, $v_O = 0$ and the voltage across $R_B = 25$ V, approximately, then $I = 25/330\ \Omega = 75$ mA. Of this current, approximately 65 mA flows through the diodes. We assume that this current raises the diode voltage drop about 0.1 V or 10 percent, and hence raises the voltage across R_E about 0.1 V or 20 percent, which results from a 20 percent or 200 mA increase of current through R_E. Thus the load current will increase to 1 A $+ 200$ mA $+ 65$ mA $= 1.265$ A when the output is shorted.

A current limiter that uses a single transistor instead of two diodes is shown in Fig. 15.16. The principle of operation of this circuit is the same as the diode circuit of Fig. 15.14. The transistor T_3 draws essentially no current until the load current that flows through R produces around a 0.5-V drop. Then the collector current I_{C3} rises sharply with load current and both I_B and I_L are held essentially constant at the predetermined value of I_L. This circuit has a somewhat sharper cutoff than the diode limiting circuit of Fig. 15.14 because I_{C3} does not flow through the load as the diode current did. In either circuit, the power dissipation in both T_2 and

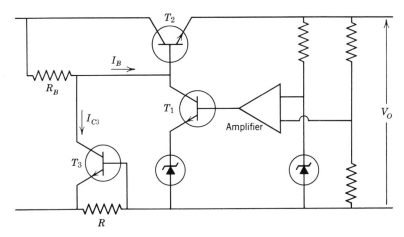

Fig. 15.16.   A transistor current limiter.

R_B is much higher than normal while the output is shorted, since nearly all the input voltage appears across these elements when the output is shorted. Therefore, the heat sink for T_2 and the dissipation rating of R_B must be adequate to prevent the destruction of these components. In either of the circuits above, the resistor R adds directly to the unregulated output resistance r'_o and is effectively reduced by the regulator to $R/(1 + \eta K_v)$, which produces a negligible increase in r_o.

The high power dissipation problem does not occur in the current limiting circuit of Fig. 15.17 because the load current is reduced to a small fraction of its maximum value as soon as the maximum value is reached. This circuit is the current limiter *only* and does not include a regulator, which could follow the limiter. In this circuit I_L flows through

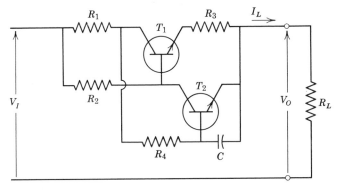

Fig. 15.17.   A current limiter that reduces load current to a small value when the current limit is exceeded.

R_1 and produces a small voltage drop (about 2 V). Resistor R_2 is less than $\beta_1 R_1$, so that transistor T_1 is normally in saturation with very small v_{CE1}. Transistor T_2 will not conduct until $v_{CE1} + I_L R_3$ rises to approximately 0.5 V (for silicon transistors). Then transistor T_2 begins to conduct and I_{C2} robs the base current of transistor T_1 and pulls it out of saturation. This action increases v_{CE1} and thus increases the forward bias of T_2. This process is cumulative, or regenerative, so that transistor T_1 is very quickly cut off and the only current that flows into the load must flow through R_2 and R_4. But these resistors normally carry only base currents and, therefore, allow only currents that are much smaller than the full load current. Transistor T_2 becomes saturated and, therefore, does not have high dissipation. However, it must safely conduct the current v_I/R_2. The circuit will not recover, even if the short is removed, until the input voltage is removed, which resets the circuit. The capacitor C is included to slow the circuit response so that the circuit will not be tripped by capacitive loads.

PROBLEM 15.8   Devise a circuit that will permit the selection of 50 mA, 250 mA, or 1.0 A as the current limiting values for the circuit of either Fig. 15.14 or 15.16. Specify resistance values.

Answer: Switch, $R = 10\,\Omega$, $R = 2\,\Omega$, $R = 0.5\,\Omega$.

PROBLEM 15.9   The transistors in Fig. 15.17 are silicon. Determine suitable values for the resistors in Fig. 15.17 if $h_{FE1} = h_{FE2} = 100$, and $I_{L\max} = 1$ A. Determine $I_{C2\max}$ if $V_I = 25$ V and the circuit is tripped. Assume $v_{CE\text{sat}} = 0.1$ V and $V_{BE} = 0.6$ V. For conduction, $V_{R2} = 2.0$ V at $I_L = 1$ A.

Answer: $R_1 = 2.5\,\Omega$, $R_2 < 200\,\Omega$, $R_3 = 0.5\,\Omega$, $R_4 < \beta R_2$, perhaps 4.7 kΩ,
$I_{C2\max} = 125$ mA.

15.5   SILICON-CONTROLLED RECTIFIERS AND TRIACS

Voltage or current regulation may also be accomplished through the use of special semiconductor devices known as thyrodes, silicon-controlled switches (SCS's), silicon-controlled rectifiers (SCR's), or Triacs. We shall first consider how these devices operate and then study several circuits that employ them.

The thyrode, or SCR, is constructed as shown in Fig. 15.18a. The action of this device can be explained by the equivalent circuit, which is shown in Fig. 15.18b. This equivalent circuit shows an n-p-n transistor and a p-n-p transistor interconnected. (In fact, the SCR action can be produced by connecting two transistors as shown in Fig. 15.18b.) If the gate is negative

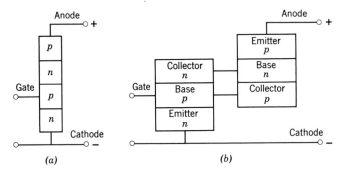

Fig. 15.18.   The silicon controlled rectifier. (a) Actual construction and (b) equivalent electrical circuit.

(or near zero volts), the n-p-n transistor will be in the cutoff condition with essentially no collector current. The collector current of the n-p-n transistor furnishes the base current for the p-n-p transistor and vice versa. Thus, if one transistor is cut off, the other transistor is also cut off, and the impedance from anode to cathode is very high.

If the bias on the gate is made positive until collector current flows in the n-p-n transistor, this collector current becomes the base current for the p-n-p transistor, which begins to conduct. The collector current of each transistor becomes the base current for the other transistor. A cumulative action is therefore initiated since an increase of current in one unit causes an increase of current in the other unit. This cumulative action culminates when both transistors are driven into saturation. When both transistors are saturated, the impedance from anode to cathode is very low. Thus, the gate is said to "trigger" the SCR "ON."

As soon as the self-regeneration action commences, the gate loses control over the action. Generally, the collector current from the p-n-p unit is much larger than the external gate current. As a result, the external gate circuit can turn the SCR ON but has difficulty turning the switch OFF. (The action is much the same as the action in the gas triode or thyratron.) To turn the SCR OFF, the gate bias must be in the reverse direction and the anode voltage must be reduced essentially to zero. In this condition, no current flows through the p-n-p unit and the gate regains control of the circuit.

If the gate is maintained at cutoff (negative potential) and the voltage on the anode is varied, the characteristics are as indicated in Fig. 15.19. As the anode voltage is made more negative, the avalanche breakdown potential of the two reverse-biased junctions is reached. (The center junction is forward biased.) At this potential, avalanche breakdown of the SCR occurs. On the other hand, as the anode voltage is made more

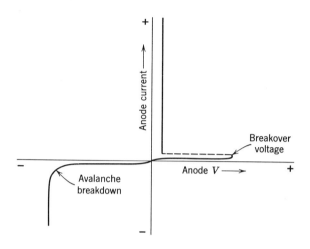

Fig. 15.19.   The current-voltage characteristics of the SCR with gate reverse biased.

positive, the center junction is reverse biased while the other two junctions are forward biased. As the center junction approaches avalanche breakdown, the avalanche current across this junction has the same polarity as a positive gate current. Thus, as avalanche breakdown approaches the SCR turns itself ON. This anode "turn-on" potential (with reverse-biased gate) is known as the *breakover voltage* of the SCR.

In normal operation, the SCR is operated with an anode potential below the breakover voltage. Then, the device is turned ON at the appropriate time by the gate. However, remember the device can be turned ON by high anode potentials. In addition, if the anode potential *changes* at a sufficiently high rate, current from the junction capacitances in the SCR may be large enough to supply the gate with sufficient current to turn the SCR ON. The symbol for the SCR is shown in Fig. 15.20*a*.

Another *p-n-p-n* device is known as a *Shockley diode*. The construction of this device is similar to an SCR without the gate lead. The Shockley diode can be turned ON by a high potential (which exceeds the breakover voltage) and turned OFF by a zero or reverse potential. The symbol for a Shockley diode is shown in Fig. 15.20*b*.

A more versatile *p-n-p-n* arrangement has a lead brought out from each area. A device with this configuration is known as a *silicon-controlled switch* (SCS). The configuration of an SCS and its symbolic representation is shown in Fig. 15.20*c*.

Two SCR's may be mounted in the same case for full-wave applications. These devices are known as *triacs* and have the anode of each device connected internally to the cathode of the other device. A single gate lead

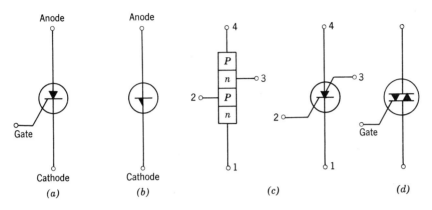

Fig. 15.20. Symbolic representations of *p-n-p-n* devices. (*a*) SCR, (*b*) Shockley diode, (*c*) the SCS, and (*d*) the triac.

(connected internally to the two gates) is provided. These devices can be switched from a blocking to a conducting state for either polarity of applied anode potential with positive or negative gate triggering. The symbol for a triac is given in Fig. 15.20*d*.

15.6   APPLICATIONS OF SCR's

The SCR (and its numerous variations) lends itself well to the control of power to various devices. For example, a simple form of over-voltage protection can be achieved by the circuit shown in Fig. 15.21. Under normal operation, the voltage across the SCR is below the anode breakover voltage. In addition, the voltage is also below the breakdown voltage of the zener diode so that the zener is nonconducting. The avalanche voltage of the zener diode is chosen so that a surge in the supply

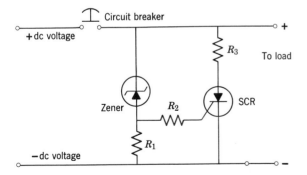

Fig. 15.21.   A "crowbar" type of overvoltage circuit protection.

voltage will cause the zener to conduct. The current flowing through the zener diode produces a positive potential on the gate of the SCR causing it to fire. Since the value of resistance in R_3 is very low, the SCR draws a heavy current from the source and opens the circuit breaker, thus protecting the load from the voltage surge. This action is similar to shorting out the supply with a crowbar whenever an overvoltage occurs. Thus the name *crowbar* type of protection. To understand the design of the circuit components, let us consider an example.

EXAMPLE 15.7   A circuit is to be constructed as shown in Fig. 15.21. The load draws 800 mA at 24 V. We shall assume that the power supply is capable of furnishing this amount of power and has a large capacitor in the output. A circuit breaker with a current rating of 1 A (slightly above the 800 mA required by the load) will be used.

An SCR is selected that will have a breakover voltage above 24 V and a current rating high enough to throw the circuit breaker. The 2N3559 will be used since it has a forward blocking voltage of 30 V and a continuous forward anode current rating of 1.6 A. The surge current rating is 18 A! Let us limit this surge current to 12 A. Then, R_3 would have a value of 24 V/12 A or 2 Ω. (Often the wiring will provide enough resistance to safely limit the surge current.) The 12 A will quickly open the 1-A circuit breaker.

The zener diode must have 24 V for its avalanche value. Let us use a 1N4749A. This silicon zener has an avalanche voltage of 24 V $\pm$ 5 percent and a power dissipation of 1 W. (If this breakdown tolerance is too great, a diode can be chosen from several diodes through the use of the curve tracer to give a more precise value of avalanche voltage.) This type of diode has a maximum reverse leakage current of 5 μA maximum. The gate of the 2N3559 will not trigger at a voltage less than 0.3 V for temperatures to 75°C. Therefore, let us choose R_1 to have a value of 0.3 V/(5×10^{-6}) = 60 kΩ. At 25°C (room temperature) the different units of 2N3559 will fire for values of gate voltage between 0.35 and 0.8 V. Thus, if the zener has an avalanche voltage of 24.0 V, the SCR will fire when the power supply voltage is increased above 24.8 V. (Some units will fire when the voltage is as low as 24.35 V.)

The resistor R_2 is included to limit the current flow to the gate of the SCR. The gate can stand a surge of 250 mA for up to 8 ms. However, the zener diode can only stand a surge of 190 mA. If the maximum expected voltage surge is 30 V, the current through R_1 will be $(30-24)$ V/60,000 Ω = 0.1 mA. Since this current is negligible compared to 190 mA, the value of R_2 is $(30-24)$ V/0.19 A = 31.6 Ω. Use a 33-Ω resistor. (The data for the 2N3559 came from the Texas Instrument handbook and the data for the 1N4749A came from the Motorola handbook.)

PROBLEM 15.10 Design a crowbar circuit like that shown in Fig. 15.21 that will protect a load of 700 mA at a voltage of 180 V. The following devices have the ratings given:

2N3562: silicon controlled rectifier
forward blocking voltage = 200 V
continuous forward anode current = 1.6 A
surge current = 18 A
minimum gate voltage to trigger = 0.3 V at 75°C
gate trigger voltage at 25°C = 0.35 to 0.8 V
maximum gate surge current = 250 mA for 8 ms

1M180ZS5: Zener diode
Zener voltage = 180 V
maximum reverse leakage current = 5 μA
maximum surge current $\simeq$ 25 mA

In the foregoing example, the circuit breaker in Fig. 15.21 removes all potential from the SCR. Thus, when the circuit is again activated, the gate maintains control of the SCR. In some applications, the power supply may have a current limiting feature. (This current limiter may be as simple as a series resistor.) In circuits of this type, the circuit breaker can be eliminated. The circuit then assumes the form shown in Fig. 15.22. In this circuit, when an overvoltage occurs, the SCR fires and effectively shorts out the output circuit. The SCR continues to conduct as long as the power is left on. Naturally, an SCR is chosen with a current rating high enough to handle the short circuit current from the power supply.

In order to reapply power to the load, the reset switch in Fig. 15.22 is closed. This reset switch shorts out the SCR element and current ceases to flow in the SCR. With no current in the SCR, the gate regains control and the SCR is turned OFF. When the reset switch (which is usually a simple

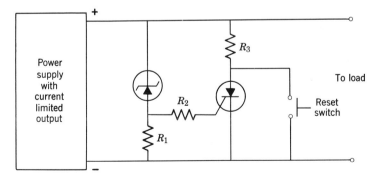

Fig. 15.22. A modified crowbar circuit.

push button switch) is released, power is again delivered to the load. As was mentioned previously, if the anode potential rises too fast, charge stored at the junction capacitances may retrigger the gate. To eliminate this effect, a capacitor may be connected across the SCR so that the anode potential will increase at a slower rate. Fortunately, in most circuits the wiring and interelectrode capacitances are sufficient to prevent transient triggering of the gate.

As the name implies, SCR devices can often be used as controlled rectifiers. For example, the SCR is widely used in a light dimmer control or for the speed control of simple series motors (such as electric saws and drills). A simple control of this type is shown in Fig. 15.23. To visualize the effect of this circuit, an analysis of the RC circuit must be considered. If the current to the gate of the SCR is assumed to be quite small, the current through the capacitor is

$$I_c = \frac{V_i}{(R_1 + R_2) + (1/j\omega C)} \tag{15.26}$$

where V_i is the input voltage. The voltage across the capacitor is $I_c Z_c$ or

$$V_c = \frac{V_i}{(R_1 + R_2) + (1/j\omega C)} \cdot \frac{1}{j\omega C} = \frac{V_i}{j\omega C (R_1 + R_2) + 1} \tag{15.27}$$

Note that if $j\omega C (R_1 + R_2)$ is much smaller than 1, V_c is essentially equal to V_i. In contrast, if $j\omega C (R_1 + R_2)$ is much larger than 1, V_c becomes approximately equal to $V_i/[j\omega C (R_1 + R_2)]$. This later value of V_c will lag V_i by approximately 90° and will have a magnitude much less than V_i. Thus, for a given value of ωC, the *magnitude* and the *phase angle* of the gate voltage

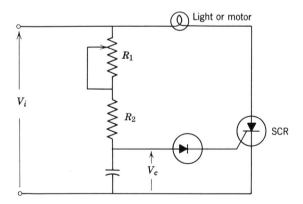

Fig. 15.23.   A light dimmer or motor speed control.

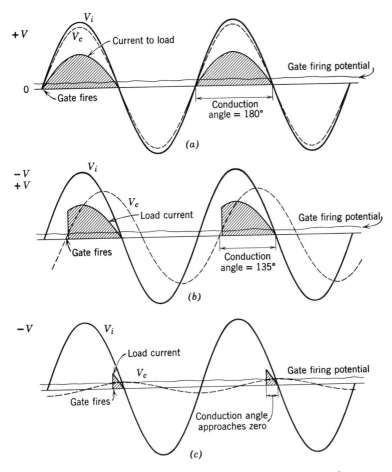

Fig. 15.24.   The effect of adjusting R_1 in Fig. 15.23 on the voltage and current waveforms. (a) $\omega C(R_1 + R_2) \ll 1$, (b) $\omega C(R_1 + R_2) \simeq 1$, and (c) $\omega C(R_1 + R_2) \gg 1$.

(V_c in this analysis) can both be adjusted by changing $R_1 + R_2$ (or really just R_1) in Fig. 15.23.

By adjusting the magnitude and the phase angle of the gate voltage, the time when the SCR fires can be controlled. This effect is illustrated in Fig. 15.24. In Fig. 15.24a, the value of $\omega C(R_1 + R_2)$ is much smaller than 1, so the gate voltage has almost the same magnitude and the same phase as the applied voltage. Since the gate potential exceeds the firing potential very early in the cycle, current flows as if the SCR were a conventional diode. At the end of the positive half cycle, the anode voltage and current are reduced to zero so that the gate regains control of the SCR. The voltages then reverse on the anode and gate. (The diode in Fig. 15.23 protects the gate against the large reverse bias voltage.) With reversed potentials, almost no current flows through the SCR. As the gate and anode voltage becomes positive, the cycle repeats.

In Fig. 15.24b, the value of $\omega C(R_1+R_2)$ is approximately equal to 1. Then, from Eq. 15.31, the gate voltage lags the applied voltage by 45°. In this case, the gate will not permit the SCR to fire until the voltage across the load and the SCR has progressed 45° through the positive half cycle. Thus, the current waveform has the shape shown in Fig. 15.24b. Notice that the average value of this current waveform is less than the average value of the current waveform shown in Fig. 15.24a. Since the power delivered to the load is I^2R, the power to the load is reduced as the conduction angle (the period when the SCR is conducting) is reduced.

In Fig. 15.24c, the value of $\omega C(R_1+R_2)$ is much greater than 1. The gate voltage now lags the applied voltage by almost 90°. In addition, the amplitude of the gate potential is reduced to such a small magnitude that the gate will not fire until the gate voltage is almost at its peak value. Thus, the gate fires when the applied voltage to the load and SCR has almost reached the end of its positive half cycle. A short pulse of current flows and then ceases as the applied voltage to the load and SCR drops to zero.

The design considerations for this type of circuit will be illustrated through the use of an example.

EXAMPLE 15.8   Let us design a light dimmer circuit to handle a 100 W light globe. We shall use the circuit shown in Fig. 15.25. The SCR must be able to handle about 1 A (current of the 100 W light) and block at least 175 V (the 110-V line may increase to 125 V with 175-V peak). Since the 2N3562 has a blocking voltage of 200 V and an average anode current of

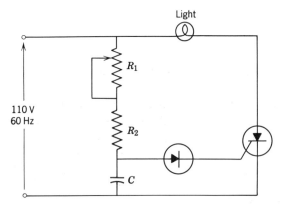

Fig. 15.25.   Light-dimmer circuit used in Example 15.7.

1 A, it will be used. The gate will have current pulses longer than 8 ms, so the average gate power dissipation of 100 mW must be used to determine the maximum allowable gate current in this circuit. Curves of gate-cathode voltage versus gate current (present in Texas Instrument manual) indicate typical 2N3562 devices have 1.3 V from the gate to cathode when the gate current is 150 mA. If the rms values of 1.3 V and 0.15 A are used, the average power dissipation would seem to be 195 mW. However, since this circuit is a half wave circuit, the actual average dissipated power is really only 98 mW, which is below the allowable gate dissipation. Maximum gate current flows when R_1 is zero. Then, the value of R_2 should limit the gate current to 0.15 A rms. The value R_2 is 110 V/0.15 = 730 Ω. To allow an extra safety margin, let us choose R_2 to be 1000 Ω.

From Eq. 15.27, we note that if R_1 is zero, $\omega C R_2 \ll 1$. Since ω is $2\pi \times 60 = 377$ rad/s, we have $C \ll 1/377 \times 10^3$ or $C \ll 2.66\,\mu F$. Let us pick a value approximately one-fifth of this value. Then $C = 0.5\,\mu F$ will be used. As a final calculation, $\omega C (R_1 + R_2) \gg 1$ when all of R_1 is in the circuit. Then, $377 \times 5.0 \times 10^{-7}(R_1 + R_2) \gg 1$ or $(R_1 + R_2) \gg 5{,}200\,\Omega$. Again, let us choose $(R_1 + R_2)$ about five times as large as this value or $R_1 = 25{,}000\,\Omega$.

The maximum reverse voltage permitted on the gate is 5 V, so the diode shown in Fig. 15.25 will be required. This diode must be able to withstand the total reverse voltage applied to the circuit as noted by Eq. 15.27. Thus, the diode must have a reverse voltage rating greater than 175 V and must be able to pass a current equal to the maximum gate current (150 mA in this example) when forward biased.

The circuit shown in Fig. 15.25 has one serious limitation. The maximum current this circuit will pass is a half-wave signal. Since a regular 117-V light globe is intended for use in a full-wave circuit, the light from the lamp in Example 15.8 will vary from essentially zero to about one-third of its brilliance in a full-wave circuit. There are two ways to achieve the full light output from the circuit shown in Fig. 15.25. These two modifications are shown in Figs. 15.26 and 15.27.

The circuit shown in Fig. 15.26 is a modification suggested by RCA.[2] In this configuration, the current to the light (or load) is controlled from zero to a half-wave value with the switch S_1 open. (The circuit is identical to that used in Example 15.8.) When S_1 is closed, the diode D_2 acts as a half-wave rectifier and the SCR with its associated circuitry can be adjusted to furnish as much of the other half of the wave as desired. Consequently, with switch S_1 open, the brilliance of the light can be adjusted from 0 to about one-third of the lamp's full brilliance. With switch S_1 closed, the brilliance can be adjusted from one-third to full brilliance of the lamp.

[2] RCA Transistor Manual, Radio Corporation of America, 1964, p. 377.

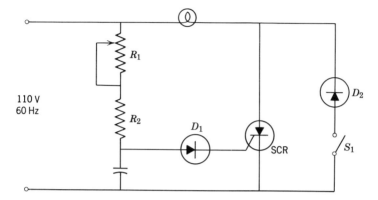

Fig. 15.26. A light-dimmer circuit with an extended range of adjustment.

The second modification of Fig. 15.25 is shown in Fig. 15.27. In this circuit, the SCR is replaced by a triac. Since the triac can switch from a blocking to a conducting state for either polarity of applied anode voltage and with either positive or negative gate triggering, the current through the load will have the waveform shown in Fig. 15.28. In this circuit, the current is controlled from zero to a full-wave value by adjusting R_1 through the proper range.

Of course, many other control circuits are possible. A large number of these circuits require the use of a *unijunction* transistor. The operating principles of the unijunction transistor will be given in Chapter 16. Then, an adjustable voltage supply that uses the unijunction device to trigger the SCR's will be considered in detail.

PROBLEM 15.11   A circuit is connected as shown in Fig. 15.25. The circuit elements are as given in Example 15.8. Sketch the input voltage V_i,

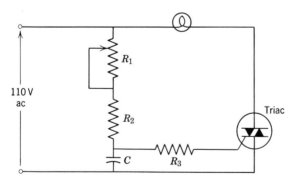

Fig. 15.27. A full-wave control circuit using a triac.

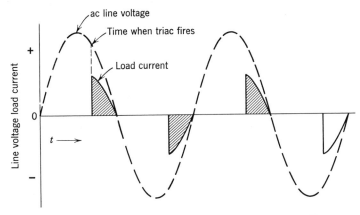

Fig. 15.28.   The voltage and current waveform for the circuit in Fig. 15.27.

the capacitor voltage V_c (if the gate loading is assumed to be negligible), and the current through the lamp i_L if R_2 is equal to 10,000 Ω. What is the conduction angle in this case?

PROBLEM 15.12   A circuit is connected as shown in Fig. 15.27. The triac is an MAC2 and has the following specifications:

  Peak blocking voltage = 200 V
  rms Conduction current = 8 A
  Peak gate power = 10 W
  Peak gate current = 2 A
  Typical ON voltage = 1 V
  Typical gate trigger voltage = 0.9 V (2 V maximum)

Find the values of each circuit element in Fig. 15.27 for control over the current range from 0 to essentially full conduction. What maximum wattage rating may the light bulb have? What is the maximum power dissipation of the triac?

CHAPTER 16

PULSE AND DIGITAL CIRCUITS

The preceding chapters of this book have dealt primarily with linear semiconductor circuits, in which the semiconductor is used mainly to enlarge or to amplify a weak input signal. The diode circuits have been an exception, of course.

In this chapter we shall consider applications in which a semiconductor behaves primarily as a switch. That is, it will normally either be forward biased so that the voltage drop across it is small, or it will be reverse biased so that very little current flows through it. The time required to switch from one of these states to the other is often of major concern.

16.1 TRANSISTORS AS SWITCHES

A simple switching circuit is shown in Fig. 16.1a and the collector characteristics of the transistor are given in Fig. 16.1b to illustrate the switching characteristics. The purpose of the transistor, with its step driving voltage, is to act as a switch in the series circuit containing V_{CC} and R_L. The transistor is not a perfect switch, however, for at least three reasons. First, a small current flows through R_L when the switch is in the OFF position. This current will be some value between I_{CO} and $(\beta+1)I_{CO}$, depending on the source resistance R_s, as was discussed in Chapter 5, where R_s was known as R_b. Therefore, this current can be minimized, and can become a negligible quantity in nearly all circuits if a silicon transistor

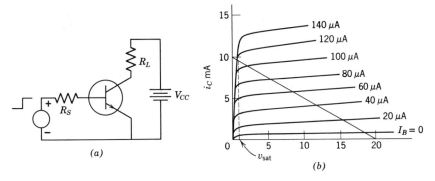

Fig. 16.1. (a) A simple transistor switching circuit and (b) its characteristics.

is used and either R_s is kept small or some reverse bias is applied to the input while the switch is in the OFF position. These measures may not generally be required.

The second shortcoming of the transistor switch is that the voltage across the transistor is v_{sat} (Fig. 16.1) while the switch is on. This v_{sat} is a function of base current as well as collector current, as shown in Fig. 16.2 where the v_{CE} scale has been expanded. Although Fig. 16.1 showed the transistor to be in saturation with $i_B = 100 \ \mu\text{A}$, Fig. 16.2 shows that the saturation voltage is reduced from 0.34 V to 0.18 V as the base current is increased from 100 μA to 140 μA and the transistor is driven farther into saturation. Any set of collector characteristics shows that v_{sat} is a function

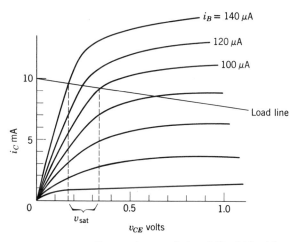

Fig. 16.2. The collector characteristics of Fig. 16.1 with an expanded v_{CE} scale to show v_{sat} versus i_B.

of collector current and becomes intolerable as the maximum current capability of the transistor is approached. Thus v_{sat} can be kept small by choosing a transistor with a maximum current rating much higher than the value to be switched and then driving the transistor well into saturation while the switch is on.

The time required to switch from one state to the other is the third limitation of the transistor switch and will be discussed in the next section. Although switching speed is listed as a limitation, the speed of semiconductor switches is highly superior to mechanical switches or relays. This advantage, coupled with the freedom from mechanical failure of the semiconductor, accounts for the wide use of the transistor switch. The FET is included as one type of transistor switch and behaves very much like the bipolar transistor except that higher input control voltage, but very little current, is required to drive the switch, and the switched current must be small in comparison with I_{DSS} in order to maintain small voltage across the FET while it is turned on.

One common application of the semiconductor switch is a modulator, in which an ac signal voltage is the collector supply voltage and a square-wave switching voltage is applied to the base, as shown in Fig. 16.3. As was previously mentioned (Chapter 4), the transistor can be used with the emitter used as the collector, and vice versa, except that the *reverse* current gains α_R and β_R are usually much lower than the *forward* current gains α_F and β_F obtained from the normal mode of operation. This gain difference does not cause appreciable dissymmetry in the output if the input switching signal is large enough to drive the transistor well into saturation in the reverse as well as the forward directions. Note that positive base potential is forward bias for either forward or reverse operation. Also note that the modulator circuit is in the emitter-follower configuration when v_I is negative and the collector is acting as the emitter. Therefore, the base switching voltage v_S must be larger than the peak input (collector) voltage v_I, and v_S must be an ac signal to insure driving the transistor into both saturation and cutoff each cycle. Generally, uniformly doped base transistors have higher β_R than graded-base transistors. Also, graded-base transistors have low V_{BE} breakdown ratings. Therefore, transistors with a uniformly doped base are most satisfactory for symmetric switching.

Another common application of the transistor switch is the chopper, which converts a slowly varying dc signal to an ac signal of fixed frequency, but with amplitude variations that match the amplitude variations of the original dc signal. This process is basically modulation and could be accomplished by the modulator circuit of Fig. 16.3. The input voltage v_I would then be the slowly varying dc voltage and the modulator would operate only in the forward, or normal, mode. However, a chopper is

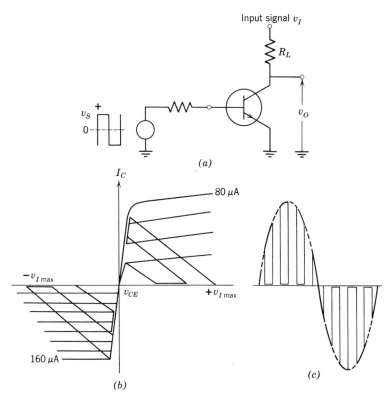

Fig. 16.3.   (a) A simple transistor modulating circuit, (b) the modulator characteristics, and (c) the output voltage.

normally used to convert a *very small* dc signal to ac, so this signal can be amplified by an RC-coupled amplifier, thereby eliminating the problem of thermally generated current (I_{CO}) becoming hopelessly mixed with the signal currents. At these very small signal levels, a problem known as voltage (or current) offset causes an annoying error in the output of the simple bipolar transistor modulator. This offset voltage is illustrated in the greatly expanded collector characteristics given in Fig. 16.4a. Because of the lack of symmetry of the transistor the collector characteristics do not converge at the point $i_C = 0$, $v_{CE} = 0$, but converge at a point $i_C = i_{Ci}$ and $v_{CE} = v_{Ci}$ instead. Only the $I_B = 0$ curve passes through the origin. Thus, the voltage v_{Ci} adds to the forward output voltage of the modulator and subtracts from the reverse output voltage. The current i_{Ci} is of the order of I_{CO} and, therefore, may be very small in silicon transistors. Thus v_{Ci} may be of the order of microvolts, which is insignificant if the signal voltage v_1 is of the order of volts, but can cause a temperature sensitive

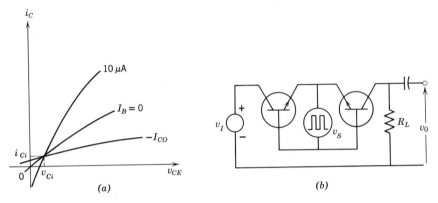

Fig. 16.4.   A symmetrical chopper (b) used to cancel the transistor offset voltage v_{Ci} shown in (a).

error that may be intolerable when the signal voltages are in the millivolt or microvolt range. This offset voltage error may be cancelled by the symmetric chopper circuit of Fig. 16.4b. Of course, the transistors must be perfectly matched and at the same temperature for complete cancellation.

A single FET can serve as a symmetrical chopper or modulator because FET's may generally have their source and drain terminals interchanged without altering their characteristics. A simple FET circuit that can be used as a switch, modulator, or chopper is given in Fig. 16.5. An example will illustrate the basic principles and design of this circuit.

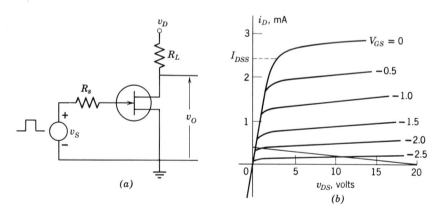

Fig. 16.5.   A switching circuit employing a FET. (a) Circuit diagram and (b) drain characteristics.

EXAMPLE 16.1   We assume that the drain-supply voltage v_D is an ac signal voltage of varying amplitude, and the switching voltage v_S is a higher frequency squarewave, so the circuit is a modulator. If the maximum drain current i_D is small compared with I_{DSS}, the voltage v_{DS} across the FET will be small compared with the pinchoff voltage V_P when the FET is switched on. In Fig. 16.5b the peak v_D is assumed to be 20 V and $R_L = 50\,k\Omega$, so $i_{Dmax} = 0.4\,mA$, which is about $I_{DSS}/7$. Smaller values of i_{Dmax} may be better, but they are more difficult to illustrate. The FET is solidly switched on when $v_S = 0$, so the resistance R_s should be large enough to limit the gate current to a small value whenever v_{GS} becomes positive. In the normal (or forward) mode of operation the maximum negative value of v_{GS} required is V_P to switch the transistor off. However, in the reverse mode when the drain and source are interchanged, the source (normally the drain) becomes negative and the base potential must become negative with respect to ground by the amount $(V_V + v_{Dmax})$ in order to switch the transistor off. In this example v_{Smax} negative must be more negative than $[-20 + (-3.0)] = -23$ V with respect to the chosen ground terminal in order to switch the FET off. Let us assume that the switching voltage v_S is symmetrical about zero and switches the transistor on with $+23$ V, with respect to ground. This would place a maximum $23\,V - (-20)$ $V = 43$ V between the gate and drain (source, while inverted) if it were not for the limiting resistance R_s. However, R_s will limit the forward bias v_{GS} or v_{GD} to approximately $+0.5$ V, so 42.5 V must be dropped across R_s. Thus, $R_s \geqslant 42.5/i_{Gmax}$, and if i_G is limited to 1 mA, $R_s \geqslant 42.5\,k\Omega$.

PROBLEM 16.1   The saturation characteristics of a transistor are sometimes characterized by a saturation resistance defined as $r_{sat} = v_{sat}/i_{Csat}$. What is the approximate value of r_{sat} for the transistor of Fig. 16.2 if (a) $I_b = 100\,\mu A$ (b) $I_b = 140\,\mu A$?   *Answer:* (a) 38 Ω, (b) 20 Ω.

PROBLEM 16.2   If the FET circuit of Fig. 16.5 is used as a chopper with $v_{DSmax} = 5$ V and the gate chopping signal is symmetrical with respect to ground, what minimum values should v_S and R_s have if $I_{Gmax} = 0.5\,mA$?   *Answer:* $v_{Smax} > 3.0\,V, R_s > 5\,k\Omega$.

PROBLEM 16.3   If the FET circuit of Fig. 16.5 is used as a modulator with the maximum signal applied to the drain $= \pm 5.0$ V and the gate chopping signal, usually called the carrier, is symmetrical with respect to ground, what minimum values should v_S and R_s have if $i_{Gmax} = 0.5\,mA$?   *Answer:* $v_S > 8\,V, R_s > 25\,k\Omega$.

16.2 SWITCHING TIMES

Despite the remarkable progress in achieving short switching times with transistor switches, there is always a need for higher speed switches. High efficiency as well as high speed may be achieved when a transistor switches its state from cutoff to saturation very quickly because power dissipation, being the product of current and voltage, is very low during cutoff because of the very low current and is also very low during saturation, providing the saturation voltage is adequately low. Thus, the dissipation is high only during the switching time. Therefore, the average dissipation may be almost proportional to the switching time, or, more accurately, the ratio of the switching time to the switching period, which is the average time for one complete cycle of operation of the switch.

The 10 percent to 90 percent rise and fall times for an amplifier output response when a square wave is applied to the input was shown in Chapter 7 to be $2.2/\omega_h$, where ω_h is the upper-cutoff frequency or bandwidth of an amplifier. A technique for either predicting the bandwidth of a given amplifier or designing an amplifier for a specified bandwidth, using a hybrid-π circuit, was also given in Chapter 7. Thus, the prediction or control of rise and fall times, or switching times, may appear to be within our present capability without further discussion. However, the techniques of Chapter 7 were developed for *linear* amplifiers and hold only approximately for a switching circuit that is driven only to the edges of the cutoff and saturation regions and therefore *not* into these regions. The main purpose of this section is to show that switching times can be reduced considerably below the values predicted for a linear amplifier if the input or switching signal is capable of driving the transistor well into the saturation and cutoff regions.

We shall first investigate the effect of driving the transistor well into saturation. The circuit diagram and the collector characteristics are given in Fig. 16.6 (*a* and *b*) for convenience. The stored charge in the transistor base and the relative circuit response times are illustrated in Fig. 16.6 (*c* and *d*). We first assume that the square-wave input v_S is just sufficient in magnitude to drive the transistor to the edge of saturation, namely $i_B = 90\,\mu\text{A}$. Then the normal charge is stored in the base region as the collector current i_C rises exponentially to almost $V_{CC}/R_L = 10\,\text{mA}$, assuming the saturation voltage is negligible. The approximately exponential rise of i_C has the time $\tau = 1/\omega_h$, as was discussed in Chapter 7, and the rise time t_{r1} is approximately $2.2\tau = 2.2/\omega_h$, as expected. When the base driving voltage v_S returns suddenly to zero, the base stored charge decreases as the collector current decreases exponentially (approximately) toward zero with a fall time equal to the rise time.

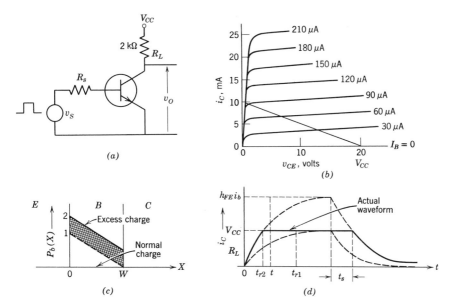

Fig. 16.6.   Illustration of the effect of base drive on switching times. (a) Circuit diagram, (b) collector characteristics, (c) stored base charge, and (d) response times.

Now let us assume that voltage v_S is increased so that the base current is doubled (180 μA) while the transistor is switched on. Then the base stored-charge rises toward point 2 (Fig. 16.6c) at the emitter junction and the collector current rises exponentially toward $h_{FE}i_B = 20$ mA with time constant τ. However, the transistor saturates at $i_C \simeq 10$ mA, which occurs at about 0.7τ, so the rise time is about 0.6τ instead of 2.2τ. Thus, the rise time is reduced by a factor of about 3.6 as a result of the excess drive. The ratio of actual base current to $V_{CC}/h_{FE}R_L$, which is the base current required to barely saturate the transistor, is usually called the *overdrive factor*. A plot of rise time as a function of overdrive factor is given in Fig. 16.7. Note that a small overdrive factor can cause a significant reduction of rise time.

EXAMPLE 16.2   Let us assume that a transistor with $h_{FE} = 100$ is to be used as a switch with $R_L = 1$ kΩ, $V_{CC} = 20$ V and $R_S = 2$ kΩ. We wish to determine the rise time if the overdrive factor is (a) 0, (b) 1, (c) 2. First we must determine the upper-cutoff frequency ω_h for the circuit as a linear amplifier, using the techniques discussed in Chapter 7. This requires additional data such as f_τ and C_{ob}. The q-point chosen for this calculation is not critical. A point near the center of the load line is appropriate. We assume that ω_h has been calculated and is 10^6 rad/s. (Of course, ω_h could

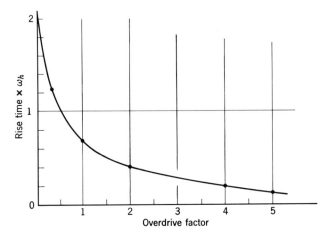

Fig. 16.7.   A plot of rise time as a function of overdrive factor.

be obtained from a frequency response curve.) Then, with a square-wave input of $i_B = 20$ V/1 kΩ(100) $= 200$ μA, $t_r = 2.2/10^{-6} = 2.2$ μs. If the base current drive is increased to 400 μA (overdrive factor $= 1$), $t_r = 0.6 \times 10^{-6}$ $= 0.6$ μs, as shown by either the preceding work or Fig. 16.7. When the base drive is increased to 600 μA, the overdrive factor is 2 and the rise time $t_r = 0.4 \times 10^{-6}$ or 0.4 μs.

The base-current overdrive, which improves the rise time, causes a problem called *storage time* in the transistor. Figure 16.6c showed that excess charge is stored in the base region when the transistor is driven into saturation. Then both junctions are forward biased, the collector current is clamped at approximately V_{CC}/R_L, and the charge in the base region must increase until the increased recombinations in the base region use up the excess base current. At the end of the input pulse the collector current cannot decrease appreciably until the *excess* stored charge is removed from the base region because the collector junction is forward biased as long as excess minority carriers are stored next to the collector junction in the base. This phenomenon was discussed in connection with diode diffusion capacitance in Chapter 3. The time required to dissipate this excess stored charge is known as *storage time* (t_s).

If the base current is just reduced to zero at the end of the input pulse, the base stored charge decreases only because of charge recombination in the base (Figs. 16.6d and 16.8). However, if reverse base current flows because of either a low-base circuit resistance or the application of a reverse bias voltage at the end of the input pulse, the stored charge is removed from the base region more quickly. First, the excess stored charge

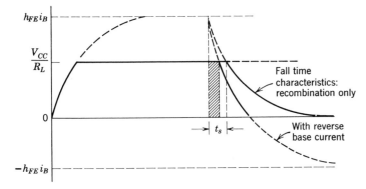

Fig. 16.8.   Illustration of the effect of reverse base drive on storage time and fall time.

is removed, shortening the storage time, and then the normal base charge is removed, shortening the fall time. Figure 16.8 illustrates the magnitude of improvement that can be expected for a given value of reverse base current $(-i_B)$. In this figure $-i_B$ was assumed to be about equal to the forward bias just required for saturation (V_{CC}/R_Lh_{FE}). Therefore, at the end of the forward-bias input signal, the negative base current $-i_B$ tends to cause the collector current to reverse, so the collector current decreases exponentially toward $-h_{FE}i_B$ rather than zero. Of course, the collector current does not actually reverse direction because the stored charge in the base becomes depleted at the same time the collector current reduces to about zero. In addition, the negative base current must also stop because the excess charge is the source for this negative base current. Thus, this reverse drive has the same effect on fall time as overdrive has on rise time. The reverse drive factor is the ratio of $-i_B$ to V_{CC}/R_Lh_{FE}. Thus, Fig. 16.7 can be used for fall time as a function of reverse-drive factor as well as rise time as a function of overdrive factor.

We can see from the foregoing discussion that the ideal base drive for a transistor switch is a source that provides significant overdrive to turn the switch ON quickly, then reduces the drive so that the *excess* stored charge in the base region is nearly zero when the switch is to be turned OFF. This action provides very small storage time. The ideal base drive should also provide reverse drive while turning off the switch to provide fast fall time. This type of drive can be provided easily when the driving source voltage is large in comparison with the Δv_{BE} required to drive the switch, which is usually the case. A resistor R_B is then placed in series with the driving source to limit the excess drive (Fig. 16.9), and a capacitor C is placed in parallel with R_B to provide overdrive at turn-on and reverse drive at turn-off. The charging current of capacitor C provides the extra drive at turn-on

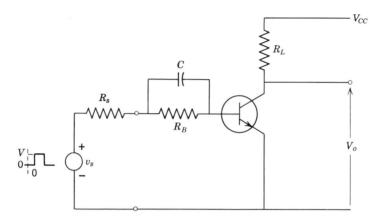

Fig. 16.9.   A typical driving circuit for a transistor switch.

and the discharging current provides the reverse drive at turn-off. Of
course, the source resistance, R_s, should be small as it limits the size of
this capacitor current.

The circuit designer must determine suitable values for R_B and C (Fig.
16.9); therefore, the equivalent circuits of Fig. 16.10 are given to provide
further insight into the circuit design. A hybrid-π circuit is first substituted
for the transistor in Fig. 16.10a. Then r_b is lumped with R_s, since the same
current must flow through both resistors, and the hybrid-π circuit is replaced by its equivalent input circuit, as was developed in Chapter 7, in
Fig. 16.10b. Next, the voltage driving source to the left of points A, B is
replaced by its equivalent current source in Fig. 16.10c. We can further
simplify this circuit only if we place some special constraints, or conditions, on it. Let us make the time constant R_BC equal to the time constant
$r_\pi C_{\text{eff}}$. Then the circuit to the right of points A, B is a balanced bridge as
shown below.

$$R_BC = r_\pi C_{\text{eff}} \tag{16.1}$$

Multiplying both sides of Eq. 16.1 by ω,

$$R_B\omega C = r_\pi \omega C_{\text{eff}} \tag{16.2}$$

Rearranging Eq. 16.2

$$\frac{R_B}{r_\pi} = \frac{\omega C_{\text{eff}}}{\omega C} = \frac{X_C}{X_{C\text{eff}}} \tag{16.3}$$

This (Eq. 16.3) is the required relationship for a balanced bridge. Then
if $R_BC = r_\pi C_{\text{eff}}$ so that the bridge is balanced, no current flows through the
conductor from point m to point n, and it can be removed without altering

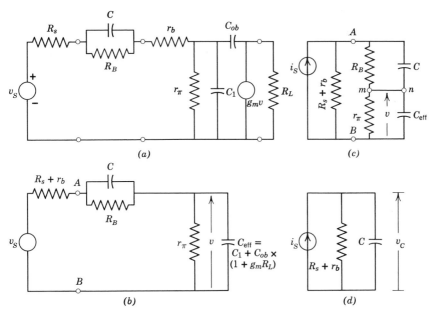

Fig. 16.10. Equivalent circuits representing the switching circuit of Fig. 16.9. (a) Equivalent circuit using hybrid-π, (b) simplified input circuit, (c) current-source version of (b), and (d) approximation of (c) when $R_B C = r C_{\text{eff}}$ and $(R_B + r_\pi) \gg (R_s = r_b)$.

the circuit performance. With this conductor removed, the resistive branch $(R_B + r_\pi)$ is directly in parallel with the modified source resistance $(R_s + r_b)$, and these resistors can be lumped together into a single equivalent resistance. Also, the capacitors C and C_{eff} are directly in series and can be lumped into an equivalent capacitance equal to $[C\, C_{\text{eff}}/(C + C_{\text{eff}})]$. This simplified equivalent circuit is given in Fig. 16.10d. However, a further simplification is indicated that may be made if our initial assumption is valid, namely that the driving source voltage is large in comparison with Δv_{BE}. Then R_B is large compared to the other resistances in the circuit, so $(R_B + r_\pi) \gg (R_s + r_b)$ and the smaller resistance branch will primarily determine the resistance of the parallel combination. Also, if R_B is predominantly large, the capacitor C in parallel with it will be much smaller than C_{eff} and the value of the series combination is approximately equal to C, the small capacitor.

The equivalent circuit of Fig. 16.10d illustrates the advantage gained from using the comparatively large drivings voltage v_S with its accompanying large value of R_B. The rise and fall times of the circuit are determined primarily by the time constant of the circuit, which is approximately

$$\tau = (R_s + r_b)C \qquad (16.4)$$

and $$t_r = t_f = 2.2(R_s + r_b)C \qquad (16.5)$$

But since R_s is normally small in comparison with R_B, the time constant τ (Eq. 16.4) is small in comparison with $(R_B C = r_\pi C_{eff})$ and the rise and fall times of the voltage v_C are small in comparison with either 2.2 $R_B C$ or 2.2 $r_\pi C_{eff}$. Since the bridge circuit (Fig. 16.10d) is balanced, the voltage v_C has precisely the same waveform and rise time as the voltage v (Fig. 16.10c) that controls the output current $g_m v$ and the output voltage $g_m v R_L$. This technique for improving rise time is precisely the same as that used in an oscilloscope probe, which converts the high capacitance of an oscilloscope input and the associated shielded input cable into a much smaller capacitance at the expense of a reduction in gain (usually a factor of 10).

The preceding technique was based on the assumption that the switching circuit can be represented by a linear equivalent circuit, which seems inappropriate and highly inaccurate. However, one requirement is that the transistor *not* be driven appreciably into saturation to avoid storage time, and if this requirement is met, the equivalent circuit is usable as a rough approximation. In fact, the time constant $r_\pi C_{eff}$ is relatively constant over a wide range of collector currents even though both r_π and C_{eff} are strongly dependent on collector current. Therefore, the equivalent circuit yields final results that are much more accurate than might be expected. The base circuit resistance R_B is chosen so that the steady-state base current during the input pulse just barely drives the transistor into saturation. Then

$$R_s + R_B = \frac{v_s - v_{BE}}{V_{CC}/h_{FE} R_L} = \frac{(v_s - v_{BE})h_{FE} R_L}{V_{CC}} \qquad (16.6)$$

and

$$R_B = \frac{(v_s - v_{BE})h_{FE} R_B}{V_{CC}} - R_s \qquad (16.7)$$

Equations 16.7 and 16.1 can be used to design the driving circuit and Eq. 16.5 can then be used to determine the rise and fall times. An example will illustrate the design procedure.

EXAMPLE 16.2   A silicon transistor with $h_{FE} = 100$, $r_b = 100\ \Omega$, $C_{ob} = 2$ pF, and $f_\tau = 300$ MHz at $I_C = 10$ mA and $V_{CE} = 5$ V is to be used as a transistor switch with $R_L = 1$ kΩ and $V_{CC} = 20$ V. The driving source has $R_s = 2$ kΩ and provides a 5.0-V rectangular pulse. Then, the value of R_B that will just drive the transistor to saturation is $[(5.0 - 0.5)/0.2\ \text{mA}] - 2\ \text{k}\Omega = 20\ \text{k}\Omega$, where V_{BE} is assumed to be 0.5 V and $V_{CC}/R_L h_{FE} = 0.2$ mA is the base current required for saturation. At $I_C = 10$ mA, which is an average value, $r_\pi = 100/0.4 = 250\ \Omega$ and $C_1 + C_{ob} = g_m/\omega_T = 212$ pF   so $C_1 = 210$ pF

(Fig. 16.10a) and $C_{\text{eff}} = 210 + 2(1 + 400) = 1012$ pF. Then, by using Eq. 16.1, $C = 250\,\Omega \times 1012$ pF/20 kΩ = 12.6 pF. A 13-pF capacitor would probably be used in parallel with the 20 kΩ resistor R_B. This completes the design. The rise and fall times for the switch can now be calculated with the aid of Eq. 16.5, which gives $t_r = t_f = 2.2 \times 2.1 \times 13 \times 10^{-9} = 6.0 \times 10^{-8}$ s or 60 ns.

We shall now check the validity of the approximations used in Eq. 16.5. The actual shunt resistance was 2.1 kΩ in parallel with 20.25 kΩ, or about 1.9 kΩ, and the actual capacitance is 13 pF in series with 1012 pF, which is very nearly 13 pF. Therefore, the rise time calculated above is about 10 percent high, which is within usually acceptable limits. You may observe that the addition of capacitor C improved the rise time approximately by the ratio $R_B/(R_s \| R_B) = (R_s + R_B)/R_s$.

16.3   BISTABLE MULTIVIBRATORS

A set of devices known as multivibrators are very useful as pulse-generating, storing, and counting circuits. They are discussed here as examples of switching-type circuits. Multivibrators are basically two-stage amplifiers with positive feedback from the output of the second amplifier to the input of the first.

A *bistable* multivibrator is shown in Fig. 16.11. We first assume that transistor T_1 is cutoff or nonconducting so that the base current of transistor T_2 is $(V_{CC} - V_{BE})/(R_{L1} + R_{B2})$. If R_{B2} is small enough, the base current is large enough to saturate T_2. Then the saturation voltage of T_2 is smaller than the turn on voltage of T_1; thus, essentially no base current flows through R_{B1} and transistor T_1 is cut off, as we assumed. The multivibrator remains in this stable state until a switching pulse is applied some

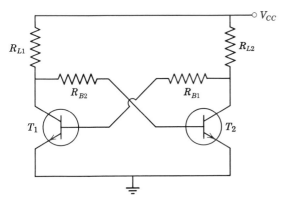

Fig. 16.11.   A basic multivibrator circuit.

place in the circuit to turn transistor T_1 on. Then T_1 will saturate and its saturation voltage will be insufficient to provide forward bias to T_2, so T_2 will be turned off. The multivibrator will then remain in this second stable state until another switching pulse is applied, which will start T_2 conducting again and T_1 will cutoff. Thus the multivibrator has two stable states and is therefore bistable. It is also known as a *flip-flop* because a trigger pulse will make it flip, or flop, from one state to the other. The trigger pulse need only start conduction in the *off* transistor and the circuit regeneration or feedback will carry through with the change of state because, as seen in Fig. 16.11, if T_1 begins to conduct, its collector voltage decreases and T_2 is pulled out of saturation. The collector voltage of T_2 then rises and forward bias is applied through R_{B1} to reinforce the initial trigger. This effect is cumulative and the change of state is completed. Note that the change of state can be initiated with a very small pulse if T_2 is just barely saturated, but if T_2 is heavily saturated the trigger pulse must be large enough and remain long enough to permit T_1 to pull T_2 out of saturation. In other words, the trigger pulse must be longer than the *storage* time.

Bistable multivibrators are often used in counting circuits, since two trigger pulses are required to cycle a flip-flop through its two stable states. A typical counting circuit is shown in Fig. 16.12. This circuit includes a triggering system consisting of C_1 and the two diodes. Also, compensating capacitors C have been added in parallel with the base circuit resistors to permit high-speed switching. These circuit features are described in the following paragraphs.

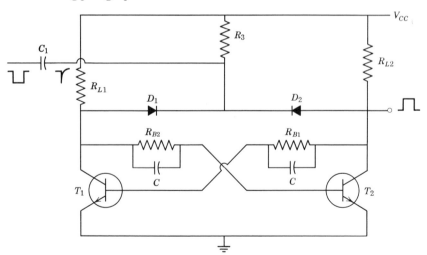

Fig. 16.12.   A binary counting circuit.

The diodes D_1 and D_2 are known as steering diodes because they alternately apply the input trigger to the collectors of T_1 and T_2. We shall assume that T_1 is not conducting, so T_2 is in saturation. Then diode D_1 is slightly reverse biased because of the base current flowing through R_{L1}, but D_2 is heavily reverse biased because of the large voltage drop across R_{L2}. Therefore, when a negative trigger pulse is applied to the junction of the diodes, D_1 is forward biased and the trigger pulse is applied to the collector of T_1. This negative voltage tends to turn transistor T_2 off and as soon as T_2 pulls out of saturation its collector voltage rises and forward biases T_1, thus completing the switching action. The next time a negative trigger pulse arrives, diode D_2 has the small reverse bias and D_1 is heavily reverse biased, so the pulse is applied through D_2 to the collector of T_2 and the base of T_1; therefore, the multivibrator changes back to its initial state. The diode currents tend to charge the capacitor C_1 to the peak value of the input pulse; therefore, resistor R_3 is required to discharge C_1 and maintain the potential of the cathodes of D_1 and D_2 at approximately V_{CC} at the beginning of each trigger pulse. The resistor R_3 is often replaced by a diode with its cathode connected to V_{CC}. The potential of the junction of the three diodes is then clamped at approximately V_{CC}.

Binary counters are often connected in cascade to produce higher countdown ratios, such as decade counters. The capacitor C_1 (Fig. 16.12) should then be small so that the coupling-circuit time constant is small compared to the duration of the rectangular input pulse. This short time-constant circuit, known as a differentiator, then produces a sharp trigger pulse at both the leading and trailing edge of the input pulse (Fig. 16.13). The positive pulses at the trailing edges of the input signal do not trigger the circuit because they increase the reverse bias of both steering diodes and are clipped (or eliminated) by the clamping diode, if one is used. When

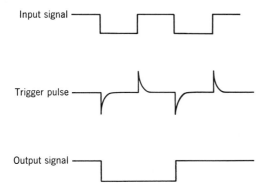

Fig. 16.13.   Input, trigger, and output voltages of a bistable multivibrator.

a clamping diode is used, the time constant of the coupling circuit is

$$\tau_c = (R_s + R_L)C_1 \qquad (16.8)$$

where R_s is the output resistance of the driving source and R_L is the approximate resistance to the trigger pulse on the right-hand side of the coupling capacitor C_1, assuming $R_B \gg R_L$. If R_3 (Fig. 16.12) is used instead of a clamping diode, the parallel combination of R_3 and R_L must be used in place of R_L.

The speedup capacitors C are the compensating capacitors discussed in the preceding section and can be calculated by the technique described there. If some doubt exists concerning their optimum value, it is preferable to err by making them too large instead of too small.

The resistors R_{B1} and R_{B2} must be small enough to insure saturation, otherwise there will be no stable states and the circuit will oscillate. When a transistor is in saturation the base and collector potentials are nearly the same. Therefore, if T_2 is assumed to be saturated (Fig. 16.12) $I_{B2}(R_L + R_{B2}) = I_{C2}R_{L2}$ and

$$R_{L1} + R_{B2} = \frac{I_{C2}}{I_{B2}}R_{L2} \qquad (16.9)$$

Transistor T_2 will be in saturation if $I_{C2}/I_{B2} \leqslant h_{FE}$ so that

$$R_{L1} + R_{B2} \leqslant h_{FE}R_{L2} \qquad (16.10)$$

or

$$R_{B2} \leqslant h_{FE}R_{L2} - R_{L1} \qquad (16.11)$$

Normally, $R_{L1} = R_{L2}$ and $R_{B1} = R_{B2}$, so the subscripts can be dropped. Then the required value of R_B for saturation is

$$R_B \leqslant R_L(h_{FE} - 1) \qquad (16.12)$$

The value of h_{FE} used in Eq. 16.12 should be the minimum expected value, from a given batch, at the minimum expected temperature.

The value of R_L is chosen in accordance with the required switching speed. As was previously discussed in this chapter and in Chapter 7, high speed, which requires short rise, fall and storage times, is possible only with high-frequency amplifiers that require high f_τ and small R_L as well as proper size speedup, or compensating capacitors. Thus values of R_L barely large enough to limit the transistor dissipation to safe values may be used in very high-speed circuits, while values of about $10\,\text{k}\Omega$ may be used in circuits not requiring high speed. The larger R_L is, the lower the power dissipation for a given V_{CC}, of course.

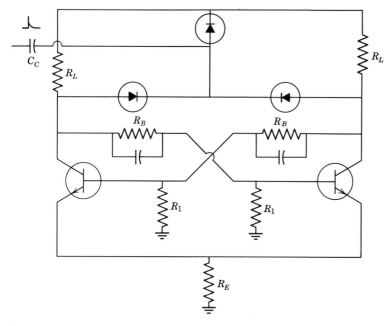

Fig. 16.14.   A bistable multivibrator with trigger desensitization.

Very often the triggering circuits are noisy. In this case, reverse bias is applied to the OFF transistor of the bistable multivibrator (Fig. 16.14), to prevent triggering on noise pulses. If the circuit is symmetrical, as shown, the current through R_E is essentially constant, being the emitter current of the ON transistor; the voltage drop $(-I_{Esat}R_E)$ across R_E reverse biases the OFF transistor. Then, the trigger pulse must be large enough to change the collector voltage of the saturated transistor by an amount greater than this reverse bias before the circuit will begin to change states. The resistors R_1 and R_2 are usually large enough so that the current through them is about the same magnitude as the base current.

EXAMPLE 16.3.   We shall now design a moderately high-speed binary counting circuit that uses a silicon transistor with $f_\tau = 300$ MHz, $C_{ob} = 2$ pF, $h_{FEmin} = 80$, $R_L = 1$ kΩ, and $V_{CC} = 20$ V. Let us choose the trigger desensitizing voltage $I_E R_E = 2$ V. Then, since the transistor is in saturation, $R_E/R_L = 2$ V/18 V and $R_E = 110$ Ω. The saturation value of $I_C = 18$ V/ 1 kΩ = 18 mA, so the minimum saturation value of $I_B = 18$ mA/80 = 225 μA. If this value of current is allowed to flow through R_1, $R_1 = 2.5$ V/225 μA = 11 kΩ. A standard value of either 10 kΩ or 12 kΩ would be adequate. Since 450 μA must flow through R_B, the maximum value of $R_B = 18$ V/ 450 μA − 1 kΩ = 39 kΩ. Since $R_1 \gg h_{ie}$, the speedup capacitors will be calculated neglecting R_1. Since the transistor, load resistance, and V_{CC} are the same in this example as in Example 16.2, we shall use the time constant $\tau = r_\pi C_{\text{eff}} = 250 \times 1012$ pF determined in Example 16.2 and $C = 250 \times 1250$ pF/39 kΩ = 8 pF. A value between 8 and 20 pF would be suitable. Let us assume that the driving source has an internal resistance $R_s = 5$ kΩ and provides a 10-μs pulse duration; we want to provide a trigger pulse of approximately 1 μs at 37 percent of its peak amplitude. Then we need $(5\,\text{k}\Omega + 1\,\text{k}\Omega)C_C = 10^{-6}$ s and $C_C = 167$ pF. The steering and clamping diodes need to be moderately fast diodes to accomplish the tasks well. This completes the design.

PROBLEM 16.4 Design a medium speed binary counter that uses silicon transistors with $h_{FE\text{min}} = 50$, $V_{CC} = 20$ V, $R_L = 10$ kΩ, $f_\tau = 100$ MHz. $C_{ob} = 5$ pF, and reverse bias = 2 V. The duration of the input pulse is about 100 μs.

 Answer: $R_E = 560$, $R_1 = 68$ kΩ, $R_B = 250$ kΩ, $C = 5$–10 pF.

16.4 MONOSTABLE AND ASTABLE MULTIVIBRATORS

The *monostable* multivibrator has only *one* stable state as the name implies. It differs from the bistable multivibrator in that one of the dc coupling circuits is replaced by an ac coupling circuit (Fig. 16.15). Transistor T_1 is normally off and T_2 is normally in saturation. Resistor R_{B2} must be small enough to maintain T_2 in saturation; therefore, $R_{B2} \leqslant h_{FE} R_{L2}$. Then T_1 remains cut off because of the lack of sufficient voltage V_{CE2} to provide forward bias. However, a trigger pulse through C_1 is coupled through C_2 and tends to pull T_2 out of saturation. When the collector voltage

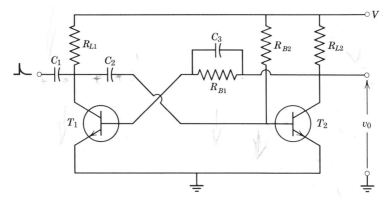

Fig. 16.15. A monostable multivibrator circuit.

of T_2 begins to rise, forward bias is applied to the base of T_1 through R_{B1} and T_1 conducts. Therefore, the collector of T_1 becomes more negative. This effect is cumulative and T_2 is cutoff while T_1 is driven into saturation, providing $R_{B1} \leqslant h_{FE} R_{L1}$.

As transistor T_1 suddenly switches on, the potential of the left side of coupling capacitor C_2 suddenly drops from V_{CC} to $V_{CE\text{sat}}$, which is almost zero. During this switch there is insufficient time to change the voltage across C_2 appreciably so that its right side must also drop in potential by the amount V_{CC}. Therefore, the base of T_2 is driven to a negative potential equal to $-(V_{CC}-V_{BE+})$ since it was at potential V_{BE+} immediately prior to the switching. This V_{BE2} potential change is shown in Fig. 16.16.

Immediately after the switching ($t = 0$), capacitor C_2 begins to charge through R_{B2} and r_{sat} (which is negligible) of T_1. Therefore, the potential on the right-hand side of C_2 and the base of T_2 rises toward V_{CC}, as may be seen in Fig. 16.15 and as illustrated in Fig. 16.16. However, transistor T_2 turns on again when V_{BE2} rises to V_{BE+} and collector current flows in T_2. Then the collector potential of T_2 drops, transistor T_1 loses its forward bias, and the monostable multivibrator switches back to its initial stable state. It will remain in this stable state until another trigger pulse is applied.

The time T_d during which transistor T_2 is cutoff can be easily determined from the relationship

$$V_{CC} = (2V_{CC} - V_{BE+})e^{-T_d/R_{B2}C_2} \qquad (16.13)$$

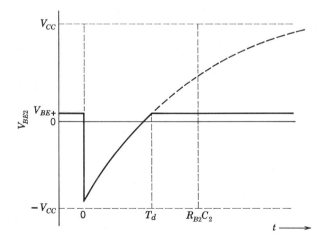

Fig. 16.16. A sketch of V_{BE2} as a function of time.

This relationship holds because the capacitor C_2 charges until its potential changes by the amount V_{CC} before it switches back to the stable state, whereas C_2 would ultimately charge until its potential changed by the amount $(2V_{CC} - V_{BE+})$ if T_2 did not again begin to conduct. If both sides of Eq. 16.13 are divided by the term $(2V_{CC} - V_{BE+})$ and then the natural log of both sides is taken,

$$\ln \left(\frac{V_{CC}}{2V_{CC} - V_{BE+}} \right) = -\frac{T_d}{R_{B2}C_2} \tag{16.14}$$

and

$$T_d = R_{B2}C_2 \ln \left(\frac{2V_{CC} - V_{BE+}}{V_{CC}} \right) \tag{16.15}$$

or

$$T_d = R_{B2}C_2 \ln \left(2 - \frac{V_{BE+}}{V_{CC}} \right) \tag{16.16}$$

Usually V_{BE+}, being about 0.5 V, is small in comparison with V_{CC}, and the term V_{BE+}/V_{CC} can be neglected. Then, since $\ln 2 = 0.69$

$$T_d \simeq 0.7 \, R_{B2}C_2 \tag{16.17}$$

The output of the monostable or *one shot* multivibrator is a rectangular pulse, often called a *gate*, the duration of which can be controlled by varying either R_{B2} or C_2, as shown in Eq. 16.15. A potentiometer is usually used as part of R_{B2} for adjustment of T_d. Large changes in T_d are sometimes made by switching C_2 values.

Graded base transistors cannot usually withstand the large reverse bias $(V_{CC} - V_{BE+})$ that occurs when transistor T_2 is switched off. Therefore, a diode should be placed in series with the base of T_2, as shown in Fig. 16.17, to prevent possible damage to the transistor and to provide the desired value of gate duration or width, whenever the maximum V_{BE} rating is less than V_{CC}.

A negative going pulse, or gate, could be obtained from the collector of T_1 but the fall time of the pulse would be seriously degraded because the relatively large base current that flows into T_2 when switching occurs at time t_d flows primarily through R_{L1} until capacitor C_2 becomes charged. This charging current causes v_{CE1} to rise exponentially toward V_{CC} at the time constant $R_{L1}C_2$, approximately. Of course, a sharp negative gate can be obtained by using *p-n-p* transistors.

An *astable* or *free-running* multivibrator can be constructed by capacitively, or ac, coupling both transistors, as shown in Fig. 16.18. Then there is no permanently stable state, and each transistor alternately saturates

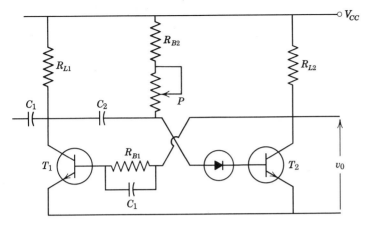

Fig. 16.17.   A monostable multivibrator with adjustable t_d and diode protection for T_2.

and cuts off. The *off time* for each transistor can be determined from the relationship $T = 0.7\,R_B C$, approximately, as was discussed for the monostable. The circuit is symmetrical if $R_{B1} = R_{B2}$ and $C_1 = C_2$, and so forth, but large disymmetry is sometimes desirable and attainable. The base circuit resistors must be chosen from the relationship $R_B \leqslant h_{FE}R_L$ for proper operation. The load resistors and transistor types are chosen in accordance with the frequency requirements as previously discussed. The rise times are determined by the coupling capacitors charging currents flowing through the load resistors, as discussed in connection with the monostable multivibrator. This charging current can be diverted from the load resistors by the addition of a resistance and a diode in each collector circuit, as shown in Fig. 16.19. When the transistor T_1 is driven into saturation, the collector voltage drops quickly and the diode D_1 is

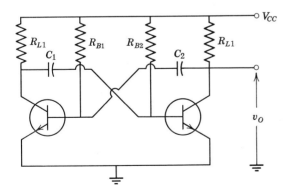

Fig. 16.18.   An astable or free-running multivibrator.

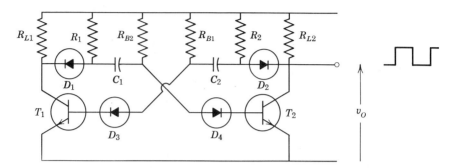

Fig. 16.19. An astable multivibrator with diodes for transistor protection and diode-resistor combinations to permit fast rise and fall times.

forward biased and therefore has little influence on the circuit operation. However, when transistor T_1 is cutoff, its collector rises very rapidly to V_{CC}, reverse biasing diode D_1, and capacitor C_1 charges through R_1. Diode D_2 and R_2 provide the same fast rise time at the collector of T_2, as shown in the sketch of the output voltage. The resistors R_1 and R_2 must be small enough to permit C_1 and C_2 to charge almost completely between the times of switching. An example will illustrate the design of an astable multivibrator.

EXAMPLE 16.4   Assume that we need a pulse generator that will produce $100\,\mu s$ pulses at the rate of 1000 pulses/s. An astable multivibrator is a suitable generator. Since the time between pulses is $900\,\mu s$, we will use $V_{CC} = 20\,V$, $R_{L1} = 4\,k\Omega$, and $R_{L2} = 10\,k\Omega$. Since R_1 is in parallel with R_{L1} when T_1 is conducting and R_2 is in parallel with R_{L2} when T_2 is conducting, we tentatively assume that $R_1 = R_{L1}$ and $R_2 = R_{L2}$ for the purpose of determining the values of R_{B1} and R_{B2}. If $h_{FE\,\min} = 80$, and by using $R_B \leqslant h_{FE}R_L$, $R_{B1} \leqslant 80 \times 2\,k\Omega = 160\,k\Omega$ and $R_{B2} \leqslant 80 \times 5\,k\Omega \leqslant 400\,k\Omega$. Then, by using Eq. 16.17, $C_1 = 9 \times 10^{-4}/(0.7 \times 4 \times 10^5) = 3.2 \times 10^{-9}\,F$ and $C_2 = 10^{-4}/(0.7 \times 1.6 \times 10^5) = 880\,pF$. The time constant $R_1C_1 = 4 \times 10^3 \times 3.2 \times 10^{-9} = 12.8\,\mu s$, which is about one-eighth of the $100\,\mu s$ during which T_1 conducts. Therefore C_1 has adequate time to charge through R_1 during the $100\,\mu s$. Also, $R_2C_2 = 10^4 \times 880\,pF = 8.8\,\mu s$, which is very small in comparison with the $900\,\mu s$ during which T_2 conducts. Thus R_2 could be much larger than $10\,k\Omega$ and R_1 could be larger than $4\,k\Omega$.

Field-effect transistors are especially useful in monostable or astable multivibrator applications that require very long time constants, as may be seen in Fig. 16.20. Since R_{G1} and R_{G2} may be very large (many megohms), the time constants can be very large compared with those of a bipolar

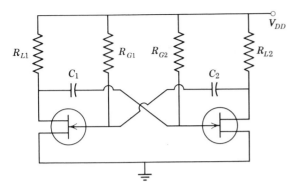

Fig. 16.20.   An astable multivibrator using FETs.

transistor. Since the FET switches on when the reverse bias is reduced to V_p, the off-period of each transistor can be calculated approximately from the following relationship (compare Eq. 16.14).

$$T_d = R_G C \ln \left(2 - \frac{V_p}{V_{DD}}\right) \tag{16.18}$$

The base-circuit or gate-circuit resistors in an astable multivibrator are usually made adjustable by the inclusion of a potentiometer, so that the time constants can be controlled.

PROBLEM 16.5   A given silicon transistor that has $h_{FEmin} = 50$, $f_\tau = 160$ MHz, and $C_{ob} = 5$ pF is used in a monostable multivibrator circuit with $V_{CC} = 20$ and $R_{L1} = R_{L2} = 10$ kΩ. Determine the other circuit components that will provide a 50 μs gating pulse. Draw a circuit diagram.
Answer: $R_B \leqslant 500$ kΩ, $C_2 = 140$ pF for $R_B = 500$ kΩ, $C_1 = 6$ pF. Diode required.

PROBLEM 16.6   Redesign the multivibrator of Example 16.4 using $R_2 = 1$ MΩ and $R_1 = 10$ kΩ to obtain the same output characteristics.
Answer: $R_{B1} = 220$ kΩ, $R_{B2} = 720$ kΩ, $C_1 = 1800$ pF, $C_2 = 610$ pF.

PROBLEM 16.7   Design a symmetrical astable multivibrator, using FET transistors with $V_{DD} = 25$ V, $V_p = -2$ V, and $g_{mo} = 2000$ μmho at $I_{DSS} = 2$ mA, that will produce a square-wave output signal with a period of 10 s.

16.5   DIGITAL INTEGRATED CIRCUITS

The semiconductor manufacturers produce a wide variety of integrated pulse (or digital) circuits. These circuits vary in complexity from units that contain 4 or 5 transistors to units with over 60 transistors. In fact, some manufacturers are now producing large-scale integrated (LSI) circuits that contain thousands of transistors. These circuits may appear to be very complex, but most of them are composed of a relatively few basic building blocks and their interconnections. Symbolic block diagrams are used to aid the designer when laying out the circuit. Consequently, digital circuits, which are so complicated that it would be almost impossible to construct them with discrete circuit elements (transistors, diodes, resistors, and capacitors), can be easily designed and fabricated with integrated circuits (IC's).

Some forms of IC's are essentially the same as circuits we have already studied. For example, the Motorola MC802 has the form shown in Fig. 16.21. This circuit is a bistable multivibrator similar to that shown in Fig. 16.11. The main difference is the parallel triggering transistors T_1 and T_4 that have been added in Fig. 16.21. To understand the behavior of this circuit, let us consider an example.

EXAMPLE 16.5 Assume that terminal 8 is at $+3$ V with reference to the ground terminal 4. Also assume that T_2 is in saturation and T_3 is cutoff. Then, terminal 7 will be at the collector saturation potential (0.2 V or less). The addition of a positive signal at terminal 1 will then have no effect on the circuit since T_2 is already drawing essentially all the current from V_{CC} that R_2 will allow. Of course, a zero or negative signal at terminal 1

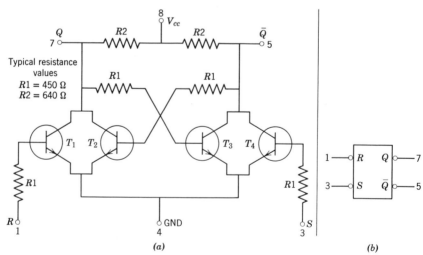

Fig. 16.21.   The RS flip-flop of the MC 802. (a) Schematic circuit and (b) symbolic block diagram.

merely drives T_1 into cutoff and in effect removes it from the circuit. If the signal on terminal 3 is zero or negative, then both T_3 and T_4 will be cutoff and the only current flowing from V_{CC} through R_2 toward terminal 5 will be the base current for transistor T_2. Then, the voltage on terminal 5 is

$$v_5 = V_{CC} - \frac{(V_{CC} - v_{BE2})R_2}{R_1 + R_2} \qquad (16.19)$$

If v_{BE2} is assumed to be about 0.6 V and R_1 and R_2 have the nominal values given in Fig. 16.21, the value of v_5 is $3 - (3 - 0.6)640/(450 + 640) \simeq 1.6$ V.

The circuit will remain in this condition until a positive signal is applied to terminal 3. Then transistor T_4 begins to draw current. This additional current through R_2 will reduce the bias on T_2 and initiate the switching action between T_2 and T_3. The manufacturer indicates that an input voltage of 0.85 V (or more) on terminal 3 will cause the circuit to switch.

After switching, terminal 3 will have a potential of about 0.2 V and terminal 7 will have a potential of about 1.6 V. The circuit will now remain in this condition until a positive pulse (0.85 V or more) is applied to terminal 1. Thus, a positive pulse on terminal 3(S) *sets* the circuit and a positive pulse on terminal 1(R) clears or *resets* the circuit. Hence it is named the *R-S flip flop*. Observe that a positive trigger pulse applied simultaneously to inputs 1 and 3, by connecting them together, will make a binary counter out of this flip-flop. Thus, the triggering transistors T_1 and T_4 perform the same function as the steering diodes of Fig. 16.12.

As was mentioned previously, the manufacturers produce a large number of different IC's. A given series of these circuits are constructed so that they can be directly interconnected. Thus, in the MC 800 series, the output signal from one IC is the proper magnitude for the input signal of another IC in the same series. For example, the output signal from the MC 802 circuit was either 0.2 V or 1.6 V. If this output signal is connected to the input of another MC 802 (or another unit in the MC 800 series), the 0.2 V signal will not trigger the second circuit but the 1.6-V signal will.

From the foregoing material we observe that there are two basic states or signals that we are working with. The transistors are either ON or OFF. Consequently, essentially all digital computer calculations are done in the *binary* (or two number) system. Actually, the basic manipulation of numbers can be accomplished in any base. The only reason our present base of 10 system has developed is because man had 10 fingers (or toes if you count that way). A comparison of the first twenty numbers as written in base 10 and binary is given in Table 16.1. With a little practice,

Table 16.1   A Comparison of Binary and Base 10 Numbers

Base 10	Binary (Base 2)
0	0
1	1
2	10
3	11
4	100
5	101
6	110
7	111
8	1000
9	1001
10	1010
11	1011
12	1100
13	1101
14	1110
15	1111
16	10000
17	10001
18	10010
19	10011
20	10100

it becomes quite easy to solve mathematical problems in the binary system. To illustrate the four basic mathematical steps, consider Example 16.6.

EXAMPLE 16.6   Let us solve the problems (a) $9+3=12$, (b) $9-3=6$, (c) $9/3=\frac{1}{3}$ and (d) $9\times3=27$ in the binary system.

(a)
$$
\begin{array}{rr}
1001 & 9 \\
+11 & +3 \\
\hline
1100 & 12
\end{array}
$$

(b)
$$
\begin{array}{rr}
1001 & 9 \\
-11 & -3 \\
\hline
110 & 6
\end{array}
$$

(c)
$$
\begin{array}{r}
11 \\
11\overline{)1001} \\
\underline{11} \\
11 \\
\underline{11} \\
0
\end{array}
\qquad
\begin{array}{r}
3 \\
3\overline{)9}
\end{array}
$$

(d)
$$
\begin{array}{rr}
1001 & 9 \\
11 & \times3 \\
\hline
1001 & 27 \\
1001 & \\
\hline
11011 &
\end{array}
$$

The main thing to remember is that $1 + 1 = 10$ conversely that $10 - 1 = 1$. It then follows that if you borrow 1 from 10 you leave 1. The rest of the rules are the same as in the base of 10 math $0 + 0 = 0$, $0 + 1 = 1$, $1 - 0 = 1$, $1 - 1 = 0$, $1 \times 0 = 0$, $1 \times 1 = 1$, $1/1 = 1$ and $0/1 = 0$.

PROBLEM 16.8   Extend Table 16.1 to 32. Is 27 correctly written as 11011 in binary?

PROBLEM 16.9   Solve the following problems using binary numbers. (a) $6 + 7$, (b) $9 - 6$, (c) $18/6$, (d) 6×3.

The designer of a system can define which state (ON or OFF) represents the 1 in his system. For this text, we shall assign a signal of 0.3 V or less as a zero (0) and a signal of 0.85 V or more as a one (1).

With the development of plastic encapsulation the cost of the IC units has been substantially reduced. The remainder of this section will use low-cost plastic encapsulated units from the MC 700P/800P series. Units of the MC 800P series will operate through the temperature range of 0°C to 75°C. The MC 700P units will operate satisfactorily for temperatures of 15° to 55°C. The MC 700P/800P series are known as RTL (resistor-transistor logic) units.

The simplest pulse circuit is known as an *inverter*. The inverter is usually a single transistor that will convert a 1 on the input to a 0 in the output, or vice versa. It is often desirable to pack several circuits in a given pack (especially if the circuits are relatively simple). For example, the MC789P/MC889P shown in Fig. 16.22 has six individual inverter circuits packaged as a single unit.

PROBLEM 16.10   Use the MC789P IC unit to construct (a) a bistable multivibrator, (b) a monostable multivibrator, and (c) an astable multivibrator. Use as few external components as possible.

A branch of mathematics known as Boolean Algebra is used to develop electronic logic circuits. It is beyond the scope of this text to develop the concepts of this approach. However, much of this logic is performed by producing a signal whenever several other signals appear (or do not appear) simultaneously. A circuit that performs this operation is commonly called a *gate*. Gates are named according to the basic Boolean algebra logic concepts they perform. For example, an AND gate produces a 1 on the output only when all input signals are also 1. If one or more of the input signals is a 0, the output signal is a 0. The OR gate produces a 1 on the output if one or more of the input signals is a 1. The output is a 0 only if all

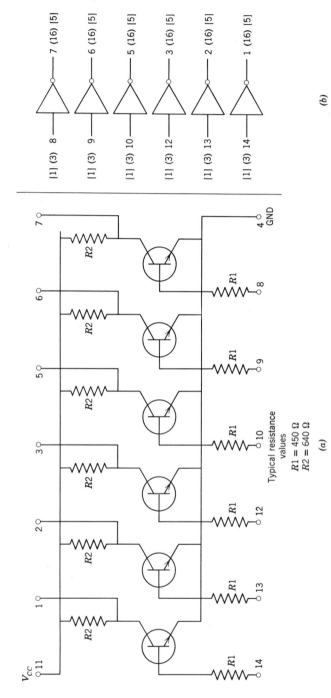

Fig. 16.22.   The MC789P/889P hex inverter circuit. (a) Schematic circuit and (b) symbolic block diagram.

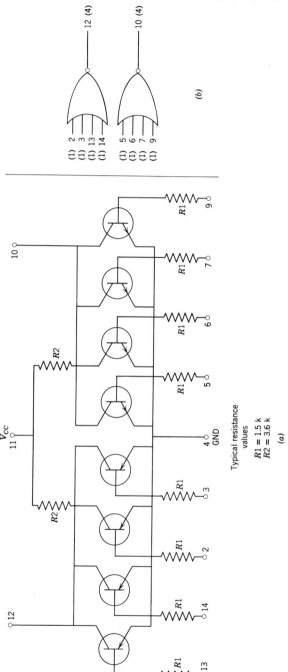

Fig. 16.23.   The MC719P/MC819P Dual four-input gates. (a) Schematic circuit and (b) symbolic block diagram.

of the input signals are 0. The NOR gate produces a 1 on the output if all of the input signals are 0. If one or more of the input signals is 1, the output is a 0. The MC719P/MC819P unit shown in Fig. 16.23 contains two separate four-input NOR gates.

Under normal operating conditions, the circuit in Fig. 16.23 will have a potential of +3 V connected to terminal 11, and terminal 4 will be connected to ground. The four transistors on the left with their associated circuitry forms one gate circuit. A 1 signal on any one of the input terminals (13, 14, 2, or 3) will drive the associated transistor into saturation to produce a 0 on the output terminal (12). If a 1 is applied to more than one input terminal, the output will still be a 0. Only if all four of the input signals are 0s will the output signal become a 1. We have a NOR gate!

PROBLEM 16.11   Show how inverter circuits could be used to convert THE MC719P (Fig. 16.23) to two four input AND gates.

To insure versatility in the circuit arrangement offered to the user, the manufacturers supply gates in a great many different package arrangements. For example, four separate two-input NOR gates are contained in the MC717P/MC817P circuit shown in Fig. 16.24.

PROBLEM 16.12   Use the MC717P circuit to construct two *RS* flip-flops (see Fig. 16.21). Use as few external circuit elements as possible.

The exact voltage on the output terminal of an IC (terminal 3 in Fig. 16.24 for example) is determined by the load connected to this terminal. If too many circuits are connected to a given output terminal, the voltage in the 1 state may not be high enough to trigger these circuits. To insure proper operation, the manufacturer provides data on the maximum number of IC's that can be connected to the output terminal and still obtain reliable operation. The information is given in terms of a *loading factor* or *fan out*. The loading factor indicates the number of "standard" input circuits that can be connected to a given output terminal. In the symbolic block diagrams, the number in parentheses near the output terminal is the loading factor for that output. In some units, the loading factors of the MC700 series and MC800 series are different. In these units, the loading factor of the MC700 series is given in parentheses, and the loading factor of the MC800 circuit is given in brackets. The actual input impedance of some circuits may be low enough to cause as much loading as two or more of the "standard" units. Thus, a loading factor (using the same notation as on the output circuits) is also given for the input circuits. For example, the MC789P (Fig. 16.22) has an input conductance equal to three "standard"

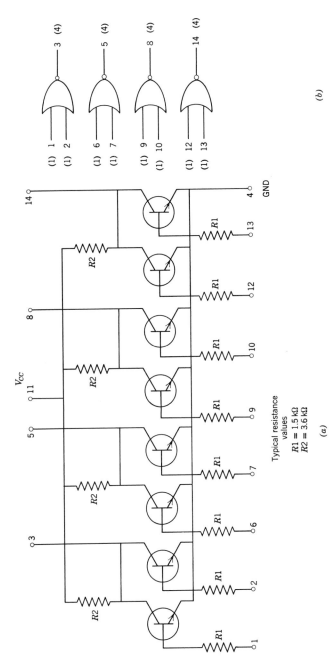

Fig. 16.24.   The MC717P/MC817P quad two-input gates. (*a*) Schematic circuit and (*b*) symbolic block diagram.

circuits. In contrast the MC889P has an input impedance equal to one "standard" circuit and can satisfactorily drive up to five "standard" circuits.

The *buffer* is designed to have a low-output impedance so that it can drive a greater number of load circuits than the basic IC. The circuit of two buffers is given in Fig. 16.25. Notice that these buffers also invert the signal.

Most simple multivibrators can be readily constructed by cross coupling inverter circuits or gate circuits. However, a more complex multivibrator configuration, known as the *JK* flip-flop is especially designed for clocked operation. (The *clock* is a unit that produces a series of evenly spaced pulses to maintain synchronization in the system.) In the *JK* flip-flop, signals (1s or 0s) are applied to the input terminals but the flip-flop in the IC is not activated until the clock pulse is applied.

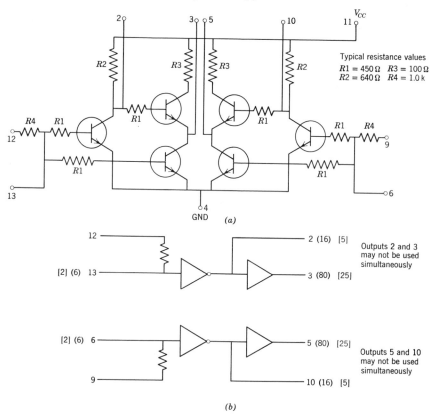

Fig. 16.25. The MC799P/MC899P dual buffer. (a) Schematic circuit and (b) symbolic block diagram.

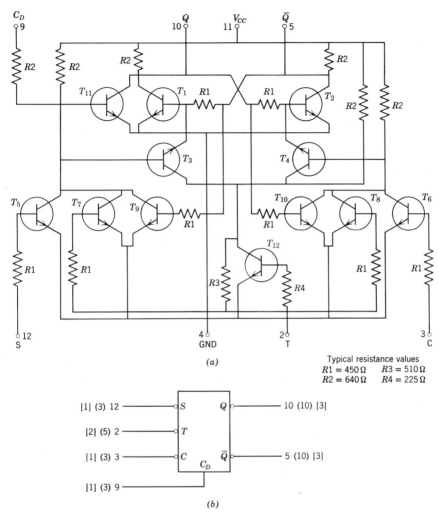

Typical resistance values
$R1 = 450\,\Omega$ $R3 = 510\,\Omega$
$R2 = 640\,\Omega$ $R4 = 225\,\Omega$

(a)

[1] (3) 12 ——— S	Q ——— 10 (10) [3]
[2] (5) 2 ——— T	
[1] (3) 3 ——— C	$\bar{Q}$ ——— 5 (10) [3]
[1] (3) 9 ——— C_D	

(b)

Fig. 16.26.   A JK flip-flop (MC723P/MC816P). (a) Schematic circuit and (b) symbolic block diagram.

The diagram of a typical *JK* flip-flop is given in Fig. 16.26. The basic flip-flop in this circuit is formed by transistors T_1 and T_2. The input C_D (terminal 9) is a *direct clear* input. Whenever a 1 is applied to this C_D terminal, the circuit will be reset so that output Q (terminal 10) has a 0 for an output. The input signals are applied to terminals 12 and 3. These inputs are known as the *set* (S) and *clear* (C) inputs. The clock or *trigger* (T) pulse is applied to terminal 2.

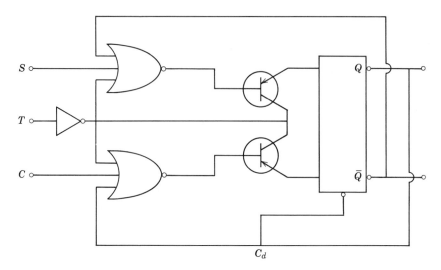

Fig. 16.27. A block diagram of the JK flip-flop in Fig. 16.26.

A careful study of the circuit in Fig. 16.26 reveals this complex circuit can be represented as an interconnection of several of the circuits already examined. For example, transistor T_{12} can be viewed as an inverter for the clock pulse. Also, notice that T_5, T_7, and T_9 are connected in parallel to form a 3-input gate. An identical configuration is formed by T_6, T_8, and T_{10}. Transistors T_3 and T_4 couple the output of these gates into the multivibrator base circuit. Consequently, the *JK* flip-flop can be represented by the block diagram given in Fig. 16.27.

Switching occurs in this circuit on the negative going portion of the clock pulse. For proper action the clock pulse fall time must be less than 100 ns. If the S input is a 1 and C input is a 0, the signal on the Q terminal (after switching) will be a 1. The signal on the $\overline{Q}$ terminal will always be opposite to the signal on the Q terminal. Consequently, $\overline{Q}$ is normally referred to as the *complement of Q* or as Q NOT. If the S input is a 0 and the C input is a 1, the signal on the Q terminal (after the clock pulse) will be a 0. If both S and C are 1s the clock pulse will not produce any change in the Q signal. In contrast, if both S and C are 0s the clock pulse will shift the output to its opposite state (a 0 to a 1 or a 1 to a 0) whenever the clock pulse appears.

Two *JK* flip-flops may be packed as a single unit. For example, the MC790P has the block diagram shown in Fig. 16.28. The circuit of each flip-flop is similar to that shown in Fig. 16.26.

To illustrate the design techniques that can be used with these IC units, let us consider an example.

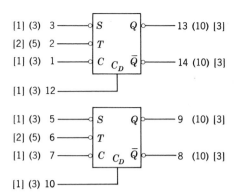

[1] (3) 3 ——o S   Q o—— 13 (10) [3]
[2] (5) 2 ——o T
[1] (3) 1 ——o C C_D $\bar{Q}$ o—— 14 (10) [3]
[1] (3) 12 ——

[1] (3) 5 ——o S   Q o—— 9 (10) [3]
[2] (5) 6 ——o T
[1] (3) 7 ——o C C_D $\bar{Q}$ o—— 8 (10) [3]
[1] (3) 10 ——

Fig. 16.28.   Block diagram of the MC790P/MC890P dual JK flip-flop.

EXAMPLE 16.7   Construct a circuit that will count to 10. When the count reaches 10, the circuit should reset itself and produce one output pulse that could be applied to a following identical unit.

Examination of Table 16.1 indicates that four different binary numbers must be used to represent the number 10. Consequently, we shall need four different flip-flops. We can use two MC790P units with two flip-flops in each unit. The counter could be constructed as shown in Fig. 16.29.

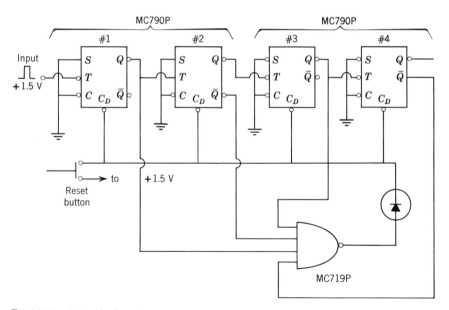

Fig. 16.29.   A simple decade counter.

Table 16.2   Truth Table for the Circuit in Fig. 16.29

| | The signal on the Q terminals | | | |
Condition	Q_4	Q_3	Q_2	Q_1
reset	0	0	0	0
after pulse 1	0	0	0	1
after pulse 2	0	0	1	0
after pulse 3	0	0	1	1
after pulse 4	0	1	0	0
after pulse 5	0	1	0	1
after pulse 6	0	1	1	0
after pulse 7	0	1	1	1
after pulse 8	1	0	0	0
after pulse 9	1	0	0	1
after pulse 10	1	0	1	0

When the reset button is pushed, a 1 signal is applied to each of the C_D inputs so that each of the flip-flops are set so that Q is 0.

Since all S and C terminals are connected to ground (0 condition), each circuit will switch whenever its clock pulse shifts from a 1 to a 0 state. For example, number 2 flip-flop switches whenever Q_1 switches from a 1 to a 0 state. Similar switching occurs in the other flip-flops. Hence, the action of the circuit can be tabulated as shown in Table 16.2.

We wish the circuit to reset itself when the tenth pulse is applied. The tenth pulse produces a 1 on Q_4, so $\overline{Q}_4$ will have a 0 for its output. Similarly, $\overline{Q}_2$ will be a 0. In addition Q_3 and Q_1 are both 0. Therefore, if we use $\overline{Q}_4, Q_3$, $\overline{Q}_2$, and Q_1 to activate one of the MC719P gates (Fig. 16.23), the output of this gate will be 1 when (and only when) the count is 10. The 1 on the output of this gate can be applied to the C_D terminals to reset all of the Q outputs back to 0. The configuration that accomplishes this switching is shown in Fig. 16.29. The diode is included in this circuit to protect the gate when the reset button is pressed.

Note that Q_4 will switch from 1 to 0 as flip-flop 4 is reset. Therefore, the signal from Q_4 can be coupled directly to the input of the next decade counter, which can be identical to that just considered.

Some of the digital IC's in the MC700P series, such as the MC796P (actually a full subtractor), have as many as 62 transistors in each package. However, even these complex units can be represented by the simple building blocks we have already discussed.

The versatility and usefulness of the digital IC units can be visualized from the foregoing material. Of course, we have just been able to introduce the basic concepts of digital IC's in this section. Nevertheless, you should be able to design some digital circuits with this background. Of course, further reading in this interesting area is strongly recommended.

PROBLEM 16.13 We wish to activate an indicator circuit from the counter in Fig. 16.29. Use buffers (MC799P) if required and MC719P gates to design a circuit that will supply a 1 signal at terminal A if and only if the count is 0, a 1 signal at terminal B if and only if the count is 1, a 1 signal at terminal C if the count is 2, etc., for the 10 positions in the counter.

16.6 FET AND MOSFET PULSE CIRCUITS

Pulse or digital circuits can also be constructed from FET or MOSFET units. These devices have the advantages of high input impedance, low power consumption, circuit simplicity, and extremely small sizes. Unfortunately, as we go to higher impedance circuits, the shunt capacitances in the circuits (as was already noted) have a greater effect on the output waveforms and switching speeds. Consequently, the FET and MOSFET circuits are slower speed circuits than their bipolar counterparts.

A typical MOSFET three input NOR gate is shown in Fig. 16.30. The transistor T_4 is used as a load resistor and is biased to a conducting state when the output terminal is more positive than the V_{DD} supply. A negative pulse on any one (or more) of the input terminals will cause the associated transistor to conduct, and the output voltage will approach zero. Only if all of the input terminals have a near zero signal will the output voltage approach $-V_{DD}$. In this case, the near zero signal is a 1. This arrangement is known as a *negative logic* system since the more negative

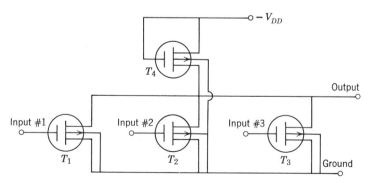

Fig. 16.30.   A typical MOSFET NOR gate.

signal is the 1 state. The circuit in Fig. 16.30 is, therefore, a NOR configuration for a negative logic system.

By using a MOSFET as a load resistance, two desirable effects are achieved. First, a MOSFET uses less area on the silicon chip than a conventional diffused resistor. Second, the effective resistance of the MOSFET is much higher than the resistance of a typical diffused resistor, as can be readily observed from Fig. 16.31. The curve for operation with $V_{GS} = V_{DS}$ (the condition of T_4 in Fig. 16.30) is shown as a solid line in Fig. 16.31. The dynamic drain resistance r_d, as discussed in Chapter 6, is defined as

$$r_d = \frac{\Delta v_{DS}}{\Delta i_S} \qquad (16.20)$$

where Δv_{DS} represents a change of drain-source voltage and Δi_D represents a change of drain current. Note in Fig. 16.31 that if operation in the shaded area is not permitted, a large change of voltage (along the $V_{GS} = V_{DS}$ line) produces only a small change of current. In fact, typical resistances with values of about 100 kΩ to 200 kΩ can be realized from a MOSFET. Compare these values with the typical resistors in the integrated bipolar transistor circuits ($R \simeq 1000\ \Omega$ or less).

The integrated MOSFET circuit can be very compact. For example, Robert Crawford[1] indicates the entire circuit of Fig. 16.30 can be constructed on an area 5.9 mils by 2.6 mils (one mil $\equiv 10^{-3}$ in.). Consequently, very complex digital circuits can be fabricated in a relatively small

[1] Robert H. Crawford, *MOSFET in Circuit Design*, McGraw-Hill Book Company, New York, 1967, p. 104.

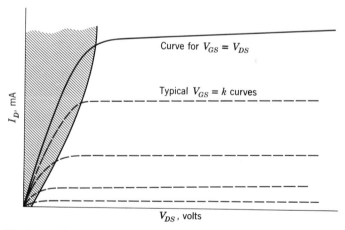

Fig. 16.31.   Operation of a MOSFET as a resistor.

Fig. 16.32.   A moderate size MOS type integrated circuit. (Courtesy of General Instruments, Salt Lake City, Utah.)

area. Thus, MOSFET devices are important contenders in the large-scale integrated (LSI) circuit area. In these LSI circuits, the digital signal processing is performed by very small MOSFET's. Larger units are used to obtain higher power levels for the output MOSFET's. A typical MOS type of LSI is given in Fig. 16.32. Over four thousand MOSFET's can be formed on a chip $\frac{1}{8}$ of an inch by $\frac{1}{8}$ of an inch!

Of course, MOSFETS can be used to construct multivibrators. For example, the configuration for an *RS* type flip-flop is given in Fig. 16.33. The similarity to the flip-flop in Fig. 16.21 is obvious. The transistors T_1 and T_2 form the load resistances for the NOR gates, which are composed of T_3 and T_4, or T_5 and T_6. By adding additional transistors, the *RS* flip-flop configuration can be converted to a *JK* type flip-flop. One proposed LSI configuration involves a large array of MOS flip-flops arranged to serve as a section of memory for a computer.

PROBLEM 16.14   Would it be possible to connect two MOSFET NOR gates (Fig. 16.30) so that they perform as an *RS* flip-flop (Fig. 16.33)? If so, give the proper configuration for this *RS* flip-flop.

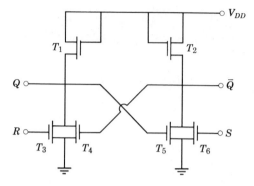

Fig. 16.33.   An RS flip-flop using MOSFETs.

16.7   THE UNIJUNCTION TRANSISTOR

The unijunction transistor is a very useful device for timing or clock pulse applications. The physical construction is essentially as shown in Fig. 16.34. A bar of lightly doped n-type semiconductor material (usually silicon) has electrical contacts or terminals connected to each end. These terminals are known as the base 1 (B_1) and base 2 (B_2) terminals. Approximately halfway along the n-type bar a region of p-type material is formed, thus producing a single junction. The terminal that is connected to this p-region is known as the emitter (E) terminal.

The symbol for a unijunction transistor as well as a typical biasing arrangement is shown in Fig. 16.35a. The electrical behavior of the circuit can be visualized with the help of the equivalent circuit shown in Fig. 16.35b. The silicon bar acts as a resistor, which is tapped where the junction is formed. The resistance from the junction to the B_2 terminal

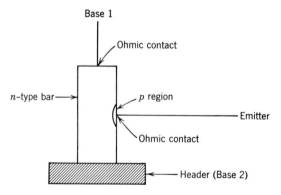

Fig. 16.34.   The physical arrangement of a unijunction transistor.

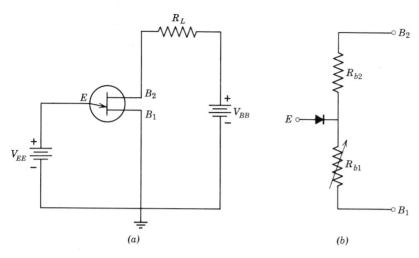

Fig. 16.35. (a) The symbol and biasing arrangement of a unijunction transistor and (b) an equivalent circuit for the unijunction transistor.

will be called R_{b2} and the resistance from the junction to the B_1 terminal will be called R_{b1}. The junction acts as a conventional p-n diode. Now, if B_1 is connected to ground and B_2 is connected to a positive potential V_{BB}, a current will flow through R_{b2} and R_{b1}. If the emitter terminal is open, a voltage V_{EO} will appear between the E and B_1 terminals. Because of the voltage divider action of R_{b2} and R_{b1}, this voltage V_{EO} will be a fraction η of the voltage on base 2, with respect to base 1.

$$V_{EO} = \eta V_{b2} \tag{16.21}$$

If the emitter is biased at a potential less than V_{EO}, the p-n junction is reverse biased and only the diode saturation current flows in the emitter circuit. However, if the voltage of the emitter is increased above V_{EO}, the junction becomes forward biased. Under these conditions, holes are injected from the p-material into the n-bar. These holes are repelled by the positive base-two end of the bar and are attracted toward the base-one end of the bar. These additional p-carriers in the emitter-to-base-one region results in a decrease of resistance for resistor R_{b1} (Fig. 16.36). The decrease of resistance R_{b1} results in a lower emitter voltage. Thus, a negative resistance effect is produced since the voltage decreases as the current increases. As more p-carriers are injected, a condition of saturation will eventually be reached, as shown in Fig. 16.36, and in the current-voltage characteristics in Fig. 16.37. The similarity of these characteristics to those of the tunnel diode is at once evident. The unijunction transistor,

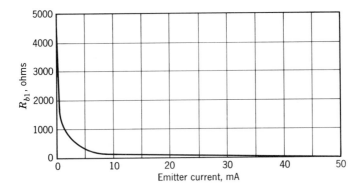

Fig. 16.36.   A plot of R_{b1} versus emitter current for a 2N492.

however, has three terminals. Therefore, to completely describe the characteristics, a family of curves (as given in Fig. 16.38) is required.

Unijunction transistors are used extensively in oscillator, pulse, and voltage sensing circuits. To help illustrate some typical applications, let us consider two examples. The first of these is a circuit known as a relaxation oscillator.

EXAMPLE 16.8   A unijunction transistor is connected as shown in Fig. 16.39. Determine the voltage waveforms on the emitter and on base 2 of the unijunction transistor. The transistor is a 2N492 (characteristics given in Fig. 16.36) with $R_{b1} + R_{b2} = R_{bb} = 7.5 \text{ k}\Omega$ and $\eta \simeq 0.67$.

At time $t = 0$, the switch S_1 is closed. The capacitor is initially uncharged, so the voltage V_{EO} is 0 at time $t = 0$. Current can flow through R_2 and the

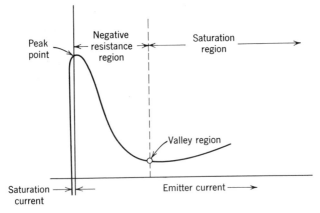

Fig. 16.37.   The emitter current-voltage characteristics of a unijunction transistor.

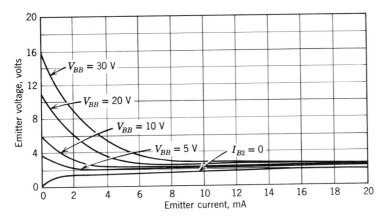

Fig. 16.38.   Static characteristic curves of a unijunction transistor.

unijunction transistor base to ground. This current will have a magnitude of

$$I_B = \frac{V_{BB}}{R_2 + R_{bb}} \tag{16.22}$$

or $12/(2.5 \text{ k}\Omega + 7.5 \text{ k}\Omega) = 1.2 \text{ mA}$.

The voltage on B_2 is $V_{BB} - I_B R_2 = 12 - (1.2 \times 10^{-3})(2.5 \times 10^3 = 9 \text{ V}$. From Eq. 16.21, $V_{EO} = 0.67 \times 9 = 6 \text{ V}$. Since the emitter voltage is initially 0, the emitter junction is reverse biased and appears as an open circuit. The circuit appears as shown in Fig. 16.40. The capacitor will charge

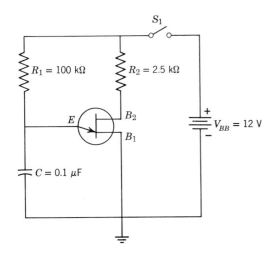

Fig. 16.39.   A relaxation oscillator.

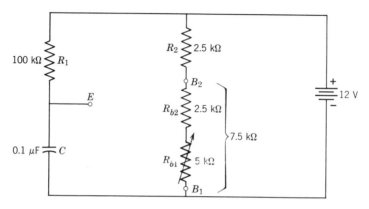

Fig. 16.40. The equivalent circuit for Fig. 16.39 when the emitter junction is reverse biased.

from 0 V toward $+12$ V with a time constant $\tau = R_1 C = 10^5 \times 10^{-7} = 10^{-2}$ s. Thus, the equation for the voltage on the emitter (while the emitter junction is reverse biased) is

$$v_E = 12(1 - e^{-t/0.01}) \qquad (16.23)$$

A sketch of v_E versus time is given in Fig. 16.41.

The voltage v_E follows Eq. 16.23 until v_E is approximately 6 V. The time required for v_E to reach 6 V can be found from Eq. 16.23. A solution of this equation with $v_E = 6$ V yields $t = 6.92$ ms. When v_E exceeds 6 V, the emitter junction becomes forward biased and p-carriers are injected into the base region. As was noted previously, these carriers reduce the resistance of R_{b1}. As the resistance of R_{b1} decreases, the voltage at the junction of R_{b1} and R_{b2} decreases. This action causes the emitter junction to become more heavily forward biased and a greater number of p-carriers are injected into the base region, further reducing the resistance of R_{b1}. In

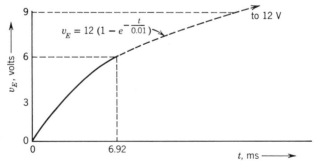

Fig. 16.41. A plot of v_E for the circuit in Fig. 16.39.

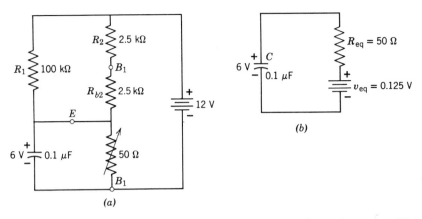

Fig. 16.42. The equivalent circuit for Fig. 16.39 an instant after v_E is equal to 6 V. (a) Equivalent circuit and (b) simplified circuit.

fact, the action is cumulative and R_{b1} rapidly approaches its minimum value of approximately 50 Ω as shown by Fig. 16.36. The forward-biased junction acts almost as a short circuit, so the equivalent circuit for Fig. 16.39 (an instant after $t = 6.92$ ms) becomes that shown in Fig. 16.42a. This circuit can be further simplified (as shown in Fig. 16.42b) by replacing all the circuitry that connects to the capacitor by a Thevenin's equivalent circuit. The equivalent resistance R_{eq} for this Thevenin's circuit is $(R_2 + R_{b2})$ in parallel with both R_1 and R_{B1}. Thus $R_{eq} = 5\,\mathrm{k\Omega} \| 100\,\mathrm{k\Omega} \| 50\,\Omega \approx 50\,\Omega$. The Thevenin's equivalent voltage is $v_{eq} = (12 \times 50)/(R_p + 50)$ where R_p is the parallel combination of $(R_2 + R_{b2})$ and R_1. Thus, $R_p = 4.76$ kΩ and $v_{eq} = (12 \times 50)/(4.76\,\mathrm{k\Omega} + 50) \approx 0.125$ V.

From Fig. 16.38b, the time constant of this circuit is $R_{eq}C = 5 \times 10^{-7}$ or 5 μs. Thus, the voltage across C (which is also v_E) will decrease from 6 V toward 0.125 V with a time constant of 5 μs. In four or five time constants (20 to 25 μs) the capacitor will be essentially discharged. As the capacitor current decreases, the p-carriers in the base 1 region decrease and R_1 begins to increase. An increase of R_1 allows less current to flow, so the action again is cumulative and the emitter junction quickly becomes reverse biased. The circuit is now in essentially the same condition as at time $t = 0$, so the cycle repeats. Consequently, the emitter voltage v_E will have the form shown in Fig. 16.43.

Since the emitter junction is essentially a short circuit while C is discharging, the lower end of R_{b2} will have essentially the same potential as v_E during this period. Thus, the lower end of R_{b2} will shift from 6 V to about 0.2 or 0.3 V with a time constant of 5 μs. As a result, the potential at the terminal B_2 will drop from 9 V (the value while the emitter junction is

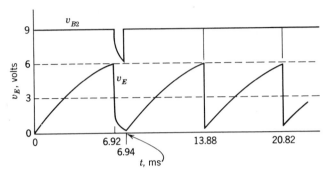

Fig. 16.43.   A plot of the voltage waveforms on the emitter and base 2 of the transistor in Fig. 16.39. The period from $t = 6.92$ ms to $t = 6.94$ ms is expanded to permit showing the detail of the waveform.

reverse biased) to a voltage v_{B2min} while the capacitor is discharging. This minimum value for v_{B2} is

$$v_{B2min} = v_{Emin} + (V_{BB} - v_{Emin})R_{b2}/(R_{b2} + R_2) \qquad (16.24)$$

where v_{Emin} is the minimum voltage on the emitter before the junction becomes reverse biased again. If we assume $v_{Emin} = 0.4$ V, $v_{B2min} = 0.4 + (12 - 0.4)2.5/(2.5 \text{ k}\Omega + 2.5 \text{ k}\Omega = 0.4 + 5.8 = 6.2$ V. The waveform for v_{B2} is also shown in Fig. 16.43.

From this example, we note that the unijunction can be used to produce a periodic trigger pulse (v_{B2}) when used as a relaxation oscillator. There are two limitations to this type of circuit that should be understood. First, if R_1 (Fig. 16.39) is too large, the current through R_1 may be no larger than the saturation current flowing through the reverse-biased emitter junction. (The leakage current through the capacitor, especially if an electrolytic capacitor is used, will add to the saturation current of the diode.) Then, the charge on the capacitor remains constant and the circuit will not operate. In the other extreme, if R_1 becomes too small, the current flowing through R_1 and the emitter junction after the circuit has "fired" may be enough to keep R_1 in its saturated (or low impedance) condition. For example, if an emitter current of 10 mA is maintained in the 2N492 (Fig. 16.39), the resistance of R_{b1} will remain at about 100 Ω. Then, the emitter of the unijunction will not be turned off and the relaxation oscillator will *not* oscillate.

PROBLEM 16.15   Repeat Example 16.8 if C is changed to 0.01 μF.

Answer: The period for one cycle = 694 μs.

PROBLEM 16.16 Modify the circuit elements in Fig. 16.39 so that the circuit can produce 1000 pulses/s.

The narrow pulses generated by the unijunction can be used to trigger other circuits such as digital IC's or SCR's. To illustrate this capability, let us consider a last example.

EXAMPLE 16.8 A phase-controlled dc power supply is connected as shown in Fig. 16.44. Analyze the performance of this circuit and determine suitable values for circuit elements to provide proper operation.

The two diodes D_1 and D_2 act as a full-wave rectifier circuit. Consequently, the voltage v_A at point A has the waveform shown in Fig. 16.45. The diode D_3 is a 12-V reference diode that acts with R_1 to limit the voltage at point B to 12 V. If D_3 is a 1-W diode, the current through D_3 should be limited to $I = W/V = 1/12 \simeq 83$ mA. In order to limit the current through D_3 to 83 mA, $R_1 = (165 - 12)\,\mathrm{V}/0.083\,\mathrm{A} = 1850\,\Omega$. We shall use a value of $2200\,\Omega$ for R_1.

The unijunction circuit operates in essentially the same fashion as the circuit in Example 16.7. However, we do wish to obtain a positive pulse to trigger the SCR's in this circuit. Therefore, a resistor R_5 is included in the

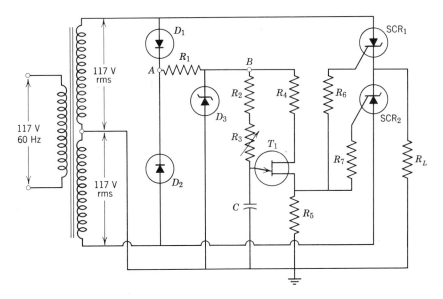

Fig. 16.44.   An adjustable dc power supply.

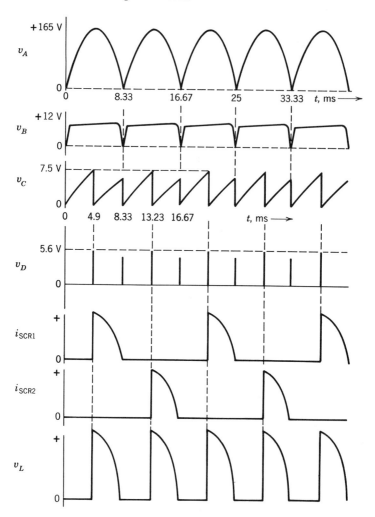

Fig. 16.45. The voltage and current waveforms of the circuit in Fig. 16.44.

base 1 circuit of the unijunction. To determine the value of R_5, the characteristics of the SCR must be considered. A 2N1597 is an industrial-type SCR with a peak-reverse voltage rating of 200 V and a forward current capacity of 1.6 A rms. We shall use this type of SCR in our circuit. The specifications for the 2N1597 indicate the most sensitive gate will not fire unless at least 0.2 V is applied. In addition, the most insensitive gate will fire if 3 V is applied to the gate. Let us use the 2N492 unijunction transistor we used in Example 16.7. In that example, a 6-V pulse was generated across R_{b1} from a 12-V supply. The value of R_{b1} was about 50 Ω while the pulse was present. Therefore, if we make R_5 equal to 50 Ω, the voltage pulse across R_5 will have a value of 3 V, which is sufficient to fire the most insensitive SCR.

To insure at least 3 V across R_5, we reduce R_4 to 500 Ω. Then, the voltage from R_{b2} to ground will be $v_{RB2} = 12(R_{bb}+R_5)/(R_{bb}+R_4+R_5) = 12(7550)/8050 = 11.25$ V. Now, the voltage pulse across R_5 will have a magnitude of 5.6 V. Since $R_5 \ll R_{bb}$, the value of $V_{EO} \simeq 0.67 \times 11.25 \simeq 7.5$ V. The time required for the voltage across $C(v_c)$ to charge to 7.5 V can be found from the equation

$$v_E = 12(1 - e^{-t/RC}) \tag{16.25}$$

where RC is $(R_2 + R_3)C$. When a value of 7.5 V is substituted into Eq. 16.23 for v_E, we have

$$7.5 = 12(1 - e^{-t/RC})$$

This equation can be solved for t to yield

$$t = 0.98\, RC \tag{16.26}$$

We would like the unijunction firing to be adjustable from 0 to 8.33 ms (the period for a half cycle). However, to protect the emitter circuit of the unijunction and to prevent the saturation of its R_{b1} we limit the current through R_2 and R_3 to 5 mA. Then when R_3 is set to zero, the value of R_2 will be $R_2 = 12\,V/5 \times 10^{-3}\,A = 2.4$ kΩ. When R_3 is all in the circuit, the firing time should be 8.33 ms. Then, from Eq. 16.26, $(R_2+R_3)C = (8.33 \times 10^{-3})/0.98 = 8.5 \times 10^{-3}$ s. We shall limit R_3 to a value of about 100 kΩ so the saturation current through the reverse biased unijunction is small compared to the minimum current through R_3. Therefore, let R_3 be a 100-kΩ potentiometer. Then, C must have a value of about 0.085 μF. We shall use a value of 0.1 μF for C.

If the resistor R_3 is adjusted so that $(R_2 + R_3)$ is 25 kΩ, the value of t from Eq. 16.24 will be $t = 0.98 \times 2.5 \times 10^4 \times 2 \times 10^{-7}\,s = 4.9$ ms. The voltage across the capacitor, v_C, will have the form shown in Fig. 16.45. At $t = 4.9$ ms, the unijunction will fire and the capacitor will discharge. The capacitor begins to charge again, but before the voltage rises to 7.5 V, the voltage v_B drops to zero. Since base 2 is connected to v_B, the voltage on base 2(v_{B2}) also decreases. As v_{B2} decreases, v_{EO} decreases. When v_{EO} is reduced to the same potential as v_C, the unijunction fires and discharges the capacitor. Thus, the capacitor always begins a charging cycle at the beginning of each half cycle of the input power. Of course, each discharge produces a positive pulse across R_5. This voltage waveform is shown in Fig. 16.45 as voltage v_D. This positive pulse is applied to the gates of the SCR's. When the positive gate potential is applied to a forward biased SCR, the SCR will fire and conduct current. The positive gate potential will have no effect on the reverse-biased SCR. Thus, one SCR will conduct for part of one half cycle and the other SCR will conduct for part of the next half cycle. The action is much the same as a full-wave rectifier except the SCR's only conduct for part of a half cycle. The current through the two SCR's will have the form shown in Fig. 16.45.

The resistors R_6 and R_7 are used to limit the current flow into the gates. The most insensitive SCR (of this type) will fire if a gate voltage of 3 V and a gate current of 10 mA is present. The actual voltage pulse v_D has a value of 5.6 V. Hence we can drop 2.6 V (with 10 mA of current) across R_6 or R_7 and still get proper operation. Thus, $R_6 = R_7 = 2.6/0.01 = 26\ \Omega$. A value of 20 Ω for these two resistors would be sufficient.

The voltage across the load will be equal to R_L times the current through R_L. This current will be the sum of the two currents through the two SCR's. Therefore, the output voltage v_L will have the form shown in Fig. 16.45. The average output current and voltage will vary as the SCR conduction time varies. Since R_3 controls the firing time of the unijunction, it also controls the average output voltage.

If the value of R_3 is reduced to a low enough value, two or more positive pulses may be produced by the unijunction transistor within one half cycle of the applied ac voltage. In this case, the first pulse fires the SCR and any following pulses will have no additional effect. The SCR is turned off only when the anode voltage drops to zero. (As was mentioned previously, a positive pulse on the gate of a reverse biased SCR will have no effect on this SCR.)

The pulse at the beginning of each half cycle may be large enough to trigger the SCR's. However, since the anode potentials are zero at this time, these pulses do not fire the SCR's. From this description, we see the conduction period of each SCR can be controlled from 0 ($R_3 \simeq 100$ kΩ) to a maximum value ($R_3 = 0$). The maximum dc voltage across the load is essentially equal to 0.636 times the peak value of one half the secondary voltage (from the center-tap to either end).

Time and space limitations have restricted our examples to a small sample of the almost-unlimited number of useful circuits. However, a fairly representative collection has been included. We hope you have an enjoyable time as you delve more deeply into the exciting field of electronics. Good luck!

PROBLEM 16.17 Determine the value of each circuit element in Fig. 16.44 if the secondary winding has 220 V rms on each side of center tap. The 2N1599 has a peak reverse voltage rating of 400 V. Otherwise, its characteristics are the same as those of the 2N1597.

SEMICONDUCTOR
DEVICE
CHARACTERISTICS

The 2N3114 is an *n-p-n* silicon planar transistor primarily designed for high-voltage, medium-power amplifier applications. This device features a guaranteed minimum LV_{CEO} of 150 V and a minimum f_τ of 40 MHz, and operates at current levels up to 100 mA.

ABSOLUTE MAXIMUM RATINGS (NOTE 1)

Maximum temperatures
Storage temperature $-65°C$ to $+200°C$
Operating junction temperature   200°C maximum
Lead temperature (soldering, 60 s time limit)   300°C maximum

Maximum power dissipation

Total dissipation at 25°C case temperature (Notes 2 and 3)   5.0 W
at 25°C free air temperature (Notes 2 and 3)   0.8 W

Maximum voltages

V_{CBO} Collector to base voltage   150 V
V_{CEO} Collector to emitter voltage (Note 4)   150 V
V_{EBO} Emitter to base voltage   5.0 V

ELECTRICAL CHARACTERISTICS (25°C free air temperature unless otherwise noted)

Symbol	Characteristic	Min.	Typ.	Max.	Units		Test Conditions
h_{FE}	dc pulse current gain (Note 5)	30	60	120		$I_C = 30$ mA	$V_{CE} = 10$ V
h_{FE}	dc current gain	15	35			$I_C = 100\ \mu$A	$V_{CE} = 10$ V
h_{FE} (−55°C)	dc pulse current gain (Note 5)	12	24			$I_C = 30$ mA	$V_{CE} = 10$ V
V_{BE} (sat)	Base saturation voltage		0.8	0.9	Volts	$I_C = 50$ mA	$I_S = 5.0$ mA
V_{CE} (sat)	Collector saturation voltage		0.3	1.0	Volts	$I_C = 50$ mA	$I_B = 5.0$ mA
I_{CBO}	Collector cutoff current		0.3	10	nA	$I_E = 0$	$V_{CB} = 100$ V
I_{CBO} (150°C)	Collector cutoff current		2.7	10	μA	$I_E = 0$	$V_{CB} = 100$ V
I_{EBO}	Emitter cutoff current			100	nA	$I_C = 0$	$V_{EB} = 4.0$ V
h_{fe}	Small-signal current gain ($f = 1.0$ kc)	25	50			$I_C = 1.0$ mA	$V_{CE} = 5.0$ V
h_{fe}	High-frequency current gain ($f = 20$ mc)	2.0	2.7			$I_C = 30$ mA	$V_{CE} = 10$ V
C_{ob}	Output capacitance		6.0	9.0	pF	$I_E = 0$	$V_{CB} = 20$ V
C_{TE}	Emitter transition capacitance			80	pF	$I_C = 0$	$V_{EB} = 0.5$ V
$R_e (h_{ie})$	Real part of input impedance ($f = 100$ mc)		70	30	ohms	$I_C = 10$ mA	$V_{CE} = 10$ V
BV_{CBO}	Collector to base breakdown voltage	150			Volts	$I_C = 100\ \mu$A	$I_E = 0$
V_{CEO} (sust)	Collector to emitter sustaining voltage (Notes 4 and 5)	150			Volts	$I_C = 30$ mA (pulsed)	$I_B = 0$
BV_{EBO}	Emitter to base breakdown voltage	5.0			Volts	$I_E = 100\ \mu$A	$I_C = 0$

NOTES:

1. These ratings are limiting values above which the serviceability of any individual semiconductor device may be impaired.
2. These are steady-state limits. The factory should be consulted on applications involving pulsed or low-duty cycle operations.
3. These ratings give a maximum junction temperature of 200°C and junction-to-case thermal resistance of 35°C/W (derating factor of 28.6 mW/°C); junction-to-ambient thermal resistance of 219°C/W (derating factor of 4.56 mW/°C).
4. Rating refers to a high current point where collector-to-emitter voltage is lowest. For more information send for Fairchild Publication APP-4.
5. Pulse conditions: Length = 300 μs; duty cycle = 1%.

Fig. 1.3A. The characteristics of the 2N3114 transistor (Copyright 1965 by Fairchild Semiconductor, a division of Fairchild Camera and Instrument Corporation.)

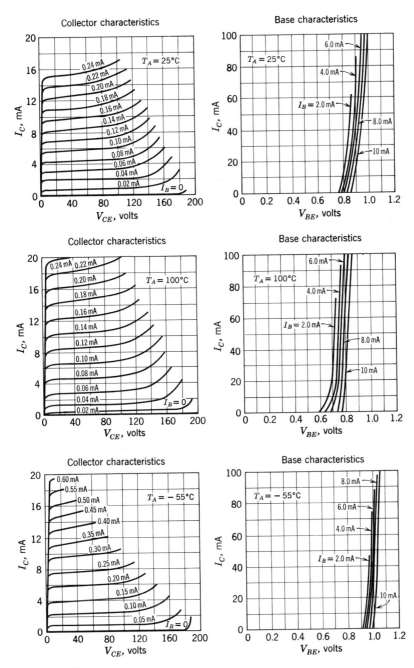

Fig. I.3B. Typical collector and base characteristics.

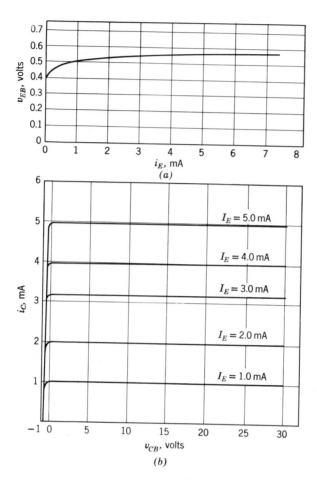

Fig. I.1. Common-base characteristics of the 2N3903 transistor. (*a*) Input characteristics and (*b*) collector characteristics.

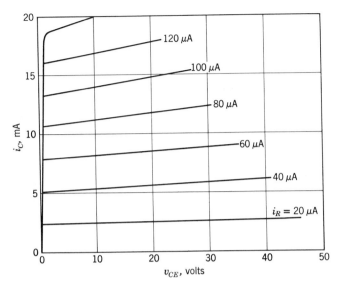

Fig. I.2.   Characteristics of the 2N3903 transistor. Maximum ratings: $T = 25°C$, $V_{CEO} = 40$ V, $I_{CO} = 0.05$ μa, $I_C = 100$ mA, $P_{diss} = 310$ mW, derate 2.81 mW/°C. Small signal characteristics: $I_C = 5$ mA, $v_{CE} = 10$ V, $f = 1$ KHx, $f_t = 250$ MHz, $C_{ob} = 2$ pF, $h_{fe} = 150$, $h_{ie} = 850$ Ω, $h_{oe} = 30$ μmhos, $h_{re} = 2 \times 10^{-4}$.

- Low Noise Figure -2.5 dB max @   60 MHz
                             5.0 dB max @ 450 MHz
- High Stable Gain in Unneutralized Amplifiers -20 dB min @   60 MHz
                                                              8 dB min @ 450 MHz
- Low Feedback Capacitance -0.5 pF max
- Guaranteed Forward AGC

ABSOLUTE MAXIMUM RATINGS (NOTE 1)

Maximum temperatures	$-55°C$ to $+200°C$
Storage temperature	$+200°C$
Operating junction temperature	
Maximum power dissipation	
Total dissipation at 25°C case temperature (Note 2)	0.3 W
at 25°C ambient temperature (Note 2)	0.2 W
Maximum voltages	
V_{CBO} Collector to base voltage	30 V
V_{CEO} Collector to emitter voltage (Note 3)	30 V
V_{EBO} Emitter to base voltage	3.0 V

ELECTRICAL CHARACTERISTICS (25°C Free Air Temperature Unless Otherwise Specified)

Symbol	Characteristic	Min.	Typ.	Max.	Units	Test Conditions
NF	Noise figure (f = 450 MHz) (Note 5)	(2N4135 only)		5.0	dB	$I_E = 1.0\,mA$ $V_{CB} = 15\,V$ $R_s \approx 130\,\Omega$
PG	Power gain (f = 450 MHz) (adjusted for min. noise figure; Note 5)	(2N4135 only) 8.0	10		dB	$I_E = 1.0\,mA$ $V_{CB} = 15\,V$
NF	Noise figure (f = 60 MHz) (Note 6)		2.0	2.5	dB	$I_E = 1.0\,mA$ $V_{CB} = 15\,V$ $R_s \approx 300\,\Omega$
PG	Power gain, neutralized (f = 60 MHz) (Adjusted for min. noise figure; Note 6)	17	21	24	dB	$I_E = 1.0\,mA$ $V_{CB} = 15\,V$
PG	Power gain, unneutralized (f = 60 MHz) (Note 7)	20	23	25	dB	$I_E = 5.0\,MA$ $V_{AGC} = 13\,V$
V_{AGC}	AGC voltage for 30 dB gain reduction (f = 60 MHz) (Note 7)	19	22	24.5	Volts	$V_{CC} = 28\,V$
$r_b C_c$	Collector-base time constant (f = 80 MHz)		2.5	5.0	ps	$I_C = 4.0\,mA$ $V_{CE} = 10\,V$
f_{max}	Maximum frequency of oscillation		3.25		GHz	$I_E = 4.0\,mA$ $V_{CB} = 15\,V$
C_{cb}	Reverse transfer capacity common emitter	0.25	0.37	0.50	pF	$I_E = 0$ $V_{CE} = 10\,V$ $f = 1.0\,MHz$ $V_{CE} = 10\,V$ (emitter & can guarded)
h_{fe}	High frequency current gain (f = 100 MHz)	(2N4135 only) 4.25		8.0		$I_C = 4.0\,mA$ $V_{CE} = 10\,V$
h_{fe}	High frequency current gain (f = 100 MHz)	(2N4134 only) 3.5		8.0		$I_C = 4.0\,mA$ $V_{CE} = 10\,V$
h_{FE}	DC pulse current gain (Note 4)	25		200		$I_C = 4.0\,mA$ $V_{CE} = 10\,V$
$h_{FE}\,(-55°C)$	DC pulse current gain (Note 4)	10				$I_C = 4.0\,mA$ $V_{CE} = 10\,V$
V_{CEO} (sust)	Collector to emitter sustaining voltage (Notes 3 and 4)	30			Volts	$I_C = 1.0\,mA$ $I_S = 0$
BV_{CBO}	Collector to base breakdown voltage	30			Volts	$I_C = 1.0\,mA$ $I_E = 0$
BV_{EBO}	Emitter to base breakdown voltage	3.0			Volts	$I_C = 0$ $I_E = 100\,\mu A$
V_{BE}(sat)	Base saturation voltage			0.92	Volt	$I_C = 10\,mA$ $I_B = 5.0\,mA$
V_{CE}(sat)	Collector saturation voltage			3.0	Volts	$I_C = 10\,mA$ $I_B = 5.0\,mA$
I_{CBO} (150°C)	Collector cutoff current			50	μA	$I_E = 0$ $V_{CB} = 10\,V$
I_{CBO}(25°C)	Collector cutoff current			50	nA	$I_E = 0$ $V_{CB} = 10\,V$

Fig. I.4A   Characteristics of the 2N4134–2N4135 transistors.

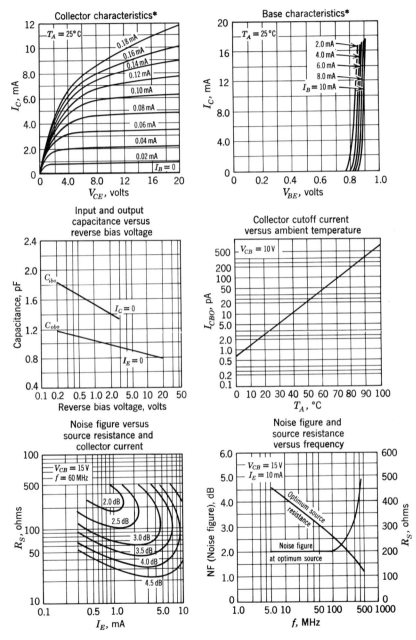

Fig. I.4B.   Typical electrical characteristics of the 2N4134–2N4135 transistors.

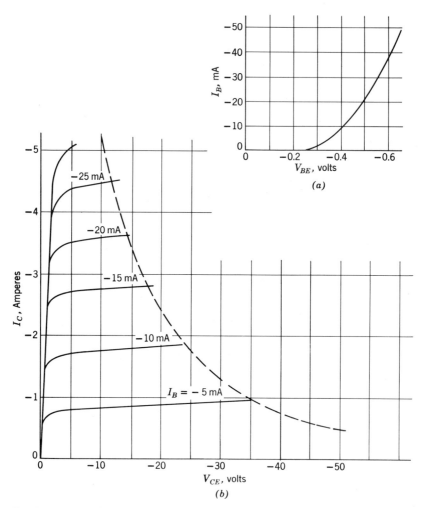

Fig. I.5.   Characteristics of the 2N2147 power transistor. (See data sheet.) (a) Input characteristics, 2N2147 and (b) collector characteristics, 2N2147.

Data Sheet, 2N2147

1. Maximum ratings:
    A. Voltage

V_{cbo}	−75 max V
V_{ceo}	−50 max V
V_{beo}	−1.5 max V
I_c	−5 max A
I_b	−1 max A
I_e	5 max A

    B. Dissipation
    (At mounting flange
        temperatures up to 81°C)   12.5 max W
    (Above 81°C)   Derate 0.66 W/C°

2. Electrical characteristics
    A. $V_{cb\ (sat)}$ ($I_c = 10$ mA, $I_e = 0$ mA)   75 min V
       $V_{ce\ (sat)}$ ($I_c = 100$ mA, $I_b = 0$ mA)   50 min V
       $V_{be\ (sat)}$ ($V_{ce} = 10$ V, $I_c = 50$ mA)   0.24 V
       I_{cbo} ($V_{cb} = 40$ V)   1 max mA
       I_{cbo} ($V_{cb} = 0.5$ V)   70 max μA
       I_{beo} ($V_{eb} = 1.5$ V)   −2.5 max mA

    B. Other data
    Thermal resistance (junction to case)   1.5 max°C/W
    H_{FE} (dc)   150
    Gain bandwidth product
    ($V_{ce} = 5$ V $I_c = 500$ mA)   4 Mc

ABSOLUTE MAXIMUM RATINGS AT 25°C

*Gate-Drain Voltage and Gate-Source Voltage (see Note on p. 1). . . . . 30 V
Gate Current (Forward Biased) . . . . 50 mA
*Total Device Dissipation at (or below) 25°C Free-Air Temperature (Note 1) . . . . 300 mW
*Storage Temperature Range. . . . . −65 to +200°C

*ELECTRICAL CHARACTERISTICS AT 25°C except as noted

Characteristic	2N2841			2N2842			2N2843			2N2844			Unit
	Min	Typ	Max	Min	Typ	Max	Min	Typ	Max	Min	Typ	Max	
I_{GSS} Gate-source cutoff current (Note 2) at: $V_{GS} = 30$ V, $V_{DS} = 0$			1			3			10			30	nA
I_{GNSS} Gate-source cutoff current at: $V_{GS} = 5$ V, $V_{DS} = 0$, $T_A = 150°C$			1			3			10			30	μA
BV_{GDS} Gate-drain breakdown voltage at: $I_G = 1$ μA, $V_{DS} = 0$	30			30			30			30			V
I_{DSS} Drain current at zero gate voltage at: $V_{DS} = -5$ V, $V_{GS} = 0$	−25	−56	−125	−65	−170	−325	−200	−500	−1000	−440	−1100	−2200	μA

V_P	Gate-source pinch-off voltage at: $V_{DS} = -5$ V $I_D = -1\ \mu A$	1.2	1.7	1.2	1.7	1.2	1.7	1.2	1.7	V
g_{fs}	Small-signal common-source forward transconductance (Note 3) at: $V_{DS} = -5$ V $V_{GS} = 0$ $f = 1$ kc	60	93	180	280	540	830	1400	1800	μmho
C_{gss}	Gate-source capacitance (Note 4) at: $V_{DS} = -5$ V $V_{GS} = 1$ V $f = 140$ kc	3.7	6	7	10	12	17	25	30	pF
NF	Noise figure at: $V_{DS} = -5$ V $V_{GS} = 0$ $f = 1$ kc $BW = 16\%$ $R_{gen} = 10\ M\Omega$ $R_{gen} = 1\ M\Omega$	0.5	3	0.5	3	0.5	3	0.5	3	dB dB

NOTES:
1. Derate linearly to $+175°C$ free-air temperature at the rate of 2 mw/C°.
2. I_{GSS} is JEDEC registered at $V_{GS} = 5$ V.
3. At 1 kc, $g_{fs} = |y_{fs}| = $ real part of y_{fs}.
4. C_{gss} is preferred subscript in lieu of C_{is}.
*JEDEC Registered Data.

Fig. I.6A.   Characteristics of the 2N2841–2N2844 FET.

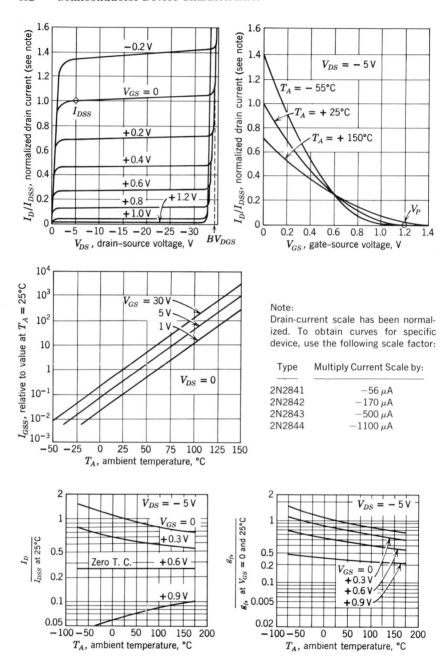

Note:
Drain-current scale has been normalized. To obtain curves for specific device, use the following scale factor:

Type	Multiply Current Scale by:
2N2841	$-56 \ \mu A$
2N2842	$-170 \ \mu A$
2N2843	$-500 \ \mu A$
2N2844	$-1100 \ \mu A$

Fig. I.6B.   Normalized typical characteristics.

Typical drain characteristics

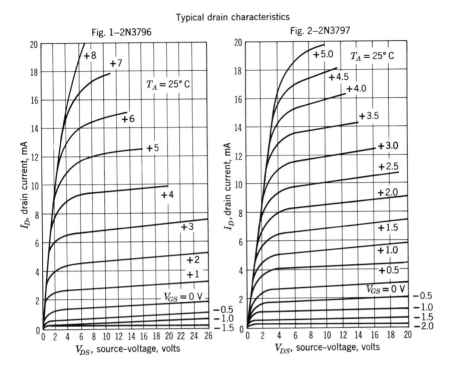

Fig. 1–2N3796   Fig. 2–2N3797

Common source transfer characteristics

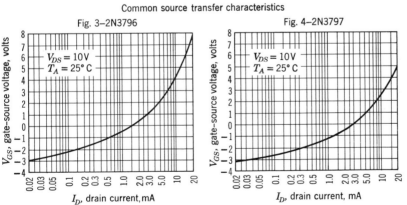

Fig. 3–2N3796   Fig. 4–2N3797

Fig. I.7B.

MAXIMUM RATINGS ($T_A = 25°$ unless otherwise noted)

Rating	Symbol	Values	Unit
Drain-source voltage	V_{DS}		Vdc
2N3796		25	
2N3797		20	
Gate-source voltage	V_{GS}	± 30	Vdc
Drain current	I_D	20	mAdc
Power dissipation at $T_A = 25°C$	P_D	300	mW
derate above 25°C		1.7	mW/°C
Operating junction temperature	T_J	$+200$	°C
Storage temperature range	T_{stg}	-65 to $+200$	°C

ELECTRICAL CHARACTERISTICS ($T_A = 25°C$ unless otherwise noted)

Characteristic		Symbol	Min	Typ	Max	Unit
Drain-source breakdown voltage		BV_{DSX}				Vdc
($V_{GS} = -4.0$ V, $I_D = 5.0$ μA)	2N3796		25	30	—	
($V_{GS} = -7.0$ V, $I_D = 5.0$ μA)	2N3797		20	25	—	
Zero-gate-voltage drain current		I_{DSS}				mAdc
($V_{DS} = 10$ V, $V_{GS} = 0$)	2N3796		0.5	1.5	3.0	
	2N3797		2.0	2.9	6.0	
Gate-source voltage cutoff		$V_{GS(off)}$				Vdc
($I_D = 2.0$ μA, $V_{DS} = 10$ V)	2N3796		—	3.0	4.0	
($I_D = 2.0$ μA, $V_{DS} = 10$ V)	2N3797		—	5.0	7.0	
"On" drain current		$I_{D(on)}$				mAdc
($V_{DS} = 10$ V, $V_{GS} = 3.5$ V)	2N3796		7.0	8.3	14	
	2N3797		9.0	14	18	
Drain-gate reverse current*		I_{DGO}*				pAdc
($V_{DG} = 10$ V, $I_S = 0$)					1.0	
Gate-reverse current*		I_{GSS}*				pAdc
($V_{GS} = 10$ V, $V_{DS} = 0$)			—	—	1.0	
($V_{GS} = -10$ V, $V_{DS} = 0$, $T_A = 150°C$)			—	—	200	
Small-signal, common-source forward transfer admittance		$|y_{fs}|$				μmhos
($V_{DS} = 10$ V, $V_{GS} = 0$, $f = 1.0$ kHz)	2N3796		900	1200	1600	
	2N3797		1500	2300	3000	
$V_{DS} = 10$ V, $V_{GS} = 0$, $f = 1.0$ MHz)	2N3796		900	—	—	
	2N3797		1500	—	—	
Small-signal, common-source, output admittance		$|y_{os}|$				μmhos
($V_{DS} = 10$ V, $V_{GS} = 0$, $f = 1.0$ kHz)	2N3796		—	12	25	
	2N3797		—	27	50	

ELECTRICAL CHARACTERISTICS (contd)

Characteristic		Symbol	Min	Typ	Max	Unit
Small-signal, common-source, input capacitance		C_{iss}				pF
($V_{DS} = 10$ V, $V_{GS} = 0, f = 1.0$ MHz)	2N3796		—	5.0	7.0	
	2N3797		—	6.0	8.0	
Small-signal, common-source, reverse transfer capacitance		C_{ras}	—			pF
($V_{DS} = 10$ V, $V_{GS} = 0, f = 1.0$ MHz)						
Noise figure		NF				dB
($V_{DS} = 10$ V, $V_{CS} = 0, f = 1.0$ kHz, $R_s = 3$ MΩ)			—	3.8	—	

*This value of current include both the FET leakage current as well as the leakage current associated with the test socket and fixture when measured under best attainable conditions.

Fig. 1.7A.   Characteristics of the 2N3796–2N3797 MOSFET transistors. Silicon N-channel MOS field-effect transistor designed for low-power applications in the audio frequency range.

INDEX